REDUCTIONISM AND CULTURAL BEING

NIJHOFF INTERNATIONAL PHILOSOPHY SERIES

VOLUME 14

For a list of other volumes in this series see final page of the volume.

Joseph Wayne Smith

Reductionism and Cultural Being

A Philosophical Critique of Sociobiological Reductionism and Physicalist Scientific Unificationism

1984 **MARTINUS NIJHOFF PUBLISHERS**
a member of the KLUWER ACADEMIC PUBLISHERS GROUP
THE HAGUE / BOSTON / LANCASTER

Distributors

for the United States and Canada: Kluwer Boston, Inc., 190 Old Derby Street, Hingham, MA 02043, USA
for all other countries: Kluwer Academic Publishers Group, Distribution Center, P.O.Box 322, 3300 AH Dordrecht, The Netherlands

Library of Congress Cataloging in Publication Data

Smith, Joseph Wayne.
Reductionism and cultural being.

(Nijhoff international philosophy series ; 14)
Includes index.
1. Sociology--Philosophy. 2. Sociobiology--Philosophy. 3. Reductionism. 4. Science--Philosophy. I. Title. II. Series: Nijhoff international philosophy series ; v. 14)
HM24.S5393 1984 301'.01 83-19399
ISBN 90-247-2884-3

PRINTED IN THE NETHERLANDS

Contents

CHAPTER 1

Statement of the argument: Philosophical orientation and the theoretical framework for critique

1.0 Introduction: The statement of the overall argument

In this Chapter I shall attempt two tasks. First I shall state the Principal Thesis (P.T.) of this work in a clear and concise form and summarize the overall argument of the work. Second I shall give some discussion and defense of the philosophical orientation of the work itself, and defend the project against some arguments which critics may advance to discredit, devalue and reject, the overall project. Such critical self-reflexivity is necessary, for only the briefest awareness of scholarly debate ensures one that many works are criticized in this fashion (especially in book reviews). The aim here is to write a strong and cogently reasoned piece of work which will convince the reader of various difficulties in a number of reviewed and criticized positions, and thus clarity in our thesis and aims is crucial.

The Principal Thesis of this work is this:

(PT) Sociobiology offers no support to the contemporary *Metaphysical Weltanschauung* of Physicalist Scientific Unificationism (PSU) because Sociobiology (and ultimately the neo-Darwinist synthesis) is biologically and philosophically flawed as a theoretical orientation. Further PSU is open to an extensive philosophical challenge with regard to all of its major tenets. Consequently it is unreasonable to believe that Culturology/ "Sociology" is theoretically reducible to, or eliminates in favour of, Sociobiology.

First, however, comment must be made with regard to the terms "Culturology", "Sociology" and "Sociobiology". I shall not attempt at this time any explicit definition of any of these terms. Indeed the desire for giving immediate explicit definitions contains its own difficulties. Wilson [1975] defines Sociobiology as "the systematic study of the biological basis of all social behaviour" [ibid., 4]. Such a definition would by definitional fiat reduce to absurdity theses such as my own. It would seem that one must, in denying the cogency of Sociobiology, embrace some form of idealism or dualism in which social behaviour does not have a "biological basis" at all. This is not my position. Such a definition then trivializes an important debate. I therefore shall not attempt any initial explication of Sociobiology until Chapter 4, where the position, which consists of a number of theories and systems of methodological principles, shall be explicated. It is hardly possible to encompass such material into any simple explict definition.

Different problems exist with our use of the terms "Culturology" and "Sociology". It is evident, as the "crisis of Sociology" argument discussed in Chapter 2 establishes that there are many models of man and society. "Sociology" if defined simply as the science of society would encompass Sociobiology, as Sociobiology does attempt to give (leaving aside the question of its adequacy), a theory of society. Under this definition my thesis is nonsense. Consequently, if an intellectually interesting argument is to be developed here a more restrictive sense of the terms "Culturology" and "Sociology" must be given.

In Chapter 8 I will atempt a more detailed account of "Culturology" as I specifically view the position. Even there, I will produce only a fragment of the explication needed to satisfactorily clarify the position. Until that chapter, the terms "Culturology" and "Sociology" will be treated as synonyms, although to reinforce my response to the trivialization argument cited above, in discussion I shall frequently write "Culturology/"Sociology"", and in situations where reference is made only to "Sociology", quotes shall occur around the world.

Now, granted this, what then is our partial explication of these terms? I propose that the following theses partially characterize my position:

(C1) There may be theoretically adequate explanations of human socio-cultural phenomena which make either sole reference to socio-cultural variables or merely trivial or minor reference to biological variables (*The Durkheimian thesis*).

(C2) Human beings are persons, that is agents which have a capacity to manipulate cultural entities and to act on the basis of held beliefs, reasons and desires. Explanations of human action with respect to such intentional variables may provide a theoretically adequate account of human action without supplementation with, or reduction to, underlying biological mechanisms (or if some form of idealism was true) or ideal mechanisms (*The Geisteswissenschaften thesis*).

Given theses (C1) and (C2) it is evident why even the gene-culture co-evolutionary theory, recently proposed by Wilson and Lumsden [1981] is rejected by Culturologists. On this position, as we shall see in our discussion and critique in Chapter 4, "pure-cultural transmission" is taken to be an evolutionary unstable strategy, open to invasion by "gene-culture transmission". Culturology in accepting (C1) and (C2) is irreconciliable with Sociobiology by definition. Wilson [1978, p. 2] has recognized this and has classified positions which conflict with his own "scientific materialism" as "mythology". I hope in this work to be able to reverse this judgement.[1]

A preoccupation with questions of reductionism and theoretical unification has marked a large proportion of twentieth-century philosophy and sociological theory. Reductionist research programmes, whether they be in the philosophy of mathematics or in the methodological individualism debate are motivated by the acceptance of the metaphysical ideal of Parsimony: that the world is ultimately simple rather than of unlimited complexity (as is accepted by a philosopher-physicalist such as David Böhm [1957] for example). For *Physicalist Scientific Unificationism* (PSU) the world is monistic, consisting of only one type of substance, the "material" or the "physical", with one type of properties and relations. Phenomena which do not fit readily into this scheme, must be shown to be in principle either reducible to the "physical", or else shown to be "illusions", fit for the ontological scrap heap. Further, the ideal of Parsimony and the ideal of the systematicity of knowledge are closely related in the PSU *Weltanschauung.* The unification and systematization of knowledge is taken to occur most plausibly through the reduction of all sciences to physics. The result of this is to produce an ontologically simple world view, described by an economical theoretical structure.

In the PSU attempt to unify the social and natural sciences, Sociobiology can be given an important role; indeed without overstatement, it is the most exciting scientific development to occur in recent years of relevance to PSU. Sociobiology seems applicable to many socio-cultural phenomena which once seemed to be the exclusive province of the socio-cultural sciences. Culture would no longer seem to escape the physicalists' net.

Of course in viewing Sociobiology in this light, I am not claiming that Sociobiology was developed solely by its constructors to support PSU. If I was, then I would be making a false claim. It is argued in Chapter 4, contrary to biologists such as Lewontin [1977], that Sociobiology is a legitimate extension of the neo-Darwinist synthesis to cope with the theoretical challenge of social behaviour, especially altruism.[2] It is no more important to attribute conscious intent to the Sociobiologists to support my claims, than it is for Sociobiologists to attribute conscious intent to their "Selfish Genes" to support their own claims. My claim is that Sociobiology is an important development for PSU, and the Australian and American "united front" of physicalists will make use of Sociobiological material with or without the blessings of the working Sociobiologist.

I have so far described the intellectual setting of the work. Now some sketch of the complex argument of the work should be given to integrate and unify the theoretical points which have been made.

In Chapter 2, I consider whether a "philosophico-Methodoligcal Reduction" of Culturology/"Sociology" to Sociobiology might be possible, that is whether there exist philosophical and methodological arguments of a nature which establish that Culturology/"Sociology" is an untenable position. If this could be done from the outset, then the project attempted here is immediately destroyed. I consider in Chapter 2 two broad sets of arguments which might achieve this. First there are the arguments of radical behaviourists and eliminative materialists against the explanatory coherence of mentalistic concepts; second there is the "crisis of Sociology" argument. The former set of arguments have been stated by Skinner [1976] and put in the context of the Sociobiology debate by Ellis [1977]. A point by point refutation of both authors is offered. The challenge of eliminative materialism is implicitly met by the combined argument of Chapter 3 which constitutes an extended critique of the central tenets of PSU.

The second argument, the "crisis of Sociology" argument is more difficult to meet, for the argument, as it is summarized in Sztompka [1979] consists of the double headed Medusa of (1) the "theoretical pluralism" and lack of "cognitive systematization" [Rescher, 1979] of the discipline as a whole and (2) specific problems in rationally justifying one orientation over another. I argue that the "crisis of Sociology" is actually a "crisis of Philosophy", and can be resolved by the building of comprehensive *Metaphysical Weltanschauungs.*

Physicalists have been building a *Metaphysical Weltanschauung* in this century, since the time of the adoption of physicalism by the logical Positivists. In Chapter 3, I isolate the central tenets of PSU as I view it and systematically argue against each. This argument is the first major prong in my defence of P.T. If PSU is an ill-defined, unjustified position, then PSU is not supported by Sociobiology.

The argument of Chapter 4 continues my offensive forward attack. There, after detailing the position of Sociobiology, and discussing its relationship to the neo-Darwinist synthesis, criticisms are advanced against both Sociobiology and neo-Darwinism. These objections are quite technical and do not make summary easy within the spatial confines here.

In Chapters 5, 6 and 7 I will continue my assault upon mechanistic biology. In Chapter 5, in response to the central criticisms made of neo-Darwinism in Chapter 4 I consider in more detail the biochemical nature of the gene, and the genetype–phenotype relation. The image of the gene as a unit of inheritance and central directing agency of the cell, and perhaps of the entire organism as Sociobiologists propose, is one which I reject. An alternative non-Mendelian anti-atomist view of inheritance is sketched. A survey is made of some contemporary alternatives to neo-Darwinism, and whilst these positions do not constitute satisfactory evolutionary theories as such, most have only recently been proposed and can be expected to undergo significant modification and development. However, they cannot be expected to develop if their ontogenesis remains only with some important fringe workers. It is the job of the philosopher of biological science to offer such critical overviews.

In Chapter 6 and 7 I will discuss two aspects of the Sociobiological debate which have not received extensive attention: sex role differences, and neuroendocrinological explanations of the "woman problem" and human and non-human ecological Sociobiology, with a consideration of Cultural Materialist theories of Culture. I wish here to offer a biological and philosophical critique of these positions. The technical nature of these arguments once more makes brief summary highly misleading and uninformative.

It is alleged on this basis that P.T. is supported. Both PSU and its associated ideas of theoretical reduction and Sociobiology are rejected as cognitively satisfactory positions. Chapter 8 of the work attempts to clarify in somewhat more detail the position of Culturology, especially with regard to that noble subject of the philosophers: human nature. It is not possible (as should be recognized from the remarks of Chapter 2 with respect to the need to produce *Metaphysical Weltanschauungs* to unify human knowledge) to produce such a unification here myself. Indeed, given the shortness of individual human life I have no confidence that any individual could single-handedly produce such a satisfactory reconstruction. Given the limitations of my medium of communication I shall attempt the most comprehensive analysis which I can within my spatial confines.

2.0 Rebuttal of methodological criticisms of the argumentative structure of the work

The structure of the argument should be sufficiently clear to enable the reader to formulate various metholodogical objections to my thesis. Before beginning the argument of the work, it is informative and reasonable to anticipate some of these objections and attempt to rebut them. These objections are now set out.

(A) (Objection (1)): an inevitable vagueness pervades much of the subject matter dealt with here. Indeed the author comes to conclude that concepts such as that of "theoretical reduction" are intolerably vague. Therefore this vagueness must reflect upon the cogency of the author's argument.

Reply to (1): the mere fact that I come to conclude that concepts such as the "physical " and "theoretical reduction" are intolerably vague does not reduce my work to absurdity, on the contrary it reduces my opponent's work to absurdity. Of course if we knew this from the outset this work would be redundant. But we do not, and must argue for this conclusion. Further, outside of rigorous formalized sciences, some degree of vagueness is inevitable because of the nature of the subject matter investigated itself. Yet the idea of vagueness is itself vague, despite advances in fuzzy logic [Goguen, 1979] so my formalist critic is self-referentially inconsistent.

(B) (Objection (2)): the overall argument of the thesis is abstract and philosophical, and its relevance to the concrete concerns of the working Sociologist and Biologist is utterly obscure: this is especially evident in the author's discussion of his position of Culturology in Chapter 8 which is incomplete and programmatic.

Reply to (2): the objection is a *petitio principii,* yet is all too frequently made to the works of authors who challenge the intellectual status quo. The argument must assume that all intellectually worthwhile endeavours must be strongly related to "concrete concerns". But why accept even that? The answer is because a concern with concrete matters will aid in the satisfaction of human interests. But do human beings not also have an interest in speculation and creative thinking? That activities such as philosophy have survived, gives some reason to believe that this is true and places an *onus probandi* upon my opponent to demonstrate otherwise. The very basis of the objection is directly challenged as well by the "crisis of Sociology" argument detailed in Chapter 2. Further, the programmatic nature of the present author's response is openly confessed. There are, however, good reasons for this given the difficulty and vagueness of the subject matter dealt with here. "Positive solutions" can only be offered after a significant slash and burning of inadequate solutions has been undertaken. It is better to be left at the end of our inquiries with the ashes of the phoenix rather than with the lotus seeds of the received and inadequate theoretical orientation. If the argument demands the rejection of assumptions cherished by the Establishment, so be it.

(C) (Objection (3)): the author, in short, rejected materialism, neo-Darwinism, received genetics and Sociobiology in the course of his argument. This seems at best an uneconomical way to support his thesis, and at worst raises the question of the competence of the author to pursue the outlined course of the argument.

Reply to (3): to continue from the reply to (2), it is the intellectual obligation of a thinker to state the truth and follow the course of the argument, no matter what prevailing dogma is challenged. Further, with regard to the question of competence, which is inevitably raised when a social scientist questions the theoretical proposals of a natural scientist, if this argument is not merely an ideological smoke screen to prevent criticism of the natural sciences, it readily undermines any of the claims which Sociobiologists have made about fields outside their specialist competence – such as the social sciences, ethics, philosophy, and theology. The critic is self-referentially inconsistent.

Space restrictions precludes a consideration of further objections. The history of thought indicates that one cannot anticipate all strong objections in many cases. One should answer what one can with what space one can afford. With no more space to afford, we commence the main argument.

CHAPTER 2

Philosophico-methodological reductionism: The alleged case against culturology

1.0 Introductory statement of the argument: Culturology and the idea of a philosophico-methodological reduction

The contemporary human Sociobiology debate offers a number of papers in which critics have attempted a "philosophical refutation" of the Sociobiological programme, either in general, or specifically with regards to the explanation and understanding of human behaviour. The epistemological and methodological criticisms advanced against both a human Sociobiology and Sociobiology in general (summarized in [Ruse, 1979]), have been advanced to establish the claim that Sociobiology is epistemologically and methodologically flawed, and that because of this the major theoretical claims either human Sociobiology or Sociobiology in general are vitiated. The discipline is taken to be untenable; I say in such a situation that a *philosophical refutation* has occurred. In a situation where a debater D supports a position P1 and advances philosophical arguments against a competing position P2 such that the conclusion that P2 is philosophically unsatisfactory is rationally justified, and P1 is itself supported by philosophical arguments indicating that a set of philosophical objections against it can be rebutted, a philosophico-methodological reduction of P2 by P1 has occurred.

There are a large number of difficulties which this idea will raise, including what actually are these "philosophical refutations". I will not attempt in the course of this chapter to deal with these questions because to deal successfully with these problems requires a theory of philosophical argument, which unfortunately I have not (nor have others in my assessment) presented in the necessary detail. But we shall take for the moment this concept as having at least sufficient initial intuitive plausibility for the initial development of the argument, for it seems that this concept is implicit in very many of the "meta-critiques" of positions to be found in philosophical discourse.

Let us now look at this matter from the side of our opponents. Are there any good reasons to believe that Culturology faces foundational problems of such a magnitude, that the position itself should be regarded as untenable? Ellis [1977] and van den Berghe [1975] [1978] both see "Sociology" as existing at present in a state of theoretical sterility, and Ellis in particular criticizes "Sociology" for its inability to find solutions to large-scale social problems such as poverty, crime and alientation. For van den Berghe, "Sociology" is a jargonistic scholastic discipline which has few claims to any merit of scientificity at all. Both authors believe that "Sociology" can only achieve scientificity by returning to its biological roots: human behaviour cannot be understood without reference to the fact that human beings are animals, and the current theory of Sociobiology presents the most promising way at present for "Sociology" to obtain scientific rebirth. We take here, by *stipulative definition*, any "Culturological Sociology", to precisely deny these theoretical claims made by Ellis and van den Berghe. Consequently the claims advanced by Ellis and van den Berghe challenge the basis of the position which I wish to defend: hence close critical attention is required.

1.1 Ellis' position: An exposition

For Ellis, the reasons for "Culturology's" alleged theoretical and practical impotence are two. First, Culturology does not seek causal explanations of sociocultural phenomena below the institutional or socio-interpersonal level, and hence remains ignorant of biological variables which may be of causal relevance to the phenomena under question. Second, and closely interrelated to the first point, Culturology upholds a general anti-reductionist stance, with its reliance upon intentionalistic and purposive explanations of human behaviour. Thus it would seem, Culturologists are ignorant of the (alleged) fact that most scientific advancements have occurred within a "reductionist framework". The physical sciences all began by attributing intentionality to physical phenomena. But their theoretical progress was nil and accelerated from zero velocity only when such explanations were discarded. The attribution of intentionality to the natural world is unsatisfactory because such attribution claims are "unfalsifiable": "purposive hypothesis are not empirically disprovable" [Ellis, 1977 p. 57]. Therefore they fail to satisfy a necessary condition for scientificity, or so Ellis maintains.

Ellis proposes that the *elimination* of "rationalistic" and "purposive" factors from social explanation is the first step for "Sociology's" recovery. A human Sociobiology will become, he predicts, an increasingly powerful problem-solver, dealing increasingly more with the major social problems facing humanity [ibid, p. 58]. "Sociology" will die a slow death, gradually disintegrating, as more and more of its potential postgraduates turn to the study of Sociobiology.

Ellis' argument can be compared quite favourably with B. F. Skinner's [1976] critique of "mentalism". Skinner allegedly conducts scientific research on the basis of the meta-scientific principle, that any scientific study of persons must eschew all intentionalistic notions as being of explanatory importance. Skinner too, writes from within a framework of a position which he believes has produced enormous results; he too claims that his position will "save mankind". With Ellis he would agree that intentionalistic explanations are the direct cause of the theoretical sterility of psychology and "Sociology". For Skinner, intentionalist explanations "distract" us from the study of behaviour [ibid, p. 17]; for Ellis, such explanations distract us from the study of biological causes.

Both object to intentionalist/mentalistic explanations on methodological grounds. Ellis' argument is that the unfalsifiable nature of all intentionalistic explanations renders them unscientific. The position is consistent, with regard to ontology, with both an eliminative materialist and epiphenomenalist view of mind, although today the former thesis is more vigorously defended. Skinner attempts to side-step the mind–body problem by advancing methodological criticisms of intentionalist explanations as inquiry-limiting, lacking explanatory power. Ellis has only one reason for this, Skinner has a number.

For Skinner, mentalistic explanations are incomplete since the mental entities referred to in such explanations will be in need of *prior* causal explanation. Skinner uses the example of an intentionalistic explanation, such as "I went to the theatre because I intended to see the play", to illustrate this point [ibid, p. 18]. He alleges that if we accept this as an ultimate explanation, we ignore questions of antecedent causal conditions which shaped this desire. Elsewhere Skinner states his position quite explicitly [Skinner, 1969, p. 240]:

> ". . . disturbance in behaviour is not explained by relating it to felt anxiety *until the anxiety has itself been explained*. An action is not explained by attributing it to expectations *until the expectations have in turn* been accounted for" [italics added]

Mentalism, Skinner alleges must embrace the thesis of the existence of an "inner agent", an autonomous *homuncule*, who provides a basic-explanation, which serves to bring to a close a possible infinite explanatory regression [Skinner, 1976, p. 19]. This *homuncule* is an initiator of its own behaviour, and is endowed with freedom of will, by which Skinner means that such behaviour is "uncaused" [ibid, p. 25]. But "autonomous man" is merely a "miraculous" god-of-the-gaps: these basic explanations are not basic explanations at all, since as we learn more about behaviour, we find increasingly more antecedental causes of our autonomous man's behaviour [ibid, p. 20]. Such an argument is consistent with Ellis' position. Here the "true causes" of human behaviour are genetically programmed strategies to maximise individual inclusive fitness under a specific environment.

A further objection made by Skinner against mentalism, is inconsistent with Ellis' sociobiological orientation; this is the objection, that mentalism involves the employment of theories [Keat, 1972 pp 61–65]. Skinner is, however, prepared to admit to the domain of the scientifically intelligible, certain private sensation data such as pain states, as a component in descriptions of contingencies of reinforcement [Skinner, 1945 p. 294]. Yet pains seem to be theoretical entities; if pains are taken to be theoretical entities because of their privacy, then there is no good reason to be critical of mental images and other members of the mentalistic tribe. Note that the question here is not the question of mind-body dualism, since Skinner in most of his works repeatedly states that his position can side-step this problem. This argument is, as Keat [1972] has shown, stated very vaguely by Skinner, and does in fact weaken Skinner's position unnecessarily. It does seem that Skinner is advancing a *theory* of behaviour and a *theory* of learning. Given the failure of a rigorous defence of the observation – theoretical distinction, in the light of the criticisms of post-empiricist philosophers of science [Suppe, (ed) 1977], Skinner's position here remains quite implausible and thus unnecessarily weakens his case.

If, however, this argument is dropped, it is possible to combine the critiques of both authors. Indeed Skinnerian behaviourism is theoretically consistent with Sociobiology. Whilst Skinner is concerned primarily with environmental causes, it does not follow that responses to environmental stimuli are not of an inheritable basis and subject to natural selection. A number of central Sociobiological concepts are behaviouristically defined, as we shall see later. However, it is not my intention here to ultimately criticize Skinnerian behaviourism, this task having already been satisfactorily performed by others. ([Efron, 1976]; [Chomsky, 1971] [1972]; [Bloomfield, 1976]; [Kitchener, 1976]; [Secord, 1971]). What must be accomplished here is a refutation of the Ellis–Skinner critique of intentional explanation. This is so, because we take here any Culturology position to be refuted, if a philosophical refutation of intentional/mentalistic explanations occurs. Hence, as part of our exposition of a Culturology position, we maintain that there exist at least some rationally justifiable intentional/mentalistic explanations of human action.

1.2 A philosophical strengthening of the Ellis–Skinner position

The Ellis–Skinner position could be greatly strengthened by the addition of two arguments from the philosophy of language and the philosophy of mind: Quine's critique of intensionality and Rorty–Quine–Feyerabend eliminative materialism. Whilst Quine's thesis is of a greater generality than Rorty's it is more convenient to discuss Rorty's position first.

Richard Rorty's [1970] (1)] position of eliminative materialism (REM) may be summarized as follows.[1]

(REM_1) Future neurophysiological research may lead us to believe that what we now call pains are in actual fact certain types of brain processes.

(REM_2) The relationship between pains and brain processes will be then the same as the relationship between a theoretical concept such as "quantity of caloric fluid" and a concept such as "mean kinetic energy of molecules": the former is eliminable without loss of explanatory or predictive power.

(REM_3) A linguistic reform may then occur whereby the observational and explanatory-predictive role performed by sensation terms may be taken over by a neurophysiological language of a specified neurophysiological or set of neurophysiological theories.

(REM_4) At this point, it could be justifiably said by the purely neurophysiological language users, that sensations do not exist, just as we now say from the perspective of theoretical physics, that caloric fluid does not exist.

To avoid the possibility of misinterpretation, it is important to point out that for Rorty weak incorrigibility is "the mark of the mental" [Rorty, (2) 1970]. Consequently it would only follow that there are in fact no entities about which we are incorrigible and hence no mental entities, and no sensations if sensations are mentalistic [Richardson, 1981 p. 476]. To avoid fruitless debates, we take "eliminative materialism" to be an abstract philosophical position, regardless of who has argued for it.

REM must be distinguished from three other forms of eliminative materialism [Cornman, 1971 p. 16]; two doctrines I will mention here, but will not discuss in detail. *Analytical behaviourism* alleges that all sentences in which mentalistic terms occur (essentially) are conceptually reducible to sentences containing no mentalistic terms, only terms referring to either bodily behaviour or dispositions to such behaviour. *Rylean behaviourism* holds with *Analytical behaviourism* that mentalistic terms are conceptually eliminable, the terms in the replacement sentences referring to either observable behaviour, or else function as open or partially open hypothetical statements supporting subjunctive conditionals, but which do not refer to unobservable phenomena. The arguments against both positions are well known: I will not readdress them.

The third form of eliminative materialism – the *strong form of eliminative materialism* (SFEM) – is the thesis that sensation and mentalistic terms refer to items, but these items do not in fact exist. Mental terms are comparable to mythological terms such as "unicorn", "Zeus", "Thor" and so on. (SFEM) is usually defended as a thesis of *postulational parsimony*, that mental terms are unnecessary items postulated by neurophysiology [Lycan and Pappas, 1972, p. 150]. This position is embraced by Quine [1960, pp. 264–266] [1966, pp. 208–214] and Paul Feyerabend [1963]. Mental entities are theoretical entities, whose postulation is justified only if they have, in Quine's words, "some indirect systematic efficacy in the development of theory" [Quine, 1960, p. 264]. There is no need to cluster our ontology with mental phenomena "The bodily states exist anyway; why add the others?" [ibid, p. 264]. Hence postulating mental entities is not justified, and neither is the claim that there are such entities.

REM does not deny that mentalistic terms denote, but maintains that what they denote are brain processes rather than "mental entities", "events" or "processes". But according to (REM_2), mental entities do not exist. Rorty illustrates his thesis with the following example [Rorty, 1970, p. 191]: natives of a primitive tribe may attribute diseases such as pneumonia to demonic visitation, and the evidential grounds upon which this is done are the observational reports of witchdoctors, who claim to see demons after taking what we would call hallucinagenic chemicals. However, there are no demons: disease is caused by all-too-material germs. The natives could abandon demon talk, without loss of both explanatory and predictive power in explaining diseases, provided certain sociopolitical changes occurred. Likewise, future neurophysiology may deny that there are sensations and other mental phenomena. (Rorty believes that sensations are more difficult for a materialist theory of mind to come to grips with; other mental phenomena such as mental images and thoughts can, he asserts, be readily dealt with.)

What relevance does SFEM and REM have to the Ellis–Skinner position? Precisely that they provide a respectible philosophical vocalization to the musings of Ellis and Skinner. The eliminative materialist Paul Churchland [1981], maintains that "folk psychology's" common-sense conception of psychological phenomena "constitutes a radically false theory, a theory so fundamentally defective that the principles and ontology of that theory will eventually be displaced, rather than smoothly reduced, by completed neuroscience" [ibid, p. 67]. Churchland is prepared to admit that the common person can, in giving intentionalistic explanations, predict and explain the behaviour of others. Yet "folk psychology" has been incapable of explaining the nature and function of sleep for example. This, it is assumed (totally without argument), is something which "folk psychology" is addressed to in the first place and raises the real possibility of folk psychology's falsity. It must also be assumed by the eliminative materialist that "folk psychological explanations" and neurophysiological explanations are incompatible, such that if one explanation is true, the other must be false. Clark [1980, pp. 8–21] illustrates this assumption perfectly, in his example of the alleged incompatability between a learning theory account of depression and a hypothesis of catecholamine transmitter substance deficiency. Now it is not part of Culturology's position to ignore neural mechanisms; rather we recognize that the causes of this neurophysiological occurrence must in turn be given, and would claim that any complete explanation here would involve reference to the subject's social environment as well as the subject *qua* agent. Neurophysiology will provide only a part of this explanation. We shall elaborate on this proposal in due course. The point obviously remains that our opponents disagree.

Churchland also criticizes "folk psychology" for its failure to progress. It is taken to be a degenerating research programme for millennia, and is fit now only for scrapping. I shall criticize Churchland's assertions in the course of this paper. In chapter three I will criticize the metaphysics underlying eliminative materialism and thus attempt to rebut part of these charges. In passing, the ideological implications of this metaphysics should be noted. Eliminative materialists claim that the mental entities of common sense do not exist, and indeed ordinary physical objects do not exist either [Guleserian, 1971, p. 282]. In this respect eliminative materialism is a form of "materialistic mysticism": all that exists in the world are the objects of "theoretical physics". It is, it may be thought, impossible to take such a position as having ideological implications, since there are no common sense entities, no ruling classes, bourgeois neuro-technocrats. However, insofar as eliminative materialism is restricted to the biochemical level explanations it provides a perfect legitimation for treating what are for the Culturologist serious socio-political problems as merely reflections of bio-chemical disorders. But "disorders" are in need of "ordering" and if phenomena such as depression, schizophrenia and so on are treated as being exclusively biochemical phenomena, it is human beings which are manipulated and their disorders corrected, rather than the sociopolitical environment examined and its "disorder" and injustices corrected.[2] This is but a sketch of the argument; I refer the reader to the work of Steven Rose for further elaboration and documentation [Rose, 1973, pp. 275–276] Rose and Rose, 1973]. (This is not to imply that the Roses are in agreement with other aspects of my position. Nor is it to imply that the position is cognitively discredited, merely by making this note.)

The second argument which would greatly strengthen the Ellis–Skinner position comes from Quine's critique of intensional concepts; the argument is perfectly captured by J. J. C. Smart's comments in 1978 on ABC radio, that all intensional and mentalistic concepts are suspect (as shown by Quine) and in need of either elimination or reduction to extensionalist concepts. Consequently, sciences which embody these concepts, especially Culturology, are problematic. To see how this argument could be developed, and the utter sceptical disasters it leads us into, let us look at Quine's critique.

Quine [1961, pp. 139–157] pointed out that one of the fundamental principles governing classical identity theory is the principle of indiscernibility of identicals: "given a true statement of identity, one of its two terms may be substituted for the other in any true statement and the result will be true" [ibid, p. 139]. Failure of substitutivity occurs in contexts which are not purely referential, such as quotation contexts, where the truth of a statement depends not merely on reference made to an object, but only the form of the name [ibid, p. 140]. Quotations are, however, only one *referentially opaque* context. Modality, belief and other epistemic contexts offer further examples. However, paradoxes lie in the very heart of these contexts, Quine believes. A representative example is given by considering this model paradox: Let "N(p)" designate the number of major planets in our solar system, then it is true that:

$$(1) \sim \Box (N(p) > 1)$$

But it also seems that the extensional identity "$N(p) = 9$" is true, so by the indiscernibility of identicals:

$$(2) \sim \Box (9 > 1)$$

which seems to be logically false, and so our argument is inconsistent. A similar argument can be constructed for belief contexts: I may believe that Tegucigalpa is in Nicaragua, but be unaware that Tully denounced Catiline, but the extension of the sentences "Tegucigalpa is in Nicaragua" and "Tully denounced Catiline" are both the truth value "true". Substitution results in the false sentence "I believe that Tully denounced Catiline".

Physicalists have in most cases taken the language of physics to be extensional, and claimed that this language is capable of providing a complete description of the world. A sentence is said to be extensional if and only if the truth value of the sentence formed by replacement in an initial sentence of any extensionally equivalent expression remains unaltered. If a sentence is a compound sentence (i.e. constructed from atomic sentences by means of logical connectives and other logical signs) then the truth value of the sentence is a function of the truth values of the atomic sentences. An extensional language is a language such that each of its sentences is extensional [Carnap, 1958, p. 133]. An intentional language by contrast is one in which there is at least one sentence which is not extensional.

We should note that the physicalist must assume that (1) the language of physics is extensionalist, and (2) intentionalist languages are untenable, to produce an argument which could support the Ellis-Skinner thesis. The second assumption could just as easily be taken as an argument for anti-reductionalism: the language of Culturology is intensionalistic; the language of physics is extensional; no plausible reduction of intentional to extensional discourse exists at present [Meyer and Routley, 1977] [Routley, 1980 (1)], therefore no plausible reduction of the language of Culturology to the language of physics can be given. The argument of Quine and Smart could be met by claiming that their conceptual elimination procedures beg the question against the intensionalist. Indeed it seems to me that good reasons have been given for rejecting the claim that physics is itself extensional, or even that mathematics is extensional [Routley, 1980 (1)] [Bunge, 1974, p. 120]. Consequently the thesis will lead to a comprehensive form of scepticism, especially insofar as a relation such as "observes" is intentional, as indeed are other epistemic relations such as "knows" and "justifies" [Routley, 1980 (1)]. Hence the argument is far too strong to be of interest to defenders of the Ellis–Skinner position. Thus we have considered this argument, only to immediately reject it.

The remaining argument which could be made from the perspective of Quine's philosophy of language, might be based upon the *indeterminacy of translation* [Quine, 1960, 1969, 1970]. Very briefly the thesis states not merely that we cannot be sure which of many translation manuals consistent with the observational evidence is correct for any two languages, but that *there is no fact of the matter* about which one can be right or wrong. The position is not concerned merely with situations of "radical translation", of the translation of the language of a hitherto unknown folk, but rather "indeterminancy begins at home", such that we cannot even determine the actual reference of terms in our own language. This has seemed to give rise to a maze of highly counterintuitive results, which Quine takes not to be the *reductio ad absurdum* of the indeterminancy thesis, but of our common sense ideas about beliefs. For example, the question as to whether someone really believes position x or y, is one which is meaningless, for there do not exist any facts of the matter to justify any opted for view [Quine, 1970, pp. 180–181]. Hence we cannot for example, really know what Quine in fact believes about the indeterminancy of translation thesis, and now it seems even what we believe; indeed the question does not make sense. Quine has attempted to avoid this consequence, but has not in my opinion (or that of others) succeeded [Thomason, 1971] [Margolis, 1973 (1)]. To further argue this point will, I hope, not be necessary on the basis of the argument strategy which I previously adopted. If Quine is right, then the problems of the "indeterminancy of translation" and "inscrutability of reference", are much wider than merely discourse dealing with the social sciences; even if physics escaped this problem, discourse in the meta-language, or the meta-meta-language of physics would not. This is not of course to answer Quine's questions, or to perform a *reductio ad absurdum* of the position, but it does show that the defender of the Ellis–Skinner thesis would thus make use of an argument which is far too general to support their claims; their own discourse about Sociobiology will immediately be faced with the above problems.

In conclusion it seems that only the position of eliminative materialism can give additional philosophical support to the Ellis-Skinner thesis. Other issues in philosophy are certainly of relevance to a detailed defence of Culturology, but I have given reason to believe that the ramifications of these questions far exceed the mere question of the plausibility of Culturology.

1.3 Van den Berghe's criticisms of Culturology

In sections 1.1 and 1.2 we considered the claims of Ellis and Skinner and sought to enquire into the contemporary philosophical literature for ways of strengthening these criticisms. Here I will attempt to do the same for Pierre van den Berghe's criticisms [van den Berghe, 1975, 1978].

For van den Berghe, Culturology can only obtain "scientificity" (a term never explicated) if a number of dogmas are abandoned. These include emphasis upon conscious motivation, anti-reductionism, and verbal behaviour; a preoccupation with value freedom and objectivity; a rejection of naive optimism about the human condition, and an abandonment of the self-imposed isolation which Culturology has adopted with respect to anthropology, biology and philosophy.

Now, apart from the question of anti-reductionism, many of these points are points well taken. They do not however establish the rational preferability of Sociobiology over Culturology. For example, if one places far too much emphasis in research upon questions of conscious motivation, social structural variables may be either ignored completely, or not examined sufficiently. Nor is even a thorough-going Culturologist such as myself committed to proposing that "intellectual isolation" for a discipline is a healthy state of affairs; human Sociobiology is untenable (as I will argue), but the Sociobiology debate has at least given Culturologists the motivation to rethink and justify assumptions which before they only tacitly accepted: indeed, to escape the "crisis of Sociology" arguments (to be detailed), it is necessary for social theory to have a much more intimate relationship with metaphysics, than it has in the past. No reasonable Culturological position should accept "intellectual isolationism".

These criticisms therefore do not substantially challenge Culturology's legitimacy at all. The only objection which seems to come close to this is the claim that Culturology is a vague scholastic tradition, which might suggest that Culturology is in fact a "pseudo-science". Andreski [1972] addressed the problem of the frequent obscurity, lack of conceptual clarity, and the uncanny tendency of "Sociologists" to make the simplest truths mysterious, alien and obscure. For him terminology is used as a "smoke screen" to hide often shallow and tautological explanations [ibid, p. 68], the work of second rate minds [ibid, pp. 85–86]. However, these criticisms are most plausibly taken as stern moral warnings to do better work rather than constituting a foundational critique of "Sociology". Nor does the claim made by van den Berghe [1978, p. 35], that "Sociologists" are involved in near-scholastic ancestor worship of the " "Saints" Marx, Durkheim, Weber, and Pareto" [ibid, p. 35] carry destructive import. It is true that biologists do not frequently work solely on *The Origin of the Species* neglecting contemporary sources; however, neither do "Sociologists", be they Marxists, Durkheimians, Weberians or Paretonians. It might *just* be the case that the theses advanced by the early modern social theorists are still illuminating, and worth considering.

In conclusion, van den Berghe's arguments and criticisms of Culturology may be welcomed; they do not in themselves constitute an argument for the untenability of Culturology. We must however attempt to state an authors case as strongly and as fairly as possible, even if this involves a rational reconstruction on our own part. Van den Berghe certainly can find such *prima facie* foundational critiques of Culturology within the sociological tradition itself: this is to be found in the "poverty of sociology" and "crisis of sociology" traditions. So let us strengthen their arguments still further.

1.4. The poverty of sociology and crisis of sociology traditions

The terms "poverty of Sociology tradition" and "crisis of Sociology tradition" are Sztompka's [1979]. The first tradition refers to the theoretical approach in which a pessimistic assessment of the state of the achievements of "Sociology" is made. This approach may vary greatly in both the degree of pessimism and the specific criticisms made of the discipline, but in *general*, the criticisms are directed against the present problem-solving capacities of the discipline, rather than against the conceptual, epistemological, ontological and politico-axiological presuppositions of the discipline. The "poverty of Sociology tradition" calls for a radical "spring-cleaning" of the discipline, abandoning certain approaches, and reorientating research to study some problems rather than others; the "crisis of Sociology tradition" calls into question the theoretical cogency of either certain parts, or indeed the total discipline itself. The former tradition is illustrated by works such as [Sorokin, 1956] and [Andreski, 1972]; the latter tradition by works such as [Gouldner, 1971], [Friedrichs, 1970, 1972], [Eisenstadt and Curelaru, 1976], [Bottomore (ed), 1975] and [Hindess, 1977].

In making this classification and citing works which I take to illustrate these two positions, it is not intended that any one author can be rigidly classified into one or other category. It also is difficult to draw a rigid line of demarcation between the two orientations. Consequently the positions should be regarded as "fussy sets" of criticisms, illustrating general critical tendencies.

If we view scientific research as beginning with problems, making use of methods to obtain results which may then be practically applied, the "poverty of Sociology tradition" may be viewed as advancing criticisms at all four levels of research. This typology is undoubtedly a very simple model, and makes the challengeable assumption that four stages of research can be distinguished in this fashion. Yet Sztompka's model is not so utterly distorted as to be of no use for a classification of these criticisms. It seems that Sociologists do in fact choose problems for research and make use of methods which produce results which are then practically applied; this in fact is the *bare minimum* of the research process. We would imagine that the problem-choosing stage is itself influenced by practical and cultural forces and by already accepted metascientific views. But insofar as at *least* four stages exist in this process, the following classification can serve to illustrate them. We offer a summary of the "poverty of Sociology" critique (see Table 2.1.).

Arguments for the "poverty of Sociology" could be constructed as a simple *modus tollens* argument, with the major premise consisting of a hypothetical statement, the antecedent of which makes an axiological or meta-scientific claim (e.g. "*if* "Sociology" was a methodologically adequate, theoretically satisfactory, or practically relevant science") and the consequent describing a non-problematic situation ("*then* problems would be chosen by proper criteria"). The minor premise denies the consequent statement, and a statement follows as a conclusion which is the negation of the antecedent. To rebut the argument requires (since the argument is formally valid) rejecting either the major or minor premise, or both. One could, for example, take the claim that "Sociology"

Table 2.1. Summary of the poverty of sociology critique (material derived from Sztompka [1979, pp. 6–14])

Research Stage 1: The problems chosen for study are improper

1. (a) "Sociology" does not in general concern itself with the crucial problems facing humanity. It does not meet the challenge of our times. Too much research is devoted to trivial problems the results being of interest only to a few academic researchers.
   (b) Many problems chosen for study are morally improper, involving ethically objectionable research, and a large quantity of research is "ideologically laden", a mere rationalization of the inevitability and naturalness of the status quo.
2. If fundamentally important problems are chosen, they are frequently selected for investigation by means of improper criteria:
   (a) the simpler problems are chosen rather than more complex and perhaps more important problems;
   (b) problems are chosen because they can be either tackled by fashionable techniques, or can be investigated with less expense than other, perhaps more important problems;
   (c) certain crucial problems may be avoided because of political and ideological ramifications; they may be of crucial importance to humankind, but their solution may involve a significant political challenge to the society of the researcher.

Research Stage 2: The inadequacy of sociological methodology

3. Sociological Methods are improperly employed by social researchers, consequently many studies are riddled with methodological flaws.
4. Methods are selected by means of improper criteria:
   (a) they are in many cases inadequate to investigate the problems to which inquiry is directed;
   (b) they are in many cases, inadequate to achieve goals of research such as unambiguous explanatory results.
5. Such methods are developed in a disproportionate fashion, with relatively well developed statistical data processing methods, but with a great underdevelopment of methods of theory construction and systematization.

Research Stage 3: The inadequacy of the research results of "Sociology"

6. The language used for the interpretation of research is imprecise, frequently obscure and meaningless and sometimes contradictory. This is due to the fact that:
   (a) "Sociology" lacks a rigorously defined technical vocabulary distinct from commonsense;
   (b) there is no general consensus on the meaning of basic sociological terms such as "society" "culture" and "institution". Thus "Sociology" is in a situation aptly described by Wittgenstein's dictum about psychology: "Sociology" has experimental methods and conceptual confusion.

Table 2.1. (continued)

Research Stage 3: The inadequacy of the research results of "Sociology"

7. The propositions of "Sociology" are unsatisfactory for both logical and empirical reasons:
   - (a) they are tautologies, or:
   - (b) fail to refer, or;
   - (c) are untestable, or;
   - (d) are not conducive to the formation of scientific laws; they are merely empirical generalizations not projectionable outside of the observed cases, or;
   - (e) such propositions are not confirmed as being true: either:
     - (1) they are false, or;
     - (2) they are not sufficiently tested.
8. Sociological propositions are insufficiently systematized;
   - (a) there is lack of systematic links established *within* homogeneous classes, such as the class of empirical findings, the class of theoretical propositions, as either:
     - (1) an extreme empiricism is adopted where "the facts" allegedly "speak for themselves", or;
     - (2) social theories are not, to a satisfactory degree, subjected to testing and falsification or confirmation by empirical research.

Research Stage 4: The inadequacy and non-utility of "Sociology's" research results

9. "Sociology" is unable to provide adequate explanations of social phenomena, contending itself primarily with descriptions.
10. "Sociology" is unable to provide successful prediction and control of social phenomena.
11. "Sociology" is unable to provide satisfactory directives for practical action.

lacks a rigorously defined technical vocabulary distinct from common sense, and meet this objection either by (1) denying this claim and illustrating what this technical vocabulary actually is, or by (2) denying that the claim in question does in fact constitute a criticism of "Sociology".

An alternative strategy would be to actually concede the criticisms, and attempt to produce a better "Sociology". The earlier considered criticism of Ellis, where it was alleged that "Sociology" failed to consider and solve major social problems, can also be met from within the sociological tradition. The answer to this criticism is implicit in that of criticism 2(c): frequently the solution to some social problem would seem to necessitate major social structural changes. Ruling classes are most unlikely to surrender their positions of power at the whim of sociologists and philosophers. Many criticisms of the alleged failure in the problem-solving capacities of "Sociology" can be met by pointing out practical difficulties in implementation. So one could in fact still concede to many criticisms made by the poverty of "Sociology" tradition without producing an argument of relevance to the Ellis–Skinner position.

Many criticisms would remain after the implementation of these strategies, questions 7 and 8 in Table 2.1 above for example. These criticisms suggest that there may in fact be something fundamentally wrong with the current state of Sociological theory. This in short is the criticism of the "crisis of Sociology tradition".

1.5 The foundational challenge of the crisis of Sociology tradition

The "crisis of Sociology tradition" is concerned with meta-scientific problems about sociological theory itself, and metaphysical and epistemological problems about the objects of social theory. The basic argument can be reduced down to two problems: (1) the problem of "theoretical pluralism" and the lack of "cognitive systematization" of the discipline as a whole; (2) specific problems in rationally justifying one orientation over another. These problems are essentially a sociological equivalent of the current debates about theory choice in the philosophy of the natural sciences. Yet all of these problems have been debated quite intensely before in the history of philosophy. To see this, some very brief rational reconstruction of Western meta-philosophy will be given. We are advancing, in what follows, a historico-conjectural articulation of a metaphysical ideal.

The metaphysical idea of *Cognitive Systematization* unites thinkers as diverse and as close as Parmenides, Plato, Aristotle, the Scholastics, Spinoza, Leibniz, Hegel, Bradley, Russel, the Logical Positivists and others. It is, in Nicholas Rescher's words "one of the great formative ideas of Western civilization" [Rescher, 1979 (1), p. 26]. Scientific knowledge itself, following Kant, is taken to be distinguishable from common sense knowledge, by its systematic unity between its knowledge claims, theories and propositions. The propositions and theories within a science must exhibit parameters of systematization such as wholeness, unity and internal integrity and be integrated into a structure which is internally consistent, coherent, comprehensive, orderly in structure and exhibiting functional simplicity, explanatory power and problem-solving capacity. It should be capable of application to phenomena not yet encountered, that is, it should have a trans-contextual aspect, and thus obtain levels of universality not possessed by commonsense knowledge.

Traditionally in the West, the very point of Philosophical and Scientific Inquiry was to discover *The Truth*. Reflect upon your existence – you find yourself in a world of things with other people. What are these particulars? Why, or how did they come to be, and where are they going? Is there a *Telos* behind this passage? You are also a member of a society, having beliefs and moral practices. Are these beliefs reasonable and are its moral practices justified, for our neighbour over the mountain does different? What then is this world, and why this world rather than some other or none at all; what is our purpose if any, and what shall we, and ought we to do in this world? As I view Western Philosophy, the attempt to systematically answer these questions by reason and evidence has been its *modus operandi*. Such a preoccupation with *The Truth* it can be argued, also underlies at least some Eastern philosophy. In Indian philosophy, for example, the fundamental epistemological question is whether illumination, as the result of Yogic-experience is possible without distortion arising from the mental addition to the act of contemplation. False awareness arises from the use of categories such as space, time, causality and substance. The sceptical traditions in Indian philosophy, such as *Mȧdhyarnika Buddhism* and *Advaita Vedȧntins* have maintained that knowledge must ultimately involve construction and hence false awareness. These latter "sceptical orientations" did accept that some knowledge of Being was possible, but only through an "Objectless Consciousness" which entails incommunicable experiences [Potter, 1975].

In Zen Buddhism, the quest for *The Truth* is not abandoned, but merely the possibility of reaching this through rational theorizing and linguistic categories [Capra, 1977, pp. 107–113]. This assumption is of course challenged by Western Rationalism, which takes Reality itself to a rationally ordered system; a coherent and integrated structure of parts, exhibiting lawfulness and uniformity of operation. The Truth must therefore be a System, and if Knowledge must reflect Truth, Knowledge must therefore be a System as well [Rescher, 1979, p. 19]. Western scepticism could then be viewed as a critique of the systematicity of *The Truth*. Hume's problem of induction, for example, seems to challenge the ontological parameter of systematization, that we are justified in believing that the world exhibits lawfulness, orderliness of operation and uniformity. Western sceptics also challenged the possibility of obtaining Knowledge by Reason, through challenging the systematicity of knowledge claims themselves. The very first of the *tropoi* of the Greek sceptics was based upon the disagreement of the philosophers. There is not a single thing which is not disputed by the philosophers, the most trivial of questions does not escape controversy. The sceptics challenge here the cognitive status of philosophical inquiry: if *The Truth* is a system and Knowledge must reflect *The Truth*, then Knowledge must be a system. But the disagreement of the philosophers precludes this. Therefore philosophy does not constitute a system and does not provide us with knowledge, and likewise for any subject, cognitively dependent upon philosophical speculation.

The first problem of the "crisis of Sociology" position is essentially a problem of the disagreement of the theoreticians. There are a large number of schools or orientations within "Sociology" which demonstrably make use of theoretical and philosophical assumptions which are mutually inconsistent; thus the discipline itself is inconsistent. If meta-theoretical relativism was true, then we could accept that the Ideal of one Orientation "accuracy corresponding to social reality" was a myth.[3] Yet it seems that the "schools", "orientations", "paradigms" or "research traditions" are involved in conflict and much time and energy is spent in smashing the foundations of opposing orientations. Marvin Harris' [1968, 1979] extensive critical adventures which has resulted in a quite negative assessment of contemporary anthropological theory (except, of course, his own research orientation of cultural materialism) are justified by him, by pointing out that following Laudan [1977, p. 120] the success of a research tradition depends upon how effective and progressive that tradition is relative to competitors. Whether Laudan is right in this allegation is open to debate,[4] but it is clear that many other social theorists believe this and are thus engaged in such warfare. Consequently "Sociology" suffers from theoretical plurality and internal conflict.

These differences have in part been over questions of first philosophy – which brings us to the second problem of the "crisis of Sociology" argument. The discipline is itself plagued by many philosophical problems which challenge the cogency of social theories themselves.[5] Apart from Sztompka's [1979] excellent work, the PHd thesis of Petryszak [1978] (c.f. also [Petryszak (1) 1980, (2) 1980]) is an extensive critique of both the contemporary élite theorists and radical social theorists, as well as classical and early modern theorists, in order to show that all these traditions accept empirically unjustified assumptions about innate human nature. Petryszak's own prejudices are towards a bio-social theory, and he presents a *prima facie* strong case to support this: if Culturologist theories are themselves committed to such assumptions, then the case against bio-social theories made by such a tradition is undermined. So with respect to the plausability of a wide range of philosophical and scientific theses, social theorists have made rarely acknowledged and more rarely defended assumptions about human nature. Yet the soundness of their arguments depends upon the truth of their premises, and if we do not know the latter, we do not know the former, and their case against bio-social theories collapses.

For Petryszak, what is needed is an empirical verification of these assumptions. However, he nowhere tells us how this can be done in a non-question begging fashion. Allegedly, ethology, behavioural genetics and Sociobiology will bring forth the answers, but I hope to establish in this work that matters are not so simple, especially since not all is well in these disciplines. What is at stake are sets of philosophical theses. Now, "Sociology" is in crisis because philosophy is in crisis, and crisis, it may be argued, is the natural state of the discipline. The reason for this, according to Rescher [1978, p. 220] is that "*philosophical issues are always such that arguments of substantial prima facie cogency can be built up for a cluster of mutually incompatible theses*". But if this is so, one wonders how *any* justification of one's adopted model of human nature can be justified, and shown to be more satisfactory than competing accounts. This is especially a problem for Culturologists and Sociobiologists alike, who have presented conflicting views of human nature and who both believe that their account presents *The Truth*. Our problem is thus a more generalized form of the problem of the rationality of theory choice debated currently in the philosophy of science, and it is a gross understatement to say that this problem is merely "difficult". But some response must now be given.

2.0 Systematizing the critique

Three major problems are isolated from the discussion in Section 1; the Ellis–Skinner critique, eliminative materialism, and the crisis of "Sociology" argument. In chapter three, the challenge of eliminative materialism will be responded to; here, however, the Ellis–Skinner argument and the crisis of Sociology argument need to be met. The most difficult, and certainly the most important argument to meet is the latter, since the central thesis of this work is that a Culturologist view of human nature is more reasonable than a Sociobiological account. In this section I will clarify the issues concerning the crisis of Sociology argument, and in the following section 3.0 respond to the critique. Following that, in section 4.0 I respond to the Ellis–Skinner critique.

There are a number of explanatory models of the diversity and pluralism within philosophy and "Sociology". These are summarized in Table 2.2.

The position which I support is *Monistic-Systemic Perspectivism*: the supporting argument proceeds by elimination. First, both forms of *Metaphilosophical/Sociological Monism* are unsatisfactory. 4(b) is a denial of the rationality of choice of such systems and merely makes S_1 a dogma. On this ground *any* system, no matter how absurd, would be acceptable. 4(a) is at best unlikely. We have not yet been presented with a philosophical (or sociological) theory which approaches completeness. On the contrary, virtually every position seems to have at least some outstanding problems. *Monistic-Systemic Perspectivism* is however as good, if not better than *Metaphilosophical/Sociological Monism*. If in fact S_1 is the best system, then from the position of *Monistic-Systemic Perspectivism* one would also find S_1, this meaning that $S_2, S_3, S_4, \ldots, S_n$, have in fact no merits worth synthesizing. The position then seems a much better bet than its monistic alternatives, as it has their merits and avoids their obvious defects.

Metaphilosophical/Sociological Scepticism represents the null hypothesis of enquiry in this domain. It is only acceptable if all of the alternative positions are untenable, for indeed this requirement must be met for the position to be justified. Note, this position might seem capable of being refuted by a self-referential argument: this position is itself a philosophical position and asserts its own untenability. This is in fact the case if this position is in fact another philosophical position. However the utter sceptic here is attempting to point beyond that which can be reasoned about, whilst then throwing the ladder

Table 2.2 Explanatory Models of Philosophical and Sociological Pluralism. (Highly Derivative from Rescher [1978])

---

1. *Metaphilosophical/Sociological Scepticism*: The diversity of opinions with respect to philosophical and socio-theoretical issues illustrates the bankruptcy of the philosophical and socio-theoretical issues under consideration, thus demonstrating their theoretical incoherence. This position with respect to philosophy has been adopted by Academic scepticism (by contrast to Pyrrhonic Scepticism, which suspends judgement upon cognitive matters); within social theory, Hindness [1977] comes close to being a socio-theoretical sceptic.
2. *Metaphilosophical/Sociological Relativism*: There are no criteria of assessment of world views independent of those internal and peculiar to the various considered philosophical and sociological theories. Hence either:
   2(a) This situation is cognitively intolerable, therefore metaphilosophical sociological scepticism is the only "rational" option. or;
   2(b) This illustrates something fundamental about the nature of Reality and Human Society; either
      (1) There is no such Reality (Ontological Nihilism); or
      (2) There is a Reality, but it consists of a multitude of "many worlds"; or
      (3) There is a Reality, but it is either unknowable, or not apprehendable by any rational means but only through art, drugs and sex, yogic consciousness, or divine relevation.
3. *Metaphilosophical/Sociological Anarchism*: The Western Rationlist ideal of *The Truth* is an utterly mistaken ideal, and the idea of a crisis of philosophy and "Sociology" is based upon a misconceived notion. We should be free to choose theoretical, metaphysical and logical principles and theories to suit our political and axiological orientations. The question of conflict of world views begs important political questions: in a free society such a pluralism would be encouraged [Feyerabend, 1975].
4. *Metaphilosophical/Sociological Monism*: There is only one correct system of philosophy or social theory and
   4(a) This system S_1, exists at present, and all competitors $S_2, S_3, \ldots, S_n$, are inadequate, this being rationally supportable; or
   4(b) S_1 is accepted fideistically as being the correct philosophical or socio-theoretic system.
5. *Metaphilosophical/Sociological Perspectivism*
   5(a) *Oriental Perspectivism*: One and only one position is right from a given orientation, but there exist a plurality of perspectives, none of which is uniquely correct. The position accepts metaphilosophical/sociological Relativism only between systems: from the perspective of some probative perspective orientation, further reasoning is possible. In Rescher [1978] these perspectives are value orientations, arising from cultural and personal intellectual experience. One can reason about these evaluative attitudes, but to do so is to already adopt an orientation, from which ones orientation is retrovalidated. Therefore the existence of pluralism in philosophy (and social theory) is inevitable. Therefore with regards to any philosophical issues there will also exist many equally reasonable incompatible solutions.

Table 2.2 (continued)

5(b) *Monistic-Systemic Perspectivism*: Whilst most positions grasp at least some of the truth about Reality, virtually all positions *at present* suffer defects. It is possible in principle to produce an ideal cognitive system which possesses all the merits of each system (consistent with various systematization constrains such as consistency) and none of the defects. It is at least possible to approach this ideal system by constructing a series of less problematic systems.

away. The position is consistent with 2(b)(3). Under pain of contradiction, this position cannot be argued for, but only experienced in abandoning the argumentative process itself. The possibility remains open that if mystical insights into the nature of Reality do occur, then they may not be in conflict with Reason. This of course, is a question which cannot be decided within the language game of philosophy.

Metaphilosophical/Sociological Anarchism presents no challenge to *Monistic-Systemic Perspectivism*, which Scepticism and Relativism do not already present. If the Anarchist attempts to rationally support his/her views by arguments, then Anarchism is just another philosophical position, and if Anarchism itself is adopted *just as a matter of aesthetic taste*, Rationalists are permitted by the Anarchist to reject the position if they so wish, without the burden of rational argument. However, Anarchists are surely noble to defend various "forms of life" against Physicalist Imperialists and other forms of intellectual totalitarianism [Feyerabend, 1975]. But dogmatism and intellectual totalitarianism should not be a part of any *consistent* Rationalist's *Weltanschauung* from the onset, and is not by definition, part of *Monistic-Systemic Perspectivism*, which recognizes the merits and defects (if any) of all positions and attempts an intellectual reconstruction from this.

Both Scepticism and Relativism are frequently supported by the *Criterion Argument*. Let C be a criterion used as a decision procedure for the determination of the truth (or even reasonableness) of propositions (or positions). Then a proposition p is true if and only if C(p), that is if p meets those specifications of truthhood of C; let these specifications be R. Then to establish the truth of p, i.e. T(p), we must produce a sound argument $C \& R(p) \rightarrow T(p)$. This argument if formally valid must also be sound. It is, questions of relevance aside, at least necessary that the premises be true, i.e. $T(C) \& T(R(p))$. The employment of C must be validated. An argument to the effect that $C \& R(C) \rightarrow T(C)$ is unacceptable as it does not give us reason to believe that C is true. But if we present now an argument $C^* \& R^*(C) \rightarrow T(C)$, either circularity or vicious infinite regression will arise; circularity insofar as self-application is made to establish $T(C^*)$, and a vicious infinite regress, insofar as to support any C_n of the infinite series $C_1, C_2, C_3, \ldots$ appeal must *first* be made to the next member of the series C_{n+1} [Rescher, 1973 (1), 1977].

This ancient argument does not in fact show that Scepticism is true, but that there can be no development of knowledge without knowledge to begin with; the knowledge process, like the process of communication, involves one in a "hermeneutic circle" [Smith 1982 (1)]. Consequently we must, if knowledge is to be possible at all, begin with paradigm cases of knowledge as well as criteria, short circuiting the criterion argument [Chisholm, 1973, pp. 12–15]. It is not possible or necessary here to produce a detailed epistemology itself. I shall attempt to say more about this in Section 3.0; here however, to rebut Scepticism and Relativism, it is sufficient to point out that for philosophical arguments to occur at all, at least some principles of logic and some facts (i.e. true propositions) must be accepted, if only provisionally, on fallible evidence. A framework of problems needs to be accepted, or the process of debates will not make sense; disagreements on absolutely every issue would destroy the very possibility of discourse including discourse about scepticism itself. Yet debate does occur and positions are rejected as being unsatisfactory; this is part of the practice of reason. This gives us some reason to believe that *Metaphilosophical/Sociological Scepticism* and *Relativism* are wrong, and at least shifts the burden of argument back upon the sceptic.[6] This is not claimed to constitute conclusive reasons for rejecting these positions. However, this will not affect the point in the debate which I wish to make, since what is denied by our opponents is the *very possibility* of giving *any reasons* in favour of any *Monism* over competing positions at all. This claim certainly has not been established.

Rescher [1978] would challenge such a claim. He claims that in matters philosophical one can present *prima facie* plausible arguments both for a thesis T and against the thesis, that is for the antithesis $\sim$ T. This does in fact seem to be true. Yet this is no challenge to *Monistic-Systemic Perspectivism* since this claim is consistent with the claim of the *de facto* correctness of one position or perhaps neither. What the *Monist* must deny is that for all theses, one can produce *de facto* sound arguments both for these theses and for the respective antitheses. If Rescher has such Protagorean commitments, it would seem that he leads us to a situation of intellectual paralysis: we could advance *de facto* sound arguments against this position as well as to any cognitive claim that could be made about the position. On the other hand, the Orientational Perspectivist's confession that there will exist equally reasonable, but incompatible orientations, seems to presuppose some transcendental standard of "reasonability" and "incompatibility" from which this meta-statement itself is made – and hence is an inconsistent claim. If the claim is made from one orientation, then the full generality of Rescher's thesis is immediately lost, since his claims are false from my perspective. All of this is utterly paradoxical and offers the best of reasons for rejection, especially insofar as no positive supporting argument has been given for this proposal by Rescher from the onset.

There seems then to be good reason to reject all competing positions to

Monistic-Systemic Perspectivism, including the sceptical position which might be taken as a null hypothesis. Yet it may well be that this position suffers internal problems of its own which would support an advocation of a suspension of judgement on all these issues. What I will attempt now is a further defence of this position, and a further clarification of how these issues relate to the problem of "the crisis of Sociology".

3.0 *Monistic-Systemic Perspectivism* and the "crisis of Sociology"

The ideal expressed by the Logical Positivists of the unification of Science had at least the merit of running counter to the "intellectual atomism" and narrow specialization which unfortunately characterizes much of contemporary intellectual life [Gal-or, 1981]. If the "crisis of Sociology" is to have any chance of resolution, then "Sociology" needs to abandon its ideal of being a self-contained domain of inquiry as indeed the Sociobiologists have pointed out (but for the wrong reasons). The "crisis of Sociology" argument demonstrates that social theories are not "metaphysically free" and their rationality is far from self-evident. We should therefore explicitly recognize this and attempt to justify our basic theoretical presuppositions. But how is this to be done? Do we for example, attempt to devise experiments and present empirical evidence, which will enable the choice between "models of man"? Such a strategy would seem from the perspective of post-empiricist philosophy of science to be unacceptable: (1) it is circular insofar as no "theory neutral" or "metaphysically neutral" observation language exists, so some perspective seems to be presupposed already, if only perhaps empiricism; (2) the Duhem-Quine and the underdetermination of theory by evidence theses, are taken to directly challenge this claim.

The issues involved here are indeed quite complex, and are matters of intense current debate. Yet the pursuit of my thesis requires now an act of bravery: to plunge head first into such cold ferocious currents of debate; but what choice does the seaker of truth have here, for the objections cannot be ignored? Such a situation does support my metaphilosophical position which views philosophical problems as standing in an intimate relationship with each other, such that their solutions are in general in clusters rather than merely one by one, or as a structureless whole. Criticisms (1) and (2) are based upon specific views about the nature of scientific theories and evidence, and the cogency of these criticisms will depend upon the cogency of these general theses. It is important to address these questions even in part since the critique of Sociobiology made here is advanced from the perspective of my own philosophy-of-science prejudices, which stand quite open to critical challenge (as undeveloped as they are).

In the next chapter, where questions of theoretical reduction are considered, the question as to what theories actually are will be addressed, but not satisfactorily resolved at all. My position (which must remain here provisional and only partially explicated) follows that of Bunge [1967 (1), 1973] in rejecting the Logical Positivist's separation of the non-logical terms and statements of a scientific theory into either "theoretical terms" on "observational terms": there are no observational terms *in this sense* within scientific theories at all. It does not follow from this, as it has been thought, that it does not make sense to appraise competing theories with respect to empirical evidence. Such a position would seem to confuse the question of the *reference* of a theory (this being the set of the referents of all referring terms) and the *evidence* for believing that the theory is true. Whether ϕ is evidence for theory T will in an intuitive sense be determined by T, just as T (often implicitly) determines what is not evidence for it. Whether in fact there is evidence ϕ or $\sim \phi$ is an empirical matter, and unless some form of theoretical idealism was true (where what theories one holds determines the ontology of the World[7]), it is a question which must be decided independently of T. If any testing experiments of T presupposed the truth of T, then vicious circularity would occur (more will be said on this point in our discussion of structuralism in the next chapter). But there are no good reasons for believing that this is the case. Even so called "incommensurable theories" (if taken on near-pains of contradiction to be "competing") can still be evidentially compared (although the results of such a comparison may well be inconclusive). Once more using metaphors, we have to look at the world as the respective theories tell us to do, and see whether "experience reads" as the respective theories tell us it should. To present an actual theory of evidence is a needed task, but one reserved for another day![8] What conclusion can we draw from this briefly presented and quite tentative proposal?

I believe then that at least in principle one can produce some empirical evidence for at least some "models of man". But the challenging problem remains as to whether empirical considerations are *sufficient* to enable reasoned theory choice. In a situation where one has competing theories $T_1, T_2, \ldots, T_n$ if it was the case that all available evidence (the set of such evidence being non-null) supported T_k, but not $T_1, T_2, \ldots, T_{k-1}, T_{k+1}, \ldots, T_n$ we could then at that time accept T_k as being the most reasonable theory, provided a set of further logical and philosophical constraints yet to be specified are met. This is an ideal situation virtually never actually achievable, if the underdetermination and Duhem–Quine theses are correct. With our available space, let us set out a response to these further difficulties.

The underdetermination of theory by data thesis is the claim that two or more theories $T_1, T_2, \ldots, T_n$ may be logically inconsistent with each other, yet consistent with all logically possible observational evidence. Theories can be logically incompatible, yet empirically equivalent [Quine, 1970, p. 179, 1975, p. 313]. Such a thesis challenges not only a "realist epistemology" [Newton-Smith, 1978], but is especially acute in the social sciences [Lukes, 1978]. But implicit in Quine's formulation of the problem are quite problematic empiricist assumptions as Lukes cogently argues [ibid, pp. 95–96]. There is one additional assumption neglected by Lukes, this being Quine's explicit equation of "scientific observational evidence" with that evidence obtained from direct sensory observation. Now this was not taken even by arch logical empiricists as providing an adequate explication of the scientific account of observability used by the working scientist, which roughly takes observables to be any quantitative magnitudes ascertainable in an experimental situation [Carnap, 1966, pp. 225–226], which indicates that Quine is implicitly accepting an empiricist account of observability rather than a scientific realist account. From the perspective of scientific realism, the underdetermination thesis can be sharply criticized, and the claim that *all* scientific theories are underdetermined by *all logically possible* evidence, demonstratable as being without cogent defence [Boyd, 1973]. However, it may well be the case that for *some* metaphysical theories, underdetermination arises once more. Examples of this are common sense realism and sceptical hypotheses which postulate that all my experiences are caused by a Cartesian demon [Slote, 1970] and the case of the theory T_1 that time is linear and history cyclic, and T_2 that time is closed [Newton-Smith, 1978, p. 79]. These are but two examples of metaphysical theories and one can produce other examples at call. How then is rational theory choice possible?

To compound our difficulties, the Duhem–Quine thesis [Harding (ed), 1976] supports the underdetermination thesis (but it is not necessarily refuted if the latter thesis is refuted). Stated very briefly, the position asserts that it is not single scientific statements which face empirical falsification, but whole scientific theories, and for Quine, the "laws of logic" and metaphysical systems themselves. The unit of empirical significance is nothing short than the whole of scientific theories.

This seems at first glance to conflict with the claim of underdetermination, since one would expect that the underdetermination thesis applies also to disconfirming evidence as well [Boyd, 1973. p. 4]. Yet this claim is consistent with Quine's rejection of the analytic/synthetic distinction. Not only is no statement immune to revision, but any statement may be held true if radical enough adjustments are made elsewhere in the system. The true/false distinction would then be applicable only to a theory as a whole. Let us agree to shelve the underdetermination problem for a while, whilst we try to give some response to these difficulties.

Quine's holism not only suffers from paradoxes [Chipman, 1974] [Quinton, 1973, pp. 216–217] but it is difficult to find anything approaching an argument for the position.[9] However, there is an undeniable difficulty facing any philosophy of science in rationally accounting for the actions one takes when theoretical predictions fail to be consistent with experimental results within a predecided error range. So a theory T and a set of auxiliary hypotheses A imply E ($T \,\&\, A \rightarrow E$), but it seems that $\sim E$. Hence either (1) $\sim (T \,\&\, A \rightarrow E)$; (2) False (T); (3) False (A); (4) False (T) & False (A) or (5) False ($\sim e$). What does one do?

"Strawmen-Popperians"[10] would wish to reject any attempts to save theories from falsification by the use of *ad hoc* measures. "Strawmen-Lakatosians" identify a hard core which is irrefutable by methodological decision, protected by a set of auxiliary hypotheses and statements of initial conditions, which are in all cases first to fall under the axe of experience. This presents us with a dilemma of either rejecting theories too soon, or making theories into unfalsifiable dogmas. My position is that either of (1) and (5) may be true, but one would have to decide this in the particular case. Nevertheless one should consider first alternative (1) or (5) (the order of which is not of importance)[11] before modifying the theory or auxiliary hypotheses. Hence it may be necessary as a matter of logic to modify one's theory [Musgrave, 1973], but it is not a matter of logic as to which alternative is a true description of the problem. It may well be that strategy (5) is correct because I was hallucinating. But to suggest as Quine has done that this can always be done, is to restate the argument of the epistemological sceptic, a thesis which can be independently attacked [Griffin and Harton, 1981].

The remaining problem confronting us which we left somewhat behind in this maze of difficulties, is the underdetermination of theory by evidence thesis. If a rational theory choice is made with heavy appeals to factual considerations [Routley, 1980 (2)] in situations of underdetermination, which would seem likely for many metaphysical theories, we would be faced with the prospects of being incapable in principle of making a reasoned choice between conflicting meta-physical theories. Perhaps one could appeal then to various pragmatic constraints such as "beauty", "simplicity", "economy", "intelligibility" "fruitfulness" and so on. However, all these concepts face semantical problems of definition and methodological problems of metricalization and scaling. Routley [1980 (2)][12] gives no actual example of optimization reasoning in his choice of non-classical ultralogic over classic logic.

My position is that if no place is made for a prior philosophy distinct from science operative in situations of underdetermination, we seem faced with relativism. If epistemology and metaphysics are to be replaced by physics, "Sociology" or hermeneutics, questions of the rationality of these disciplines must be regarded as meaningless and intellectual disputes within the discipline itself be regarded as incapable of rational resolution [Briskman, 1977]. In short, naturalized epistemologies are merely thinly disguised forms of relativism, and if philosophy is merely an underlabourer to the sciences, there is no way in which the questions of the relativist can be answered. But this is itself a philosophical position and so does not in any case free one ultimately from dreaded metaphysical and epistemological questions.

But what is the philosophical reasoning to which I refer? What methodology does philosophy follow? Answering these extremely difficult questions can only be done by articulating and defending my own metaphilosophy. It is extraordinarily difficult to say what precisely metaphysics is without appealing to concepts even vaguer [Schlesinger, 1981] and the difficulties are compounded for "philosophy" itself. The problem is perhaps one of being faced with a hermeneutically virtuous circle rather than a logically vicious one, so that our quest is to get properly into the circle rather than out. So let us take an example and address Newton-Smith's [1978] challenge to the realist, taking his own example of how to choose between a linear model of time with a cyclic history and a closed view of time, for this might give us an intuitive grasp on this difficult concept. One response which has been made to the position of closed time is that the claim is incoherent: only if there is a linear time does it make sense to speak of cyclic time [Whitrow, 1972, pp. 119–120] [Zwart, 1976, pp. 244–246]. If such an argument was sound we have immediately adequate reasons for rejecting the cyclic view of time: it is an outrightly incoherent conjecture. Such sorts of arguments illustrate philosophical reasoning at its best. But what precisely is the mode of argument?

This example illustrates but one form of reasoning which is employed constantly in philosophy but rarely in physics, an argument from self-referential inconsistency.[13] [Boyle (et al.), 1976]. I cannot and hopefully need not, outline explicitly the main forms of philosophical arguments, although this is in general a needed task at present.[14] The point which I wish to make is that philosophy seemingly can stand as a weapon of critique of various "forms of life", and indeed of itself. Considerations of consistency, coherence, the clarity of meaning of presupposed terms, the logical adequacy of advanced arguments are common considerations of philosophical practice, and as far as I can see, can also be advanced against the proposals of scientific positions from logic to physics, to biology and "Sociology". Having arrived at this point, its great tentativeness and utter incompleteness I admit fully. I will return now to the "crisis of Sociology" argument and outline a meta-methodology for its resolution, recognizing full well the limitations of my own argument as a good Critical Rationalist should.

Sztompka [1979] has produced in my opinion one of the most important pieces of work in the philosophical methodology of the social sciences in recent times: important for its argumentative strategy. His response to the various sociological dilemmas is to reject the common meta-assumptions of the debate and attempt to formulate a more satisfactory position from a synthesis of the common ground of agreement of both positions: Ramsey's maxim [Ramsey, 1931, pp. 115–116] (or more fairly "Hegel's maxim"). Sztompka's use of this one methodological maxim has had much success (it is maintained) and suggests a general strategy which one might adopt in attempting to resolve the "crisis of Sociology". "Sociology" needs to return to grand systems building on a scale much more embracing than that suggested by Talcott Parsons. An embracing World-System is needed, a *Principia Philosophica* encompassing a cosmology, metaphysics, epistemology, philosophy of logic and language, and so on. Such a System given the shortness of our lives, probably could not be constructed but at best approximated. Yet there seems that there are no good reasons to believe that this task is *a priori* impossible, and even with a very incomplete sketch of such a system some entire social orientations could be subject to foundational appraisal. This has already been conducted by the "new social realists" who are critical of "positivist" and "hermeneutic" orientations for their explicit acceptance of a logical empiricist philosophy of the natural sciences [Keat & Urry, 1975]. I suggest a more extensive critical enterprise should be undertaken encompassing the totality of human knowledge. Insofar as such a systematic world view could be presented, the "crisis of Sociology" argument could be overcome, for this "crisis" is nothing more than a "crisis" of philosophy, in differing metaphysical perspectives.

This "master-builder" view of philosophy will be found by many, especially Sociologists to be utterly unpalatable. Yet I believe that even those that profess the naturalization of epistemology such as Quine, are still, (despite themselves) engaged in the traditional grand philosophical quest; the problems with accounts of reference, modality and justification which Quine has raised hardly seem capable of resolution by physics, psychology, anthropology or logic itself. And Rorty [1979] who has produced the most challenging critique of the position of philosophy advocated here, seems to be ultimately involved in a self-refuting circle from which he does not escape. He wishes philosophy to be a therapeutic, edifying endeavour, which does not attempt to say what there is, why it is, and the philosopher certainly should not engage in imperialist activities such as criticizing the scientists or the politicians.

A paradox presents itself. Rorty's conclusions only support another view of philosophy as a sophisticated positivism. It is assumed that the empirical sciences are pure, metaphysically unpolluted domains of inquiry. Yet many professional workers in these sciences have seen that their fields of inquiry accept implicit metaphysical assumptions which when modified would substantially alter the face of their sciences – Routley in logic [Routley, 1980 (1)], Böhm in physics [Böhm, 1980], Webster and Goodwin in biology [Webster and Goodwin, 1982]. The problem becomes particularly acute with regards to the social and political sciences which are also value-impregnated fields of inquiry [Thomas, 1979]. Rorty's proposals would leave us here in a state of crisis, even though, from within this form of life there is an increasing tendency to abandon neo-positivist assumptions of metaphysical freedom. These are excellent reasons, in his own terms, for abandoning his position, for he has in his own words no reason or right to impose his own metaphilosophical prejudices upon Metaphysicians.

Perhaps analytic critiques of Rorty [Hacking, 1980] [Kim, 1980] [Pompa, 1981] beg the question at issue, since Rorty is not attempting to state meta-theoretical truths about the incoherence and "ugliness" of systematic philosophy (a task which would be in itself incoherent and "ugly") but is himself in his book "edifying". But if so it is difficult to make sense of Rorty's general strategy of *argument* which most certainly seems to make truth claims. Hermeneutics, for example, is taken to be inconsistent with systematic philosophy. This seems to be an alleged truth claim: if so Rorty is inconsistent. If not, it is not the case that hermeneutics and systematic philosophy are inconsistent with each other, and so his argument collapses. Further it seems that some form of a hermeneutic circle is involved as a necessary condition of any act of inquiry [Bhaskar, 1979, pp. 195–203],[15] therefore Rorty's assumption of incompatibility is false. Needless to say, the relationship between Systematic Philosophy and hermeneutics requires statement, and here we can hardly even address the problem of detailed exposition of the methods of systematic philosophy and the other "sciences". Even so, I would hope, despite the very general and abstract level of the argument here, to have given some initial response to the "crisis of Sociology" argument insofar as it is taken to strengthen the Ellis–Skinner position. Further, I have explicitly assumed that Rorty is *arguing for a philosophical position here*, and in the course of the argument of the next chapter will consider a possible defence which Rorty and other eliminative materialists make (may make), which will involve a rejection of this assumption. It is more convenient to consider that objection in the course of the argument of the next chapter.

Let us briefly examine the state of the argument. In this section I have given a Monistic-Systemic Perspectivist response to "the crisis of Sociology" (and philosophy argument). I have said that to avoid relativism a concern with prior philosophy must become an increasingly important supplement to factual considerations. In particular a concern must be addressed to *Weltanschauung* building. If the diagnosis of the problem given in earlier sections is correct this conclusion seems inescapable. It is not however trivial. Rorty's position and the strong naturalization of epistemology and socialization of epistemology challenge this. Here I have given a strategy of how my position would cope with these difficulties but do not claim on this basis to have presented anything resembling a much needed "philosophical refutation" of these positions. What has not been done, I have repeatably stressed, is to produce in full explicit detail a study of what precisely this domain of philosophical reasoning is, as needed, and as difficult as this task is. This remains one of the most urgent problems of philosophy.

Two arguments against "Sociology"/Culturology remain unrefuted. This is the Ellis–Skinner argument and the eliminative materialist argument. I will respond to the Ellis–Skinner argument in the next section. The challenge of physicalists, theoretical reductionists and eliminative materialists will be addressed in the next chapter, where I consider the cognitive plausibility of the Physicalist Scientific Unificationist *Weltanschauung.* Whilst this work hardly makes a substantial contribution to the philosophy of mind, my initial and tentative response to the difficulties facing my own position advanced by my physicalist opponents, will be to attempt to undermine some of their central metaphysical and metascientific assumptions; especially the "unity of science by micro-reduction" principle, which is especially relevant to my thesis.

4.0 A response to the Ellis–Skinner critique of the fundamental assumption of Culturology

In this section I will respond to the critiques of Culturology of Ellis and Skinner outlined in Section 1.1. My interests here are to simply meet and beat these arguments; the most economical and satisfactory way to do this, is to formally isolate the arguments and give a point by point refutation. The main point of concern is with the epistemological arguments. The argument from the failure of "Sociology"/Culturology to solve the central problems facing humanity thus necessitating a replacement by Sociobiology is a fallacious argument. Neither Ellis, or any other sociobiologist has demonstrated that Sociobiology has an "emancipatory interest", and a case will be made out later that human Sociobiology has strong ideological implications. Moreover, the failure of "Sociology" to *solve* the central problems facing humanity can be readily explained from a sociological perspective. Answers to many problems such as world poverty, underdevelopment and the ecological crisis, would seem to involve not merely radical redistributions of income, but radical changes in social structure and modes of living. It is widely recognized by both "radical theorists" and the "elite theorists", that various powerful "ruling groups" would most certainly oppose, and indeed have opposed any attempts to radically alter society, such that "private property" would be eliminated, and vast individual (be this individual person, company or nation) accumulations of wealth, outlawed and distributed in an egalitarian fashion. If Sociobiologists recommended merely changes in social structure to solve social problems, they would tell us nothing more than what the sociological tradition has already said. On the other hand, Sociobiology can rationalize the legitimacy of private property based social systems, by claiming that by endowing one's offspring with wealth one is maximising ones own inclusive fitness [Hartung, 1976]. No reasons exist for believing that Sociobiology has any emancipatory interest, and there are good reasons against this. We turn immediately to the epistemological arguments.

(1) Epistemological argument No. 1: The unfalsifiability claim

The argument here can be summarized as follows:

(1) All scientifically adequate explanations are empirically falsifiable

(2) No intentionalistic/mentalistic explanation is empirically falsifiable

Therefore,

(3) No intentionalistic/mentalistic explanation is scientifically adequate.

This formally valid argument can be criticized by criticizing either premise (1) or premise (2). The easiest premise to criticize is premise (2). This premise is false because at least some intentionalistic explanations are false, or are capable of being shown to be false. If we were never (even fallibly) capable of ascertaining the falsity of any intentionalistic explanations, the highly co-ordinated and regimented social life of industrial capitalist societies could only be explained through appeal to a "cosmic coincidence". It seems that we do falsify many intentionalistic explanations constantly in day to day social practices such as working, cooking meals (e.g. we assume that if someone says that they are hungry, and we have no reason to believe otherwise, then it is a fact that they are hungry) and driving cars (e.g. we assume that the other drivers intend to obey the road laws, and soon find out via road accidents when they do not). One could no doubt challenge these claims by defending solipsism. However, even if one was driven to such extremes, it seems that at least some explanations of ones' own behaviour with reference to ones' intentions, are in principle falsifiable. It might seem to me at time t_1 that I intended to do act A on the basis of reason R_1, whilst latter at time t_n, upon reflection, I may come to believe that R_1 was not my real reason for A at all because of evidence E; I take this to be a common aspect of experience. It seems then that premise (2) is false, and the argument unsound.[16]

(2) *Epistemological argument No. 2: The abandonment of intentional explanations is necessary for Socio-Scientific progress*

Both Ellis and Skinner claim that intentionalistic explanations must be abandoned in the Social Sciences as they allegedly have been in the natural sciences for scientific progress to occur. The following argument seems to be made:

(1) Progress in the natural sciences has occurred.
(2) Rejection of the explanatory role of mentalistic items such as intentions, purposes and beliefs, had occurred early in the history of the natural sciences.
(3) (1) has occurred because of (2).
Therefore
(4) Progress in all sciences will occur if the explanatory role of mentalistic items is rejected.
(5) Progress in the Social Sciences has not occurred.
(6) Rejection of the explanatory role of mentalistic items has not occurred in the Social Sciences.
(7) The Social Sciences are sciences
Therefore
(8) Progress in the Social Sciences will occur if the rejection of the explanatory role of mental items occurs.

One difficulty with this argument is premise (4). It is not supported by premises (1), (2) and (3). It may well be that a rejection of mentalistic items was necessary for progress to occur in the "natural sciences" such as physics, because such items do not describe any reality physics deals with. Consequently the inference from (1), (2) and (3) to (4) is fallacious. A second difficulty with this argument, is also premise (8). It is false to claim that no previous social-scientific research programme has rejected the explanatory role of mentalistic items. Radical behaviourists have attempted to this. Yet there is, as I pointed out earlier, good reason to believe that Radical Behaviourism is not only a degenerating research programme, but one which has already died many times over.

Skinner, as I pointed out in section 1.1, has an argument to support the claim that mentalistic explanations are inquiry limiting. The argument can be summarized as follows:

(1) An action is not explicable by reference to mental attributes such as intentions, until the intentions have been given *prior* explanation with respect to their causal antecedents.

(2) "Mentalism" does not explain the causal antecedents of intentions and other mental items, as such items are not given prior causal explanation.

Therefore,

(3) "Mentalism" does not explain action.

The problem with this argument is premise (1). If we are to accept that an action is not explained by reference to mental attributes such as intentions until the antecedents have been in turn explained by their causal antecedents, we have, from this claim about the nature of causal explanation, landed ourselves in a vicious infinite regression. If no causal explanation is acceptable without prior explanation of its own referred to generative mechanisms, then we are committed now to rejecting all causal explanations, since those mechanisms must be explained, and those again, and so on, such that we find that we must give an infinity of explanations to explain even one initial phenomena [Smith, 1982 (2)]. If we accept the law of universal causality, that every event has a cause, as well as the premise that if an event E is caused by an event B, E cannot be the cause of any other event, we are led into inconsistency. For as E could be any event from the set of all events, E being caused by B means that B in turn must be caused (by assumption of universal causality). But if B is caused then by our second assumption, B cannot be the cause of E. Therefore E is not caused by B, which is inconsistent with the proposal that E is caused by B. As E and B are any events, it follows that there are no causal chains [Flew, 1978, pp. 142–143]. I conclude that Skinner's argument cannot be accepted unless we are prepared to accept a strong form of scepticism about causal relations, and since Skinner is not a causal sceptic, this alternative is unacceptable, and the argument collapses.

This concludes my response to the Ellis-Skinner critique: none of the proposed arguments are sound. Therefore this position fails to establish a case for a philosophico-methodological reduction of Culturology to Sociobiology.

5.0 Conclusion: The state of the argument

In this chapter I have examined the most plausible lines of argument in developing what I called a philosophico-methodological reduction of Culturology to

Sociobiology. Here my concern has been to outline and criticize those arguments which would constitute a philosophical refutation of my own position. Many philosophers and sociologists would wish to claim that a (human) Sociobiology is untenable, myself included. However, it seems unreasonable to criticize my opponents on various grounds if my own position was untenable. So I have sought to isolate such critical arguments, and give a response to them.

Three arguments have been isolated in this chapter but only two have been addressed; the remaining criticism from the perspective of eliminative materialism will be addressed in the following chapter. The other two arguments are the Ellis–Skinner argument and the crisis of Sociology argument. Whilst the Ellis–Skinner argument was stated first, it was addressed last, for a further discussion was presupposed. This presupposition, and what was in fact my response to the crisis of Sociology argument, was to suggest the legitimacy and the need of a prior philosophy. A question of a purely factual choice between various "models of man" seems to be challenged by the underdetermination of theory by evidence thesis, although I do not accept that this necessarily occurs for all competing theories. But I know of no good arguments to exclude the real possibility of it occurring *at all*. The problem increases in magnitude if my interpretation of the crisis of "Sociology" as being a crisis in philosophy is correct. This is so because the underdetermination of theory by evidence seems especially common in metaphysics.

Problems were earlier presented by the crisis of "Sociology" argument, in the claim that the mass of mutually disputing and seemingly inadequate orientations/paradigms/research programmes rendered the cognitive systematization of the Social Sciences impossible. My advanced solution to these problems has been to argue that a World-System is needed. Various views on the nature of philosophical debates were considered whereby a process of critical elimination, one remaining position *Monistic-Systemic Perspectivism* was opted for, and consequently defended. I have been necessarily vague about what this system is, for I do not have it at hand. It is at least in principle capable of construction and the crisis of Sociology argument rebutted. It is another matter whether this in actual fact will occur. Lehrer [1976] has given a mathematical argument based upon Markov transition matrix properties, that it is demonstratably irrational for a group sharing the same information and respecting each other's opinions, to disagree. Continual disagreement may well then not indicate a problem with our methods [Rabinow and Sullivan, 1979, p. 4] but with the present institutionalization and professionalization of knowledge. This is a topic recommended for a sociology of knowledge examination; it does not affect the argument here.

CHAPTER 3

Theoretical reductionism and physicalist scientific unificationism: The case against

1.0 Introductory statement of the argument

To place diverse and prolific social theorists and philosophers within the constraints of some classifactory scheme is a task which always leads to major difficulties: virtually all of these thinkers will break one's neatly drawn constraints in some respects. Nevertheless, to fail to do this has much worse consequences: one is condemned to an atomistic methodology which to be sure enables a clear exposition of the point of view of one theorist (or tradition), while failing to expose what the similarities there are between seemingly different view points. This, from the perspective of the metamethodological comments of the previous chapter is objectionable because one must then ignore insights which may well arise from such broad comparative analyses, and is self-defeating given the metascientific aim to understand as much about the World as possible. Oversimplifications are a small price to pay for this possibility.

Given these qualifications let us get on with the business of canvassing the central argument. One problem which has divided theorists for generations, and in fact, I hope to show still does, is the question of the cogency of either a *Naturalistic* or *Anti-Naturalistic* basis of Sociology, and the other social sciences. This is a many faceted problem, and it is best that we attempt to clearly distinguish this question from others to which it bears intimate theoretical relationships, but to which it remains logically distinct.

The Naturalism versus Anti-Naturalism debate is logically distinguishable from the various positions which could be taken on the question of the ontological status of social entities, although such questions are raised in the context of the former debate. It also is distinguishable from the so-called "Methodological Individualist-Social Holist" debate. The latter debate itself encompasses as many issues as the debate which I am attempting to explicate here, ranging over questions of methodology, ontology and theoretical reduction. However, this much can be stated without producing unjustifiable complications at this point:

one important part of the Methodological Individualist-Social Holist debate with regard to metascientific matters of theoretical explanation is whether sociological theories are reducible (in respect to various models of theory reduction) to psychological theories or whether psychological theories are reducible to sociological theories. Once more, we should note the possibility that the resolution of this debate would have important implications for the Naturalism versus Anti-Naturalism debate, but it is not an accurate characterization of this debate. I will attempt to support these remarks in more detail shortly.

What then is the debate? The issue being debated here is metascientific, concerning "the *relationship* between the natural and social sciences as disciplines". This statement is exceedingly and deliberately vague: we can in fact distinguish two main debates which have been frequently confused by many parties at the dispute. The first debate is concerned with the question of whether there is a *methodological unity* between the social and natural sciences, and if so then in what respects. The second debate is concerned with the question of the theoretical autonomy of "Sociology" with respect to the natural sciences: are substantial sociological theories reducible (in specified senses) to specified natural scientific theories? It is the second question which is of principal concern to this work, although we should note once more the possibility that any attempt to answer this question will possibly have some implications for answers which might be given to a number of other previously cited questions. To clarify the second question, consideration of the implications of the first, will be given.

It is notoriously difficult to explicate what precisely a "methodology" is. In the sense specified by researchers in the philosophy of science, a methodology M outlines criteria for the evaluation and assessment of theories, such that theory T is rationally preferable to theory T* if and only if T satisfies M where as T* does not. This itself is still vague: some specification of the aspects of preference needs to be detailed – but this account will suffice for the present argument.[1] In speaking of the methodological unification of the natural and social sciences, what then is alleged is that there is no qualitative difference in the methodologies for evaluation of natural and social scientific theories; that is, if there exists an M satisfactory for the evaluation for natural scientific theories, then M will be satisfactory for social scientific theories. Naturalists accept this claim: Anti-Naturalists reject it.

As one would expect from such a broad classification diverse thinkers find themselves companions of other thinkers whose positions they would be in general unsympathetic to in other respects. Naturalists would include, for example, physicalists such as J. J. C. Smart [1963 (1)] and D. M. Armstrong [1968] at one extreme, and an anti-physicalist (although "emergent materialist") such as Bhaskar [1979] at the other; Anti-Naturalists would include such diverse thinkers as J. Habermas [1972], C. Taylor [1971, 1980], Winch [1958] and Skjervheim [1974]. In terms of traditions, rather than individuals, among the Anti-Naturalist orientations we may classify "Critical Theory", "Ethnomethodology", "Symbolic Interactionism", "Goffman's Dramaturgy", "Ethogenetics" and a variety of "Phenomenological", "Dialectical-Hermeneutics" and "Ordinary Language" orientations (see for a further exposition and defence [Sztompka, 1979, Ch. 2], [Radnitzky, 1970] [Fay and Moon, 1977]).

Central to the Anti-Naturalist position is the critique of "Positivism", especially in the form of "Logical Empiricism" [Radnitzky, 1970] [Suppe (ed.), 1977]. Here one path taken has accepted the metascience of Logical Empiricism as satisfactory for the natural sciences and has argued that this view was inapplicable for the human sciences, for reasons such as an alleged logical incongruity between reason-based explanations and causal explanations [Melden, 1961]. This particular mode of argument came under attack initially by Davidson [1963] and increasingly criticized throughout the seventies by the "neo-Realists" [Keat and Urry, 1975] [Bhaskar, 1978 (1), 1979] arguing that the Anti-Naturalists case was based upon the explicit acceptance of the Logical Empiricist metascience of the natural sciences. At the same time Logical Empiricism came under heavy fire from historically orientated approaches championed by Kuhn [1970], Feyerabend [1975] and others (cf. [Laudan, 1979]) with an increasing recognition of the social processes of science. From this orientation, Naturalism has also been defended [Thomas, 1979] and the claim that all sciences are hermeneutic has been made [Hesse, 1972].

The question must now be asked: what are the implications of the present situation in sociological methodology for my central thesis; do we find any methodological/metascientific short cuts to lessen our work load? That is to say, does the current state of this debate suggest any possible meta-critiques which might be made against Sociobiological reductionism? Having led the reader through a thicket of conceptual issues already, we face now the bitter dregs of disappointment. Sociobiological reductionism hardly seems capable of being refuted merely by pointing out that human beings are *persons* and *agents*, or that communicative cognitive interests are irreducible to empirical-analytic interests. With regards to to the first point Sociobiologists who accept the *ontological* irreducibility of the person, would argue, that the principal causal determinant of human behaviour is gene-related biochemical activity. With regards to the second point, it might well be conceded that at a methodological level communicative cognitive interests are irreducible to empirical-analytic interests: nevertheless communication is an aspect of human behaviour readily open to Sociobiology. The point to be made here, and which has been conceded already, is that arguments against Sociobiology from a sociological perspective are quite weak, and indeed for the Sociobiologist, in fact question begging, especially for the Sociobiologist with eliminative materialist sympathies.

In this chapter Sociobiology is located within the framework of the *Metaphysical Weltanschauung* of *Physicalist Scientific Unificationism* (PSU). This is not to say that the working sociobiologist recognizes himself-herself readily in such an intellectual environment; rather, this is a method for producing the strongest possible challenge to Culturology as we saw in the last chapter. However, it is not to maintain that such a construction is an utter perversion of the spirit of the Sociobiological research programme: textual evidence from Wilson [1975] and especially Ellis [1977] is readily citable and quite obvious.[2] The argument of this chapter is that PSU is untenable. I hope by this strategy to destroy the remaining eliminative materialist argument against Culturology outstanding from the previous chapter by suggesting that the metaphysical assumptions of this orientation are unsatisfactory. Moreover I hope to give some *metaphysical rationale* to Culturologist claims, which PSUist would regard as outrightly question-begging. This is an ambitious task, one which one would hope to avoid if one was searching for easy marks, confortable positions and smiles from the opposition. I however will seek the Truth, and live dangerously.

2.0 The Weltanschauung of Physicalist Scientific Unificationism

As was stated in the previous chapter, the ideal of systemicity runs throughout the history of human thought; Physicalist Scientific Unificationism is but one response to the diversity of the World and the schemes of human thought. As a *Weltanschauung* it encompasses metaphysical, logical, epistemological and methodological commitments, but as was pointed out at the outset of this chapter, the richness of the thought of any one of the to-be-cited thinkers breaks the bounds of this, and possibly all such classifications.[3] But if we are to bring order to our thoughts, some classification, despite its limitations must be imposed.

2.1 Metaphysical commitments of PSU

The first metaphysical thesis of PSU is a thesis about the ontology of the world: the world is monistic, consisting of only one "substance" with only one sort of properties and relations. For PSU, by contrast to philosophical cosmologies such as Machian neutral monism and Berkeleyian subjective idealism, the world is "material". There are a variety of explications of this notion of "materiality", but the most frequently adopted account, is the view that the world contains nothing more than the entities recognized by physics [Smart, 1963 (2)] [Armstrong, 1978 (1), p. 268] [Levin, 1979, pp. 60–63].[4] The ontological thesis of "physicalism" or "materialism", is most frequently justified by an epistemological foundationalist thesis about the fundamentality of physics, which I will detail shortly.

Physicalism itself provides a neat solution to both *Eddington's two table problem* and *Sellars' problem*. For Eddington [1958], despite appearance to the contrary, the desk at which I am now writing is not "solid". In fact, microphysics tells us that the table is in fact a "swarm" of entities which possess quantum mechanical properties, which no common sense table could possess. Given that the universe is a "homogeneous", "one-layer world" [Radnitzky, 1970, p. 73], it seems that one is faced with ascribing mutually incompatible properties to the one particular. PSU provides a simple resolution to this philosophical problem: there are only the microphysical entities.[5]

Not only does Eddington's two table problem raise the problem of reconciling the seeming heterogeneity and multi-layerness of the world with a monistic metaphysics, but it also raises a problem of the causal relationship of "levels": what, for example is to prevent situations where macroscopic objects violate microphysical laws? [Cartwright, 1979, p. 85]. The thesis of "Universal Mechanism" [Bantz, 1977, p. 6] is an attempt to escape this problem: everything is composed of physical entities and obeys the laws of physics, and a description of the parts of a thing with reference to such laws, provides an adequate understanding of the thing and its interactions. Consequently no inconsistency can arise, the PSUist alleges.[6]

Sellars' Problem is closely related to Eddington's two table problem. In a classic paper, Wilfred Sellars [1963 (1)] drew a distinction between two images of human beings: the *Manifest image* and the *Scientific image.* The Manifest image is a specific modification of the "original standpoint" of Man-in-the-midst-of-the-World (with slight distortion, what "phenomenologists" have called the *Lebenswelt*[7]) in both an *empirical* and a *categorical* sense. With respect to the empirical refinement, Sellars refers to the refinement of the *Lebenswelt* through the addition of techniques of "correlational induction" [ibid., p. 7]. In this sense the Sellarsian Manifest image is both "disciplined and critical", and a part of the Scientific world view. However, the Manifest image does not permit modes of postulational induction involving the postulation of imperceptible law-governed entities, to explain phenomena in the world: the basic objects of the Manifest image are *Persons* [ibid., pp. 9–10], and by a "categorical refinement" Sellars means that today some entities are not classified as persons which once were (e.g. trees, and other parts of the physical environment by some societies), whilst some "entities" are now classified as persons which once were not (e.g. "slaves") [ibid., p. 10]. However the category of a "person" remains primary, despite these modifications in its domain of reference.

The conflict which arises between the Manifest and Scientific images of humanity, Sellars himself compares to the two table problem [ibid., pp. 35–36], how in fact to reconcile "*the ultimate homogeneity of the manifest image with the ultimate non-homogeneity of the system of scientific objects*" [ibid., p. 36]. For Sellars, a reconciliation is possible, insofar as the Manifest image is primary from the perspective of *practical reason* in viewing human beings as part of a community of intentional agents, but from a theoretical standpoint, it is the Scientific image which is the measure of the real.[8]

A distinction equivalent to Sellars' distinction between the Manifest and

Scientific images of humanity and the world, is found in the work of both identity theorist PSUists (i.e. the mental is identical (in a specified sense) to the physical; the Manifest world is identical to the Scientific world) and eliminative materialist PSU'ists (i.e. the mental, like phlogiston does not exist, nor are there mental properties, there are only "physical objects" with "physical properties"; the Manifest world does not exist, there is only the Scientific world). For the identity theorist Wilson [1979], the distinction is between the OA (ordinary "animist") model and the PO (physicalist objectivist) model; the OA model and PO model are equated with Sellars' Manifest and Scientific images respectively [ibid., pp. 29–30]. For the eliminative materialist, Paul Churchland [1981], the distinction is between "folk psychology" and (physicalistic) neurophysiology. In both authors, the Manifest image is considered to be radically defective.

Some further crucial points underlying the PSU *Weltanschauung* can now be explicitly stated. The Scientific world view is essentially that of the natural sciences. It would seem that an analogous problem to that of the causal relationship of the levels would arise if "Science" was not restricted to that of the natural sciences. For example, are societies to be particulars which we admit into our ontology, along with alleged socio-historical laws, and if so how are they causally related to material reality? The identity theorist solves this problem by identifying societies with some permissible (so they believe) objects such as sets of human organisms, and attempts to reduce socio-historical laws to physical laws through a prior reduction of the former laws to biological laws; the eliminative materialist does so by simply eliminating societies. Not only is it assumed throughout this debate that there is "this thing called science", but a demarcation between the natural and social sciences must be drawn.[9]

2.2 Logical orientation of PSU

I shall not deal very extensively with the logical commitments of PSU, first for the reasons to be cited, and second, whilst such a study is of interest in itself, it will not contribute to the arguments to be advanced in criticism later, and indeed a satisfactory critique already exists [Routley, 1980 (1)]. The PSU programme of Oppenheim and Putnam [1958] stressed that a unified science required a unified language for all scientific disciplines. This requires the conceptual reduction of the terms of one science to those of physics, excluding those of logic and mathematics (the status of philosophy is unclear: one way PSU has dealt with this, is to pronounce the "death of philosophy" and "naturalize epistemology"; we considered and rejected this move in the last chapter). This is an extremely strong requirement, and is not frequently advocated today. Identity theorists usually permit a dualism of languages, pointing out that it is the identity of the reference of mentalistic terms, for example, with physicalistic terms' referents, which are of interest to them. Eliminative materialists accept a duality of language, admitting for example that mentalistic terms either do not denote at all, or denote entities, but these entities do not exist.

Within PSU, there is also a strong preference for extensionalistic logic especially among the followers of Quine, but on the other hand possible world semantics is quite popular with PSUists in America and Australia nowadays. However, there seems to be wide acceptance of the position to put it vaguely, that the "natural sciences are extensional", whilst "psychology and sociology are intensional". The failure of extensionalist reduction programmes [Meyer and Routley, 1977] raises problems for physicalists, as the mental and the social seem to escape through the net of the natural sciences [Wilkes, 1978, ch. 2]. This exit, taken by some non-reductionists (e.g. [Margolis, 1978]) is closed for me, given the recognition that physics is non-extensional [Routley, 1980 (1)].

To sum up, there are no further points that can be made here, which will generate argument of direct relevance to my critical interests here.

2.3 Epistemological orientation of PSU

Two remarks can be made about the PSU orientation under this heading. First, all PSUists are scientific realists with respect to the natural sciences at least: non-logical and non-syncategorematic terms of such scientific theories typically refer and the theories in which they feature in are approximately true [Putnam, 1978, pp. 20, 21] [Laudan, 1980, 1981]). Some such as Smart [1963 (1), pp. 50–61] and some eliminative materialists would deny that sciences "from biology up" are genuine sciences, because these lack, they believe, authentic scientific laws. Most, however, take a realist account of all empirical sciences, so that the ideal of unification of science by micro-reduction is operative.

Second, it is believed by PSUists that there is a natural taxonomy of the sciences based upon some intuitively plausible "levels" in the world: elementary particles, atoms, molecules, cells, multicellular living things, and social groups is the list given by Oppenheim and Putnam [1958]. Given their mereological commitments, there are no unique particulars corresponding to psychology's universe of discourse; that is, the usual ranking – physics, chemistry, biology, psychology and "Sociology" is not in accordance with Oppenheim and Putnam's schema. But from our discussion of Sellars, this problem is easily solved: psychology's universe of discourse is that of *persons*, and their properties and "Sociology's", systems of persons and their properties and relations.

The fundamentality of physics and chemistry is asserted in both a causal-ontological and an epistemological sense. As was pointed out above, given the universality of natural laws, physics and chemistry are taken to be unrestrictively true (if true at all): everything in the universe "obeys" the laws of physics because everything in the universe is *physical*. A field F_1 may be taken to be more fundamental than a field F_2, if and only if F_1 explains everything F_2 describes and more, in either greater detail, or a greater domain of phenomena, perhaps themselves in greater detail as well [Dresden, 1974, p. 137]. In this respect, PSU will regard physics as more fundamental than any other field of study.

Physics is also taken to be fundamental in an epistemological sense: physics is a Science *par excellence* (at least an *empirical* science *par excellence*, logic and mathematics are still hot favourites, as we shall see, for providing models of the structure of scientific discourse). *If* there is such a thing as "scientific progress", we are more likely to find it in physics, than in a discipline such as "Sociology" which, like philosophy, seems to dwell in a constant state of epistemological crisis – or so we are told.

Finally, physics is epistemologically fundamental insofar that if current and foreseeable physical theories run into severe foundational problems, such as the occurrence of a multitude of antinomies, or major conflicts between various theories of this discipline, then all of human empirical science is thrown into disarray. This seems to be a consequence of the PSU claim that all empirical science can be unified by micro-reduction to physics.[10]

2.4 Methodological orientation of PSU

As was stated previously, the central motivating force of PSU is to unify and systematize human knowledge. The principal way in which this is to be done is through *theoretical reduction.* Here I will not discuss the various accounts of this reduction in detail: further consideration must be given to this question shortly, since there exists no widely accepted unproblematic account of the relation of theoretical reduction (and those which do exist are problematic in my opinion). Not only this, but matters become worse when we attempt to precisely specify what in fact the entities which we are reducing, that is, "scientific theories" and "disciplines", are, and how such entities are individuated. In this work a fully worked-out response to this question, important as it is, cannot be conducted; indeed these difficult questions stand on the frontlines of metascientific research. But the problems, which I will in turn canvass, are those which any systematic thinking upon these matters must raise, and the answers which I shall give, will be exceedingly negative.

3.0 Systematizing the argument: Sociobiology and PSU

The material to date has consisted in a lengthy exposition of the PSU metaphysical *Weltanschauung.* A critic, patient throughout the previous section, at the present moment will either find the material the incoherent bumblings of the metaphysician, or else accept that placing Sociobiology within a *Weltanschauung* is of course necessary for making the point of a thorough-going foundational demolition, but will dispute that PSU is in fact the correct world view to locate Sociobiology within. For instance, it is quite possible that the Sociobiologist, whilst wishing for a unification of the social and natural sciences, may well hold that biology is not reducible (in senses to be specified) to chemistry and physics.

Now in reply to this criticism, it is conceded that as a matter of logic no Sociobiologist need embrace PSU at all. However, such a criticism is based upon an ignorance of our already clearly stated methodological aims: to see whether *in principle* a reduction of Culturology to Sociobiology is possible, and thus to examine the tenability of the philosophical foundations of Culturology (for if this occurs, the position is by virtue of its rugged anti-reductionism absurd). This methodological aim can, I believe, be instrumented, regardless of whether any living thinker can as a matter of textual fact be precisely classified as adhering to PSU.

However, my argument does not as yet adequately respond to the claim that a Sociobiologist may believe in the reductibility of Culturology, and yet deny that biology and Sociobiology are reducible to physics and chemistry. My reply here is that this is in fact a *prima facie* implausible position given the frequently made claims of PSU that the remaining programme resistant phenomena are those of the social sciences. It is especially implausible in the light of the frequently made claims by PSU, that "genetics" has been reduced to molecular biology – implausible because Sociobiology is constituted to a significant degree by genetics, especially population genetics [Wilson, 1975].

It is true that discrediting a number of central tenets of PSU *will not* constitute an adequate critique of Sociobiology. On the other hand the current critiques of Sociobiology from the perspective of the social sciences, seem ultimately question begging, and utterly impotent against the philosophical critique which can be made against them in turn, which we discussed in detail in the previous chapter. It is no coincidence in my opinion that such a critique seems to presuppose the metaphysical *Weltanschauung* of PSU.[11] In facing PSU head-on, one in effect tackles the strongest possible critique which PSU could make upon my own socio-theoretical position. But in doing this, we must not, as my first critic cautioned, restrict our considerations purely to speculative metaphysical matters; indeed in this work I will formulate a strategy for conducting a metaphysical critique of Sociobiology itself. But first PSU.

This chapter consists of two remaining sections. In section 4 I will examine the prospects of a Culturology-Sociobiology theoretical reduction and the multitude of conceptual and theoretical problems this involves. In section 5 I will advance criticisms which tell against most of those assumptions of PSU, and through such criticism rebut the outstanding challenge made against Culturology by the eliminative materialist. This is not the place to develop a comprehensive metaphysics and philosophy of the sciences, and it would be opening my work to easy demolition by the critics to claim this. However, I hope to be able, in the space available here, to show that there are good prospects for developing an alternative metaphysical *Weltanschauung* to PSU, consistent with the central claims of Culturology.

4.0 Culturology, Sociobiology and theoretical reduction

In this section I will concentrate my critical fire power upon the concept of theoretical reduction. I will argue after criticizing most of the strongest accounts of this relation, that we should abandon further attempts at reconstruction and that the research programme of "theoretical reduction" is, as a research programme, most degenerative. My attitude towards this concept is thus contrastable with that of the anti-realist towards the concept of verisimilitude: it is internally problematic, out of touch with the history of science, and its elimination will produce no tears of regret from the non-PSUist working scientist. Such is my attitude towards the concept of theoretical reduction. One should recognize that there are few cases of (intuitively regardable) complete "reductions", even within physics [Bunge, 1973] [Popper, 1974]; yet presumably science and philosophy of science have survived, so the centre of our theoretical world hardly collapses by this proposal. Reductionists have an extremely weak inductively supported record, contrary to their public relations work, in the introductory and concluding sections of their books and articles.[12]

The main problems facing a theoretical reduction are these: (1) to outline philosophically satisfactory accounts of theoretical reduction; (2) to analyse what it means to say that one discipline has been "reduced" to another (i.e. is discipline reduction merely a generalized form of theory reduction, because disciplines are, nothing but sets of theories?); (3) to perform the reduction and put the results on paper for all to see. Here we will not get past steps (1) and (2), given the difficulties facing these steps. Nevertheless, the explicitly detailed work done to support (3), especially involving social theories, is hardly as voluminous as one would be inclined to believe from PSU rhetoric.

Indeed, in the past twenty years the theoretical reduction programme has been viewed with increasing critical awareness [Wimsatt, 1979]. This has come through an extensive revolt against the logical empiricist view of science. Criticisms have been that this metascience faces internal conceptual problems and analyses of basic concepts such as reduction, confirmation and so on, bear little coherence with the history of science, for on closer analysis the paradigm cases of theoretical reduction reveal surprising complexities past research did not investigate [Hooker, pt. I, 1981, pt. II, 1981, pt. III, 1981] [Wimsatt, 1979].

Philosophy of science, or more accurately, "meta-science", has taken a number of directions over this period. With regards to the concept of reduction there are only two trends. First, it is either outrightly abandoned, as I advocate, or, second, some rational reconstruction is attempted. Some have attempted to rework and defend the logical empiricist account of reduction [Yoshida, 1977]

[Ruse, 1976] [Goosens, 1978]. The structuralist accounts of scientific theories, which we shall review later, have also attempted to define a formal relation of theoretical reduction.

In such a situation of "epistemological crisis", it is best to attempt a precise yet comprehensive survey critique of the various approaches, to at least inform the reader of where and why I take the stand which I do.[13] Whilst such an approach is necessarily selective, one can still survey a sufficient number of *strong representatives* of a wide variety of orientations, and by undermining basic assumptions of such orientations, establish a more general, and for the PSUist, depressing claim. We shall consider first the early accounts of reduction proposed by Nagel, Kemeny/Oppenheim/Putnam and Adams/Suppes and Quine. First exposition, then criticism.

4.1 Early accounts of theoretical reduction: Nagel, Kemeny/Oppenheim/Putnam and Adams/Suppes and Quine

4.1.1 Nagel's account in The structure of Science

For Nagel [1961] theoretical reduction consisted in the following:

> "... the experimental laws of the secondary science [i.e. the theory to be reduced] (and if it has an adequate theory, its theory as well) are shown to be the logical consequences of the theoretical assumptions (inclusive of the coordinating definitions) of the primary science" [ibid., p. 352]

Nagel distinguishes between *homogeneous* and *inhomogeneous* reductions. In homogeneous reductions, the vocabulary of the secondary science does not contain any descriptive terms which do not appear with approximately the same meanings as those terms of the primary science. In inhomogeneous reductions, the secondary science has descriptive predicates in its vocabulary which do not occur in the vocabulary of the primary science. Reduction now, is a more complex matter, involving both formal and non-formal conditions for reduction. These are summarized in Table 3.1. (Criticism will follow after the exposition of other accounts.)

4.1.2 The Kemeny/Oppenheim/Putnam account

The Kemeny/Oppenheim/Putnam account of reduction [Kemeny and Oppenheim, 1956] [Oppenheim and Putnam, 1958] may be summarized as follows [ibid., p. 5]:

(KOP/D_1) For theories T_1 and T_2, T_2 reduces to T_1 = df (1) The vocabulary of T_1, and (2) Any observational data explicable by T_2 is explicable by T_1, and (3) T_1 is at least as well 'systematized" as T_2.

The reduction of a branch of science B_2 by another B_1 is defined as follows:

(KOP/D_2) B_2 is reduced to B_1 at time t = df there exists a theory T_1 in B_1 at t such that T_1 reduces T_2.

One should note that this account presupposes (1) that the observation-theoretical

Table 3.1. Summary of Nagel's [1961] conditions for theory reduction

(A) Formal conditions for reduction

(1) *The requirement of maximum explictness*

The terms of the descriptive vocabulary of both sciences must have their meanings fixed unambiguously by explicitly stateable rules of usage. These will include:

(a) theoretical postulates and theorems derivable from them;
(b) the experimental laws;
(c) observation states and
(d) "borrowed" laws (governing the operation of experimental instruments but not explained within the theory in question [ibid., pp. 345–349].

(2) *The requirement of analysis*

Every term of both the primary and secondary sciences "can be analyzed as a linguistic structure, compounded out of more elementary expressions in accordance with tacit or explicit rules of construction [ibid., p. 349]. There must exist in each science a set of primitives sufficient to explicate, by means of logic, the meanings of all other expressions of the science.

(3) *The condition of meaning invariance*

Expressions belonging to a particular science must possess meanings fixed by the rules and procedures of explication of the science in question, such that the expression can be understood in a sense associated with it in that science, regardless of whether a reduction has occurred.

(4) *Of inhomogeneous reductions*

If the secondary science contains terms $t_1, t_2, \ldots, t_n$ not contained in the primary science T_1, then the following are necessary formal conditions for the theoretical reduction of T_2 to T_1:

(a) *The condition of connectability (bridge principles)*

Assumptions are introduced for each theoretical term of the secondary science $t_1, t_2, \ldots, t_n$ not in the vocabulary of the primary science, which establish relations between the designata of $t_1, t_2, \ldots, t_n$ and phenomena designated by the theoretical terms of the primary science.

(b) *The condition of derivability*

On the satisfaction of the condition of connectivity, all the laws of the secondary science must be logically deducible from the primary science's theoretical premises and associated coordinating definitions.

(c) *The condition of empirical support*

There must be adequate empirical support for these postulated linkages.

(B) Non-formal conditions for reduction

(1) ". . . theoretical assumptions of the primary science [must] be supported by empirical evidence possessing some degree of probative force" [ibid., p. 358].

Table 3.1 (continued)

(B) Non-formal conditions for reduction
- (2) If the secondary science is not prepared at some stage of its development to operate effectively with the primary theory, theoretical reduction may result in little increased knowledge.
- (3) Theoretical reduction involves the deduction of statements, not properties, as logical connections hold between statements, not properties or phenomena.

distinction can be drawn, and (2) that the two branches have the same observational vocabulary. Explanation on the KOP account involves the deducation of the observational data of one theory from that of another; T is said to explain the set O of observational data if and only if $O_1 \cup O_2 = O$ and O_1 and O_2 are disjoint and $T \,\&\, O_1 \rightarrow O_2$.

4.1.3 The Quine, Adams/Suppes accounts

Finally, three further early accounts of theoretical reduction will be summarized: that of Quine [1966], Adams [1959] and Suppes [1967].

Quine's principle discussion of theoretical reduction [Quine, 1966], occurs in a paper primarily concerned with questions of ontological reduction in the philosophy of mathematics, including the onotological significance of the Löwenheim-Skolem theorem.[15] It seems, however that Quine is advancing a general theory of reduction; this is at least how he has been interpreted in the secondary literature [Schaffner, 1967], a charge which he has never denied. He has this to say:

(QDR) "The standard of reduction of a theory θ to a theory θ' can now be put as follows. We specify a function, not necessarily in the notation of θ or θ', which admits as arguments all objects in the universe of θ and takes values in the universe of θ'. This is the proxy function. Then to each n-place primitive predicate of θ, for each n, we effectively associate an open sentence of θ' in n free variables, in such a way that the predicate is fulfilled by an n-tuple of arguments of the proxy function always and only when the open sentence is fulfilled by the corresponding n-tuple of values" [ibid., p. 205].

Quine hopes to avoid the Pythagoreanization allegedly produced by the Löwenheim-Skolem theorem, as now we must in fact specify an actual proxy function which performs the mapping specified by (QDR). However, as Massey [1973, p. 55] points out, we can in fact show that such proxy functions exist, and the world is reduced to numbers! This is surely absolutely counter-intuitive, and a strong reason for rejecting Quine's account of theoretical reduction. There is, however, as I shall cite shortly, a further criticism which can be upheld against both Quine's position and the remaining accounts of Adams and Suppes.

Whilst Adams [1959] is cited in the secondary literature as having produced an account of theoretical reduction [Schaffner, 1967], close study of his paper reveals that this is not the case. However, he, along with Suppes, had laid the foundations for the recent structuralist view of theories, which will be considered shortly. Let us consider in any case Adams' alleged proposals.

For Adams, a theory is a set-theoretical entity: $T = \langle C, I \rangle$, where C is the set of all entities satisfying some specified axioms and I is the set of intended models. Adams lays down two conditions for reduction:

Condition A: Let $T_1 = \langle C_1, I_1 \rangle$ and $T_2 = \langle C_2, I_2 \rangle$ be two theories such that T_2 is reduced to T_1 by relation R. Then for all i_2 in I_2 there exists i_1 in I_1 such that $i_2 R i_1$.

Condition B: Let $T_1 = \langle C_1, I_1 \rangle$ and $T_2 = \langle C_2, I_2 \rangle$ be two theories such that T_2 is reduced to T_1 by relation R. Then for all c_1 and c_2 if c_1 is in C_1 and $C_2 R c_1$ then c_2 is in C_2.

Clearly this cannot be a definition of "theoretical reduction", since it presupposes this very relation. Nor, as Adams admits, do these conditions *define* the relation R, but at best show that if T_2 is reduced to T_1, if T_1 is correct, then T_2 is correct. However Adams' reduction relation is so weak that it allows itself to be trivialized and reduced to absurdity by defining a reduction relation for any two "correct" theories, which may well be utterly unrelated in subject matter.

Suppes [1967] seems in basic agreement with Adams, that scientific theories are set-theoretical entities. If M_1 is a model of the reducing theory T_1 and M_2 a model of the reduced theory T_2, then T_2 reduces to T_1 if and only if, there is a model M_1^* such that M_1^* is isomorphic to M_2.[16] On such an account for example, a sociological theory would be reduced to a sociobiological theory, if for any model of the former, it was possible to construct an isomorphic model within the latter.

4.1.4 Difficulties facing the early models

The difficulty for both the Suppes, Adams and Quine accounts of theoretical reduction is that one can perform absurd reductions, especially in the light of the Löwenheim-Skolem theorem. Indeed on Suppes' account, nothing prevents us reducing T_2 to some absurd theory T_1 which meets the specified model-theoretical requirement. This is surely contrary to "our" philosophical-reductionist intuitions and suggests in itself no reason for taking these intuitions to be flawed. Further, no specification excludes the possibility of mutual reductibility, on either of these accounts. But our reductionist intuitions tell us that any non-trivial reduction relation is asymmetric, at least if the PSU position is to be upheld.

This latter difficulty, although it has not been widely advertised, faces the KOP account as well; this is so even if two competing mutually inconsistent theories T_1 and T_2 are considered, provided (KOP/D_1) is met: from a formal perspective nothing is specified to preclude this. Next, it is surely far too weak a requirement to claim that discipline reduction occurs if at least *one* theory of the to-be-reduced-discipline is reduced, as this requirement is consistent with quite a robust anti-reductionism. Moreover, this explication does not avoid the trivialization claim, since mutual reducibility of disciplines is not ruled out by the formal definition itself.[17] This violation of asymmetry also faces Nagel's account [Kuipers, 1982, pp. 107–108].

Both the Nagel and KOP accounts have faced a critical onslaught, through the widespread criticism of the logical empiricist meta-science. For example, with respect to the KOP model as was stated earlier, if one rejects the observation-theoretical distinction, denying that there are any purely observational data [Achinstein, 1968] [Suppe (ed.), 1977, pp. 80–86], the model collapses.

The Nagel model has suffered extensive critical fire, especially from Feyerabend [1962]. First the logical empiricist account of theoretical reduction fails to accurately depict the recent history of science, ignoring the existence of the outright replacement of theories. Successive theories are taken to be *incommensurable* with regard to both the sense and reference of theoretical terms, since both sense and reference are, for Feyerabend, determined contextually by the whole theory itself. Consequently, Feyerabend argues, Nagel's condition of meaning invariance is violated. Second, the condition of derivability fails, insofar as T_1 and T_2 are frequently found, despite "partial derivability" to be logically incompatible. Finally, scepticism about the plausibility of the deductive-nomological account of explanation itself (see also [Harré, 1970] [Bhaskar, 1978] [Suppe (ed.), 1977]) further weakens the Nagel model.

Whilst Feyerabend's "incommensurability thesis" has been widely criticized, and most researchers recognize that the task of presenting a descriptively adequate theory of meaning remains unfinished, there has been widespread acceptance that Nagel's condition of derivability fails, indeed for some *paradigm cases* of theoretical reduction [Smart, 1965, pp. 159–160] [Glymour, 1970, p. 351] [Sklar, 1967, pp. 112–113] [Nickles, 1973, p. 186, 188]. Yoshida [1977] has argued against the "incompatibility argument", but his arguments have in my opinion been effectively met by Hooker [1979, pp. 93–94].[18]

One popular strategy is to modify the logical empiricist account of reduction, and argue that "it is the approximate truth of the reduced theory. . . that is deduced from the reducing theory" [Putnam, 1965, p. 206]. Here an approximation theory T^* is taken to be deducible from T_1 and T_2 i.e. $(T_1 \rightarrow T^*)$ and $(T_2 \rightarrow T^*)$. However ignoring problems in clarifying what precisely T^* is [Gaa, 1975, pp. 353, 354], we cannot say that T_2 reduces to T_1 without also justifying the counter-intuitive claim that T_1 reduces to T_2.[19]

It seems then that these early attempts are highly problematic and there is no need to consider whether these analyses would be satisfactory for an understanding of the relationship between "Sociology" and Sociobiology, since they fail, even with regards to natural scientific theories.

4.2 *More recent accounts of theoretical reduction*

Throughout the 1970's there has been an extensive production of accounts of theoretical reduction. Not all can be surveyed here, but frankly speaking this is not necessary, since virtually all proposed accounts have been subjected to extensive critique. Here I will select those accounts which I intuitively judge to be the most developed and criticize them. We begin.

4.2.1 *Causey's Unity of Science and Hooker's development*

Robert Causey's text Causey [1977] continues the PSU tradition of taking the appropriate method for the unification of knowledge to be through micro-reduction. We are not given any metatheoretical defense of this claim; Causey only argues for the thesis that if T_2 can be micro-reduced to T_1 then the most satisfactory method of unification is by micro-reduction [ibid., pp. 121–131]. This is a trivial consequence of the antecedental assumption and not a justification of his programme at all, by contrast to some alternative method of unification. Further, Causey's account of reduction is very restrictive, making unified science a quite "remote ideal at the present time" [ibid., p. 1], thus leaving us with a contemporary disunified science, and a nadir of empirical support for his thesis.

Causey restates the PSU goal of unifying all branches of science by successive micro-reductions to a "unified physics". What this "unfied physics" actually is (and what precisely is the reference of this term "science") is unclear, since as I will point out later, there does not exist at present any such entity (and indeed if logic is a science and is inexistential, then on Causey's strongly ontological account of reduction, general unification of knowledge is logically impossible). Let us outline first his account of theoretical reduction.

He proposes that the reduction of a theory T_2 to a theory T_1 involves the explanation of the laws of T_2 by the laws of T_1, where "explanation" follows the deductive-nomological account.[20] A scientific theory $T = F \cup I \cup D$, where F is a set of fundamental law sentences of T, I is a set of true identity sentences and D is a set of derivative law sentences of T. No law sentence in F is explicable within T (although it may be in another theory T*) and each law sentence in D is explicable by $F \cup I$. T consists of non-logical predicates – the familiar thing – and attribute predicates.

Let "Dom_1" designate the non-null set of objects in the domain of T_1 and "Dom_2" designate the non-null set of objects in the domain of T_2, then the necessary condition for a micro-reduction of T_2 to T_1 is that the elements of Dom_2 are identified with the elements of Dom_1. A theory T_1 has elements of two sorts, *basic* ("Bas_1") and *compound* ("$Comp_1$") elements, such that $Dom_1 = Bas_1 \cup Comp_1$. The compound elements are structured wholes whose parts are elements of Bas_1 and which can be completely understood in terms of Bas_1.[21]

Causey's major contribution to the reduction debate comes from his view about what the bridge principles in the Nagelian account of theoretical reduction must be: these are all identity statements, both thing identity and property identity statements. Such statements' designata do not require explanation, since identities are not subject to causal explanation, yet because they are *synthetic statements*, they are in need of *justification*. In short, the ontology of T_2 must be contained within the ontology of T_1 and the set of laws stated by T_2, contained in the set of laws stated by T_1 [ibid., pp. 91–92], if the reduction of T_2 to T_1 is to occur.

This position might be thought to have merit in avoiding outrightly the mutuality of reducibility criticism: if this did occur then $T_1 = T_2$. However, on closer reflection it seems that in such a situation we can just as rightly say that T_2 micro-reduces to T_1 as T_1 micro-reduces to T_2. Nevertheless this outcome is not in my opinion a strong criticism of Causey's position even though it told against the other considered positions. The condition of sameness on ontology prevents the reduction of theories which possess logically incompatible ontologies.

Nevertheless, whilst avoiding this criticism, Causey must reject alleged cases of reduction where ontologies are logically incompatible. This would exclude, as was pointed out earlier, many paradigm cases of theoretical reduction such as Newtonian mechanics to Special Relativity (where STR attributes to objects the property, that their masses vary with their velocities, a property not attributed to objects in Newtonian mechanics). If we are to have a consistent ontology, then Causey-micro-reductions are impossible in such cases. But if this is so, unification cannot be made primarily through micro-reductions as Causey conceives them. Therefore his methodological claim, the principle motivation for his programme, collapses.

A further major difficulty for Causey's position hinges around the construction of bridge statements as identity statements. Ager, Aron and Weingard [1974] argued that Causey's identity criterion – a statement will be a contingent identity statement if and only if the designated phenomena do not require a causal explanation – is circular, for if "you conjoin a postulated identity with a reducing theory, you cannot use the reducing theory to support the identity" [ibid., p. 131]. R. N. McCauley [1981, p. 223] has argued in reply that the circularity claim of Ager, Aronson and Weingard fails because it is based upon a confusion between Causey's explicit attempt to supply a semantical criterion for distinguishing identities from synthetic co-extensionalities, and an epistemic criterion of support. It is true that Ager, Aronson and Weingard do not explicitly make this distinction. Now if the criterion in question is epistemological, then the criterion does seem outrightly circular, since to decide whether or not an explanation is needed, must be decided independently of identity claims. But if Causey's criterion is taken as a semantic criterion, it also seems to fail. There may well be, if we have independent reasons for accepting some form of dualistic interactionist account of the mind–body relation (e.g. all alternative positions are absurd and it can be successfully defended itself), synthetic co-extensionalities which are not in need of explanation, although in need of justification. If "identities" are not in need of explanations, then there is no reason to assume that "differences" are.[22]

Hooker [1981, pt. I] has argued that theoretical reduction involves a spectrum of relations from straightout theory replacement, through to his own development of Causey's account of reduction. To summarize very roughly, but not falsely, T reduces to T′ if and only if in the reducing theory T′, theoretical models O′ of the basic objects O of the to-be-reduced theory T are constructable such that theoretically basic properties and entities of T are identifiable with properties and basic objects of T′. On his spectrum view Hooker can avoid the severe difficulties facing Causey's attempt to unify science solely through micro-reduction, but at the expense of not producing a unified account of reduction. Nevertheless, his own account of reduction faces problems of its own.

Let us examine Hooker's definitions of thing and property identities. They are as follows:

(DH_1) Two thing terms H_1 and H_2 have the same reference = df the same set of properties are attributable to each referent by virtue of the "theoretical roles" of their designating terms.

(DH_2) Two predicate terms F_1 and F_2 designate the same property = df (1) they "have the same applications, instances or extensions considering their extensional occurrences in predicative position" [ibid., p. 213]; (2) "the designata of the predicates instantiate the same properties, considering occurrences of the predicates in subject position" [ibid., p. 213].

On the face of it, (DH_2) seems circular, since via clause (2), to ascertain whether the designata of the predicate terms F_1 and F_2 instantiate the same properties, presupposes the establishment of the fact that F_1 and F_2 designate the same property. Further, to settle the question of extensions via (DH_2) (1) requires settling the question of property identity via (DH_1), leading us into vicious circularity.

Hooker seems to recognize this dilemma, but attempts to escape it by maintaining that contingent identifications must be introduced in "wholes". This claim is allegedly supported by the reduction of macroscopic objects to microscopic structured aggregates, one of "the great aims of science", which sadly "has not yet been satisfactorily achieved" [ibid., p. 214], words which hence indicate that Hooker has not supported his case. But he then admits that identifying each macro-object straightforwardly (holistically?) with such microstructures may well be "ontologically incoherent" [ibid., p. 213]. This leads him back to abandoning his holistic identification programme, and introducing a variety of contingent identifications. Thus he effectively abandons his holistic solution (which I fail to see how it solves the problem of circularity anyway), and falls back into vicious circularity.

Neither Causey's, nor Hooker's accounts of theoretical reduction is satisfactory.

4.2.2 Reduction and the structuralist view of theories

The so-called "Semantical View of Scientific Theories" [Suppe (ed.), 1977, pp. 221–230] [Suppe, 1979] takes scientific theories to be some form of mathematical structure which stands in a representational relationship to the objects of scientific inquiry. This approach is adopted by Beth [1948, 1949, 1961], van Fraassen [1970, 1972, 1974] and Suppe [1967, 1972, 1973, 1974], but there remains considerable disagreement as to the nature of the ascribed mathematical structure. The "Structuralist" approach, associated with work of Sneed [1971], Stegmüller [1976, 1979] and others,[23] takes the mathematical structure in question to be a set-theoretical structure. Since this approach has been preoccupied with providing an account of theoretical reduction, and since the other "semantical approaches" have not significantly shared in this preoccupation, my exclusive attention must be directed towards the Structuralists.

For the Structuralists, the logical empiricist's problem of theoretical term is still a matter of concern and the role of theoretical terms unclear. To clarify this role, a distinction is made between two dichotomies: the "observable–nonobservable" and theoretical–nontheoretical" dichotomies. What is called "T-theoretical" concepts are, roughly, those whose values cannot be calculated without recourse to the theory T. More precisely, a function f_i is T-theoretical if and only if in every application of T, f_i is measured in a T-dependent way: f_i is measured in a T-dependent way if and only if there is an individual $x_0 \in (D_i \cap D_1(f_i))$, where "$D_i$" is the domain of the i-th application of the theory T and "$D_1(f_i)$" the domain of f_i, such that the application i of theory T contains no description of a method of measuring $f_i(x_0)$ which does not presuppose that some application of T is successful [Stegmüller, 1976, p. 45] [Sneed, 1971, p. 31]. The mass function in classical particle mechanics is cited as an example of a T-dependent function, allegedly because determination of mass ratios presupposes that classical particle mechanics must apply to the physical systems used to make the measurements, whilst they allege, this is not the case for the position function [Sneed, 1971, p. 32].[24]

The difficulty which this view in general produces, is that theoretical functions seem to have a circular character [Stegmüller, 1976, p. 29] – worse, if some correct application of T is required *before* measurements can be made, it becomes logically impossible to ascertain the correctness of the initial applications for by the definition of T-theoretical we would require some further correct application, hence leading to a vicious infinite regression, or else the correctness of the initial application of the theory is ascertained in a T-non-theoretical manner, leading to contradiction. To avoid this, appeal is made to the so-called "Ramsey-sentences". This question leads us to the question of how in fact the Structuralists view scientific theories in the first place – a question which must be answered prior to answering the question of what their account of theoretical reduction is.

Theories, or what the Structuralists call "theory elements" are *set-theoretical entities* of the form $T = \langle K, I \rangle$, where I is a set of intended applications, and $K = \langle M_p, M_{pp}, M, C \rangle$. M_p, the set of partial models, is an m + k matrix, i.e. M_p is a non-empty set and for m, k $\in$ non-negative integers such that $m > o$:

$$(1) \quad (\forall x)(\exists n_1)(\exists n_2)\ldots(\exists n_m)(\exists t_1)(\exists t_2)\ldots (\exists t_k)(x = \langle n_1, \ldots, n_m, t_1, \ldots, t_k \rangle)$$

holds. M_{pp}, the set of partial possible models if defined as follows:

$$(2) \quad M_{pp} = \text{df.} \{\langle n_1, \ldots, n_m \rangle | (\forall t_1)(\forall t_2)\ldots(\forall t_k) (\langle n_1, \ldots, n_m, t_1, \ldots, t_k \rangle \in M_p)\}$$

The set of models M of the theory T is a set D (such that M = D) and $D \subseteq M_p$. The constraint C of T is defined as follows:

$$(3) \quad C \subseteq B^1(M_p)^2 \;\&\; (C \neq \emptyset) \;\&\; (\emptyset \notin C) \;\&\; (\forall x)((x \in M_p) \subseteq (\{x\} \in C).$$

whereby recursive definition $B^0(M) = M$ and $B^{n+1}(M) = P(B^n(M))$, where "P(M)" is the power set of M [Stegmüller, 1979, pp. 89–90]. For $T = \langle K, I \rangle$, the empirical claim of T is the sentence $I \subseteq r^2(B^1(M) \cap C)$, where by recursive definition $r^0(\langle n_1, \ldots, n_m, t_1, \ldots, t_k \rangle) = \langle n_1, \ldots, n_m \rangle)$ and $r^{i+1}(X) = \{r^i(Y) | Y \in X\}$ for $X \in B^{i+1}(M_p)$.

T-theoretical terms are taken to be *Ramsey-eliminable* such that one "need only investigate certain non-theoretical entities, thereby evading the restriction of producing values of the theoretical functions" [Stegmüller, 1978, p. 44]. Empirical claims of the form:

(1) c is an S

are by Ramsey-substitutions (i.e. replacing the T-theoretical functional symbols by variables and existentially quantifying) transformable into Ramsey-substitute sentences, thus allegedly solving the problem of theoretical terms.[25]

Our principal matter of interest is the Structuralist's account of theoretical reduction. Let Q, Q′ be non-null sets and $T = \langle M_p, M_{pp}, M, C, I \rangle$ and $T' = \langle M'_p, M'_{pp}, M', C', I' \rangle$. Then a *reductive correspondence* rd(R, Q′, Q) exists between Q′ and Q if and only if:

(1) $R \subseteq Q' \times Q$;

(2) $D_1(R) = Q'$;

(3) $\tilde{R}: D_{11}(R) \rightarrow Q'$.

On this basis the following definition of theoretical reduction can be given: (SDR) R reduces T′ to T = df. (1) rd (R', M'_{pp}, M_{pp});

$$(2) \quad (\forall X)(\forall X')[(\langle X', X \rangle \in B^1(M'_{pp})) \times B^1(M_{pp})(X \subseteq r^2(B^1(M) \cap C) \;\&\; (X' \notin \emptyset) \;\&\; ((\langle X', X \rangle \in \tilde{R}) \rightarrow (X' \subseteq r^2(B^1(M') \cap C')))]$$

where $\tilde{R} = df. \{\langle X, Y\rangle \in B^1(Q') \times B^1(Q) | (\exists c)((X \leftrightarrow Y) \& (\forall x)((x \in X) \& \langle x, c(x)\rangle \in R))\}$

(3) $(\forall x')(\forall X')[(x' \in X') \& (X' \in I') \rightarrow (\exists X)(\exists x)((x \in X) \& (X \in I) \& \langle x', x\rangle \in R)]$

It is possible to define other reduction relations such as "strong reducibility", "weak reducibility" and "incomplete reducibility", but this will not be necessary here.

For those critical of the "theoretical-non-theoretical dichotomy, the philosophical significance of Ramsey-sentences, and Kuhn's philosophy of science (reconstructed by Stegmüller [1976]), structuralism will be viewed unfavourably. Here is not the place for an epistemological critique of structuralism as interesting and needed as this would be. Let us restrict our scope to the question of the plausibility of SDR. The major difficulty which exists here, I feel, is that the reduction relation is so weak as to allow reduction relations to hold between intuitively regardable, unrelated theories, especially in the light of the Löwenheim-Skolem Theorem. This was a problem facing the accounts of Adams and Suppes and seems applicable to virtually all "formalist" accounts of reduction which I know of. The great generality of logic and set theory virtually guarantees this.

Part of this difficulty must lie with the abstract set-theoretical style of definition practiced by the Structuralists. The point to be made, as David Salt [1971] stated in a similar context, is that physical (or social) theories are *always* interpreted theories, *not abstract mathematical structures*. The Structuralist position leads to some absurd results from a realist and non-Pythagorean position: consider for example Sneed's [1971, p. 114] definition (D7) of particle mechanics, which makes no reference in the definiens to physical particles at all! It is easily modelled in the domain of positive integers in the light of the Löwenheim-Skolem Theorem. On the realist account of axiomatization, championed by Bunge [1967 (2)], since scientific theories are interpreted theories they have no models (other than the true physical interpretation itself), although the signs of the syntax may be given any number of interpretations.[26]

The conclusion which I am led to, is that the structuralist account of reduction does not avoid the major difficulties facing alternative accounts.

4.3 *The abandonment of reduction*

The argument from sections 4.1 to 4.2 had led us to an attitude of scepticism about the satisfactoriness of some of the central and most developed accounts of theoretical reduction. This leaves of course one further account, viz. what Schaffner [1967, p. 138] has called the "Popper-Feyerabend-Kuhn account of reduction", being the outright replacement of theories. Yet this position can also be viewed as the denial of the possibility of theoretical reductions at all. This is in fact the claim which Paul Feyerabend made in his classic paper

"Explanation, Reduction and Empiricism" [Feyerabend, 1962], when he stated that "a formal account of reduction and explanation is impossible for general theories" [ibid., p. 28]. Here I go beyond Feyerabend's claim (only with regards to reduction) and maintain that developing accounts of theoretical reduction is simply a waste of intellectual resources, and that neither our science nor meta-science, will suffer from its elimination.

A more fruitful strategy of inquiry for the philosopher to investigate is the idea of a "better explanation"; to outline, explicate and contrary to Feyerabend, justify the general features of rational theory choice and theory criticism, as part of a global pattern of metaphysical inquiry and *Weltanschauung* building. I will attempt to expand on this point in section 5: the point to be made here is that some alternative to unification by reduction must be sought, and "world picture making" readily suggests itself as a mode of the unification of all human knowledge. This was classically the philosopher's role, and one which must once more be fulfilled.

Given that I am advocating the elimination of the programme of theoretical reduction, this does not reduce to absurdity my own work here, since I began my inquiry attempting to discover whether a theoretical reduction of "Sociology"/Culturology to Sociobiology was plausible. Yet we have not even met first base here in considering what such a reduction would be like, since it has been argued that the principle accounts of theoretical reduction are flawed. How in fact can the contemporary Sociobiology debate be understood?

One should note that initial formulations of questions of inquiry may well come to be rejected in the course of the argument, but this does not lead me into vicious circularity at all. Rather, what I wish to say is that "Sociology"/ Culturology and Sociobiology are simply competing orientations and that what is at issue in this debate is the "*explanatory rationality*" of these respective orientations. In the next chapter, when examining the question of the respective "explanatory reduction", but I am not intending this to explicate the general idea of theoretical reduction, for I see no reason to believe that my own analysis will succeed where others have failed. These remarks are applicable also to the concept of "reduction" introduced in the previous chapter. It is rather a low grade aesthetic device, implemented to add continuity throughout the work, given that we do not end this work now on the note of our present sceptical conclusions, but that a further discussion of other related issues follows.

Let us quickly get our sense of direction in this argument. The most plausible metaphysical *Weltanschauung* which Sociobiology could be located in, is the PSU Culturology could be subjected to a metaphysical critique. I have attacked one support of the PSU programme, and in section 5 will continue this attack in the hope of resolving outstanding problems. Further, if I am to follow my own recommendations, Culturology itself needs its own metaphysical *Weltanschauung*, the articulation of which is an even more difficult matter than the initial clarification of the position itself.

5.0 Unification without reduction; Philosophy without physicalism

In this concluding section I will directly attack some of the remaining assumptions of the PSU orientation, and state briefly an alternative world hypothesis which will be presupposed throughout my foundational critique of Sociobiology. Here a recipe for a critique of the PSU metaphysical *Weltanschauung* will be canvassed in two broad steps. First, an internal critique will be presented, centred around internal difficulties which the PSU programme faces. Second an external critique will be sketched in broad outline, where an alternative model of the unification of human knowledge will be given. In doing so, I hope to present a strong reply to the eliminative materialist critique of Culturology, and thus put my own philosophical house in shape for the critique of Sociobiology which will follow.

5.1 Systematizing the internal critique

Virtually all of the major assumptions of the PSU metaphysical *Weltanschauung* are open to debate, although explicit critiques of these assumptions are lacking in the literature. Not all of these points can be covered here, as condensed and concise as the discussion will be. The major points which warrant immediate critique are these: (1) the coherence of the PSU ontology, especially the idea of the "physical" or the "material"; (2) the epistemological-theoretical thesis of the explanatory fundamentality of physics (or in general the natural sciences) and the plausibility of the enterprise of taking such sciences as model(s) for the unification of human knowledge. Discussion of the last point will immediately lead us back to the question of how in fact human knowledge should be unified, and provide a response to the eliminative materialist challenge to Culturology.

5.2 Against the "physical" in physicalism

Despite the intense debate generated in the past twenty-five years in the philosophy of mind over the plausibility of various forms of "materialism" and "physicalism", there has been little systematic thinking about what the "physical" and the "material" actually are. Previous attempts to actually define "materialism" popular in the Marxist tradition, such as the attempt of Lenin [1970, p. 100], take "materialism" to be the doctrine which denies the ontological reduction of the world to mental experience. This position is, however, more accurately described as the position of "Realism" [Rubin, 1977, pp. 4–5]; in any case the definition seems to necessarily presuppose dualism (especially evident in statements such as "the material is that which exists independently of mind") insofar as ineliminable reference is made to the "mental" in the definiens of the definition, that is, the "material" is defined in terms of it.

The principal strategy adopted by the "physicalist" in explicating the nature of the "physical", is to maintain that physical object's properties and relations are precisely those postulated by physics [Smart, 1963 (2), p. 651] [Armstrong, 1978 (1), p. 268] [Feigl, 1963, p. 242] [Levin, 1979, pp. 60–63]. The central motivation for this appears to be dependent upon the thesis of the explanatory fundamentality of physics, a thesis which will be criticized shortly.

Armstrong and Feigl are vague about what precisely this thing "physics" is. Levin wishes to relativize the concept of "matter" and the "physical" to that currently appealed to by the "physics community". Smart in a later paper [Smart, 1978 (1)] adopts the "two-bob-each-way" strategy of keeping for the purposes of the mind–body identity theory an account of physicalism relativized to contemporary physics. For the purposes of general metaphysics, however, he states: "I can tie my definition of physicalism to present day physics, and a twenty-first century physicalist can tie it to twenty-first century physics" [ibid., p. 340].

The major difficulty with this strategy of explication, is that it leads to intuitively absurd results. (We gave an argument against this position previously (cf. note 10) and summarize our argument once more and consider some further exit strategies). If the physics community is one's judge, then by an arbitrary social revolution, physics in a society may become dominated by "scientists" who refer to "anti-physicalist" entities such as God in physical explanations, especially with regard to cosmology. "Physicists" of the Scientific-Creationist movement already do this. It is no good to argue that there is no good reason to believe in God, since the decisions of the physics community is your standard of ontological commitment and further, no constraint has been placed upon the rationality of this community. To place such a constraint requires an independent characterization of a reasonable metaphysics that epistemic naturalists such as Smart do not supply. So according to Smart, God now is a "physical" object. This I take to constitute a *reductio ad absurdum*.

Further if we restrict our attention to the "physics" of today, we find further difficulties. The physicalist's criterion of the "physical" is based upon an acceptance of a realist conception of quantum mechanics, and there are well known difficulties facing realist interpretations [D'Espagnat, 1971] [Mermin, 1981]. Some physicists have been led to regard mentality as a principle of central importance in our understanding of the quantum domain [Wigner, 1967, 1970] [Cochran, 1971] whilst others – perhaps the majority – hold to some variant on the Copenhagen position (and instrumentalism *is* a position still defended [Cornman, 1971]). It is utterly unclear what ontology "physics" accepts at present (if we are not merely to appeal to arational methods such as asking for a show of hands!). Therefore it is unclear what ontology physicalism accepts. So the position is theoretically bankrupt.

Levin [1979, pp. 60–61] argues that physicists such as Wigner and Cochran cannot be right because mentality is a property of complex organisms. The point is both question begging and irrelevant. Pan-psychists [Globus, 1976] would immediately reject Levin's claim. The point is irrelevant insofar as to consider the adequacy of an explication it is sufficient to appeal to thought experiments to derive the logical consequences of a position. It is sufficient to point out that on the relativistic account of the physical under consideration here many entities would be "physical" which are usually taken as not being so.

A retreat strategy is to divorce "physicalism" from "materialism". Smart [1978 (2)] has done this to avoid the difficulties which the abstract entities of mathematics pose for physicalists who are frustrated with the progress of mathematical nominalism, and are reluctant Platonists. Since physics presupposes these entities, and no non-circular physicalistic reduction seems possible, they are accepted to be "physical" but not "material" objects. This, apart from being, for the non-converted, an *ad hoc* move, renders physicalism demonstratably inconsistent with the thesis of naturalism. Naturalism is "the doctrine that reality consists of nothing but a single all-embracing spatio-temporal system" [Armstrong, 1978 (1), p. 261]. Armstrong, who has addressed this problem in more detail than Smart, recognizes that abstract entities are a paradigm case of "transcendental entities" outside the naturalistic sphere of influence [ibid., p. 263]. To bring them within the framework of naturalism he attributes causal powers to them [ibid., p. 267] thus contradicting his own naturalism, since we have an example of an entity which exists according to a causal criterion of existence which is not part of the naturalists' spatio-temporal frame.[27] Physicalism and Naturalism are hence inconsistent. Physicalists should therefore find Smart's *ad hoc* move quite unpalatable.

An alternative physicalist strategy, adopted by Ellis [1976, 1979, p. 46], takes physical entities to be any entity which was energy, so that neutrinos are "physical", whereas abstract entities are not. Physical properties are then taken to be any properties of physical entities which "make a [causal] difference" to some physical process involving that entity. Physical events are any changes in the energy distribution in the universe, and physical processes are any causally connected sequence of physical events [Ellis, 1976, pp. 179–180].

On this account forces and energy itself remain, as Ellis recognizes, problematic, being either "non-physical" particulars or attributes, whatever the case may be. Likewise for space: if an absolute view of space is adopted [Nerlich, 1976] where space is a particular, such a particular is non-physical (as space has no causal powers [Bunge, 1977, pp. 328–329]), or if a relationist position is taken, then spatial properties are non-physical properties. Yet from an intuitive point of view, space is surely a "physical" phenomena. Worse yet, a Cartesian mind, interacting with the brain, constitutes a physical event, and such a Cartesian mind is then, according to Ellis, a *physical* property of the brain, which is outrightly incoherent.[28]

The final account which I will consider here is presented by Cornman [1971, pp. 13–14]. Cornman recognizes both the extreme importance and the difficulties in characterizing the "physical" and rejects himself a number of accounts (especially [Sellars, 1965]) for their circularity. He finally settles on this definition [Cornman, 1971, p. 14]:

(CDP) O is a physical object = df O has physical properties, and O has no properties that a spatio-temporal individual which is living, or is a part of a living individual, would have only if it, or what it is part of, were living.

Physical properties and relations, of and between individuals, are either properties of, or relations between, spatio-temporal individuals, which are neither living individuals nor attached parts of such. They are not properties or relations which such individuals would have if it, or what it is a part of, were living.

The difficulty with this definition, as has been recognized by D. H. Mellor [1973, pp. 107–109], is that the concept of "living" is not adequate for the analysis of the concept of the "physical". Apart from suffering from clear counter-examples, such as artificially sustained brains in dead bodies [ibid., p. 109], the analysis cannot work since any plausible contemporary account of "the living" (e.g. [Bunge, 1979, pp. 79–80]) will make reference to various biochemical concepts, and the same fate which Cornman [1971, pp. 10–11] presented for Sellars' account will meet his own: it will be *circular* since at least some of these bio-chemical concepts will be taken to be "physical" concepts.

Our conclusion is brief: PSU has yet to produce a satisfactory explication of the "physical" or "material" and consequently physicalism is an unsatisfactory, ill-defined philosophical position. Consequently one of the major metaphysical weapons which is used against the Culturologist is disarmed: it cannot be argued that because cultural entities have properties which are seemingly "non-physical" that they are thus ontologically suspect, for it is the PSU idea of "the physical" which is suspect. Let us turn now to the third major thesis of the PSU programme.

5.3 PSU and explanatory fundamentality

Does physics, or any other "natural science" for that matter, occupy an epistemically privileged and explanatory fundamental place in the scheme of human knowledge? The answer which I will give to this question is once more negative. Consequently, physicalist responses to problems such as the "secondary qualities", are problematic insofar as the title of physics (or any other natural science) as the judge of what there is, is wrestled away. If the secondary qualities for example, are inexplicable within the framework of physics, so much the worse for physics.

I should remark at the outset that there are well-known attempts to reconcile the PSU ontology with a general anti-theoretical reductionist orientation, the most notable being Donald Davidson's *anomalous monism* [Davidson, 1970] and the functionalist-tokenism of Jerry Fodor [1974, 1981]. These positions have come under very intense, and I believe successful, critical attack and I refer the reader to the appropriate literature.[29]

Let us return to the main course of the argument. There are two questions which we need to tackle: the question of the epistemological primacy of physics or the natural sciences in general and the question of the explanatory fundamentality of physics or the natural sciences. I will now address the former question.

The first task is to clarify our object of inquiry. Talk of the epistemological primacy of physics and the other natural sciences seems to encompass two questions. First is the claim that physics, or the natural sciences, allegedly give us ideal models of the structure of knowledge and rational inference itself. Second is the claim that if current and foreseeable physical and natural scientific theories ran into foundational difficulties, then all of human knowledge is thrown into disarray: this is a direct consequence of the PSU claim that the sciences are unified by micro-reduction.

With regards to the first question, it is undoubtably true for all but those who embrace epistemological scepticism, that the natural sciences have supplied us with much knowledge. Yet one can, as epistemological sceptics have done, not only challenge whether so-called scientific knowledge is really *knowledge*, but also challenge some metaphysical presuppositions of science itself (e.g. that an external world exists or that observers exist). The attempts of the positivists to reject such a question as being cognitively meaningless are widely conceded to have failed. Therefore the natural sciences are not epistemologically primary; rather it is First Philosophy which takes such a title. Therefore this PSU claim collapses.

Let us address the second part of the question. It is true that severe foundational crisis within the natural sciences would leave human knowledge in disarray, and there seems adequate material for a contemporary Sextus Empiricus to present a critique of the major fields of the natural sciences.[30] Yet if physics "crumbled", it is not at all clear how this would affect "Sociology", and whilst the effects of a widespread foundational crisis in physics may have *some* impact upon "Sociology", that it would have *the* impact which the PSUist dreads, requires a demonstration of the unification of science by reduction thesis. The fact that there might well be radical changes in physics and biology without resultant radical changes in "Sociology", gives some reason to doubt the truth of this PSU claim. In any case since this epistemological thesis presupposes the reduction thesis, it cannot be cited in support of the latter reduction thesis without circular argument.

To continue a thesis which has run throughout this work, what would produce a total disarray of human knowledge is the failure to resolve philosophical problems such as Hume's problem of induction, the challenge of scepticism and solipsism, to name but a few. If for example, I had no reason to believe anything at all as the sceptic proposes, then I necessarily have no reason to believe anything of physics and "Sociology". Thus, in this sense, it is philosophy, not physics, biology or "Sociology" which is "epistemologically basic".

We turn now to the question of the explanatory fundamentality of physics and the other natural sciences. Our immediate task is to clarify the meaning of this question, so let us review the account of fundamentality given in Section 2.3 here:

(DF) A field F_1 is more fundamental than a field F_2 = df F_1 explains everything F_2 describes and more, in either greater detail or a greater domain of phenomena, perhaps themselves in greater detail as well.

On this account, physics or any other natural science is *not* explanatory more fundamental than "Sociology", unless it can be shown that the former fields can *at least* explain all socio-historical phenomena, and this presupposes the truth of the reductionist programme. Hence ontological considerations, most usually of a nihilistic mode, justified by appeals to the fundamentality of physics (for example with regards to the *sui generis* nature (or lack of) of the secondary qualities) are circularly defended.

The PSUist would no doubt feel that something is drastically wrong with (DF). Still, we undoubtably have an intuition that the natural sciences are in some way "deeper" than the social sciences. The next explication is this: a field F_1 is more fundamental than a field F_2 if and only if the entities of F_2 obey all the laws of F_2. Whilst all the phenomena in the universe "obey the laws of physics", a restricted domain of phenomena obey biological laws, and less still "Sociological" laws.

However, even if we grant that all phenomena are governed by the "law of physics", it does not follow, and indeed is explicitly recognized in the above statement, that there are no further laws than those of physics. This gives no support to the PSU programme. What it in fact shows is that physical descriptions are necessary, but not alone sufficient, for the explanation and description of phenomena in the universe [Grene, 1971, pp. 20–21] [Polten, 1973, pp. 231–232].

Indeed the above explication of fundamentality, taken literally, is false. It is not true that all biological phenomena, for example, obey *all* the laws of physics: genes for example, do not obey the gas laws because genes are not gases, and thus, the gas laws are inapplicable. As a second example, some cultural entities such as numbers, propositions, arguments and theories do not seem to be capable of being brought under the sway of physical laws: indeed the mathematical formalism of contemporary physics indicates that it is physics which *presupposes* mathematics, not the other way around. So if we are not to speak nonsense, an explication of fundamentality something like this must be accepted: a field F_1 is more fundamental than a field F_2 if and only if the entities of F_2 obey all the *applicable* laws of F_1, but it is not necessary that all the entities of F_1 obey all the *applicable* laws of F_2. This account suffers further difficulties (such as: if the laws of F_2 are *applicable*, then it is self-contradictory to suppose that F_1 entities do not obey them, as this is precisely what "applicability" involves), but the central point to be made here is that *if* some account such as the above is accepted, the thesis of "universal mechanism" must be rejected, and this is a crucial assumption of the PSU programme. True, one might say, but uninteresting. What the PSUist is surely talking about are some very basic laws of nature, which all phenomena must obey. If this claim was rejected, then it would seem, as an aside, that one has a decisive answer to the mechanist: if biological phenomena are *nothing but* physico-chemical phenomena, then biological phenomena obey all the *basic* laws of physics and chemistry; but if biological phenomena do not obey all the *basic* laws of physics and chemistry, then biological phenomena are *nothing but* physico-chemical phenomena.

I need not defend such a controversial thesis here to make my point: the latter claim whilst quite plausible, does nothing to support the PSU programme; it is consistent both with "emergentism" and forms of dualism, that is to say, with competing metaphysical research programmes. However, I wish further to draw two items of attention to the reader's notice before passing on to my conclusion:

Item 1: If a *strictly* "Realist" interpretation of quantum mechanics is abandoned, then it is not metaphysically self-defeating to suppose than quantum mechanics is not applicable to the phenomena of life and consciousness [Wigner, 1969, 1970] [Cochran, 1971].

Item 2: Interpreting the second law of thermodynamics raises many perplexing problems [Popper, 1965] [Prigogine, 1973] [Petersen, 1975], especially with regards to living systems and their possible actions as Maxwellian Demons [McClare, 1971, 1972]. The philosopher-physicist P. J. Zwart [1976] has cut the Gordian knot and maintained that if the second law is valid at all, then it is not universally valid for all biological processes. (He defends this thesis against the usual counter-attacks such as "the open-systems argument" [ibid., pp. 129–130]).

These items establish that the point under consideration is at least a matter of controversy, at least in need of rational investigation rather than dogmatic dismissal. Hence we must accept at least the logical possibility, that what for the PSUist, are causal antinomies, *may* well exist. Whether they *do* is, however, another matter for inquiry.

We pass now to a conclusion. We have found no reason to regard the natural sciences and in particular, physics as epistemologically basic and explanatory primary in any interesting sense for PSU. Consequently, the essential foundations of the PSU *Weltanschauung* are rejected – and hence its presupposed system of metaphysics is rejected. But this leaves us with one question: what is the alternative? This is always the moment the critic dreads, when the spot light of the burden of proof is shifted from centre ring and trained upon him/her.

5.4 An alternative: Systemic Emergent Realism

Let us be clear, before we are lost in the clouds of metaphysical reconstruction why the issues which are being pursued here are of relevance to the central concerns of this project. What we have attempted to do in this chapter is to criticize the metaphysical system which I propose most plausibly underlines Sociobiology. Now I wish to sketch in broad outline some alternative direction from which both my critique of Sociobiology and response to the eliminative materialists critique of "Culturology" will proceed. The material to follow must be considered as quite tentative footnotes to a much more comprehensive research project.

The root metaphor underlying this work is the ideal of *Systematicity* – and indeed it was the seeming failure of the satisfaction of this ideal by the social sciences, which necessitated the long and torturous metaphysical and meta-scientific journey already taken. But why accept this ideal at all? The answer which was given in the previous chapter was that "The Truth" itself we postulate, exhibits systematicity: the Truth is a system, Knowledge is a statement of the Truth, therefore Knowledge is a system [Rescher, 1979, p. 19]. But what does it mean to say that "the Truth" is a system? Nothing more than that the World is a world of systems. But what are these systems? How does one individuate systems from other particulars in the world? Or are all particulars systems, in which case our position is vacuous and intellectually uninteresting? Clearly some general account of a system is needed.

When we turn to General Systems Theory for aid in such explications we face frustration, for neither the "founding father" Ludwig von Bertalanffy [1968] nor one of its chief philosophical expounders Ervin Laszlo [1969, 1972 (1), 1972 (2)] have presented a satisfactory general account of the concept of a *system*. This failure, and the often confused use of this term has stood as one central line of criticism of General Systems Theory [Sadovsky, 1974] [Esposito, 1976] [Berlinksi, 1976]. There are two important attempts to deal with this problem: that of Marchal [1975] and Bunge [1979], both of which suffer a major metaphysical difficulty, but one which can be repaired.

Marchal [1975] surveys a number of previously advanced definitions of the concept of a system and rejects them because they violate one or more of these material conditions of adequacy for such definitions:

(MCA) A satisfactory definition of a system must:

(1) make clear the basis for identifying and distinguishing between *different* systems;

(2) make clear the basis for identifying and distinguishing between different *kinds* of systems;

(3) will not result in *everything* becoming a system [ibid., p. 452].

His own formal explication is as follows:

(MDS) S is a system only if $S = \langle E, R \rangle$ where:

(i) E is an element set and

(ii) $R = \langle R_1, R_2, \ldots, R_n \rangle$ is a relation set, i.e. $R_1, R_2, \ldots, R_n$ are relations holding among the elements of E [ibid., pp. 462–463].

On this position, the *type* of system, be this a physical, mathematical or philosophical system, is specified by the conditions placed upon (MCA)(1) and (2). Major metaphysical difficulties arise from Marchal's attempt to meet (MCA)(3): to exclude things such as chairs and people as systems he argues that "physical objects. . . are not even sets, let alone the sort of set a system is" [ibid., p. 465]. This is an extremely counterintuitive consequence, since we would surely wish to regard concrete systems as real objects rather than constructs. The same difficulty faces Bunge's position.

Bunge [1979] produces both a general and a more specific account of a system. In general a system $\Sigma = \langle C, E, S \rangle$ over a non-null set T, where $((C \subset T) \& (E \subset T)) \& (C \cap E = \emptyset)$ and where S is a non-null set of relations on $C \cup E$. A conceptual system is defined where T is a set of conceptual particulars; a concrete system is defined if T is a set of concrete particulars. However this account is but a schema of a definition, and we need to know how C, E and S are interpreted. My principal concern here is with the concept of a concrete system.

A precise set-theoretical account can be given through defining the concepts of *A-composition*, *A-environment* and *A-structure* of a concrete particular x and a class A of particulars. Let "$\sqsubset$" designate the relation "is a part of" such that "$y \sqsubset x$" reads in English "y is a part of x". Then the A-composition of a particular x, $C_A(x)$, is the set of its A-parts, i.e.

$$(DB_1)\ C_A(x) = df\ \{y \in A \mid y \sqsubset x\}$$

Let "$\triangleright$" designate the relation "acts upon" such that "$a \triangleright b$" reads in English "a acts upon b", then a and b are *bonded* if and only if $((a \triangleright b) \vee (b \triangleright a))$. The set of relations amongst the components of x consist of bonding relations B_A and non-bonding relations $\bar{B}_A$.

The A-Environment $E_A(x)$ of a particular x, with A-composition $C_A(X)$ is defined as follows:

$$(DB_2)\ E_A(x) = df\ \{y \in A \mid y \notin C_A(x) \& (\exists z)((z \sqsubset x) \& ((y \triangleright z) \vee (z \triangleright y)))\}$$

Finally the A-structure $S_A(x)$ of a thing x, with A-composition $C_A(x)$ and A-environment $E_A(x)$ is defined as follows:

$$(DB_3)\ S_A(x) = df\ \{R_i \in B_A(x) \cup \bar{B}_A(x) \mid (B_A(x) \neq \phi) \& (1 \leqslant i \leqslant n)\}$$

We can now define the concept of a concrete system Σ at t:

(DB_4) A concrete system Σ at time $t = \langle C_A(\Sigma, T), E_A(\Sigma, t), S_A(\Sigma, t) \rangle$, where:

$$(1)\ C_A(\Sigma, t) = df\ \{x \in A \mid x \sqsubset \Sigma\}$$

$$(2)\ E_A(\Sigma, t) = df\ \{x \in A \mid x \notin C_A(\Sigma, t)) \& (\exists y)((y \in C_A(\Sigma, t)) \& ((x \triangleright y) \vee (y \triangleright x)))\}$$

$$(3)\ S_A(\Sigma, t) = df\ \{R_i \in B_A(\Sigma, t) \vee \bar{B}_A(\Sigma, t) \mid B_A(\Sigma, t) \neq \phi) \& 1 \leqslant i \leqslant n)\}$$

Two further concepts are of importance here: that of *level* and that of *emergence*. The so-called "levels" or "strata" of the world are not particulars for Bunge, but sets – hence levels cannot act upon each other. One level is said to *precede* ("$<$") another if and only if all the particulars of the latter level, have as their composition, particulars of some or all of the particulars of the former. A particular belongs to given level if and only if it is composed of some, or all, particulars of preceding levels. Formally:

(DB_5) Let $L = \{L_i \mid 1 \leqslant i \leqslant n\}$ be a set of non-empty sets of concrete particulars. Then:

$$(1)\ L_i < L_j = \mathrm{df}\ (\forall x)[(x \in L_j) \rightarrow (\exists y)((y \in L_i)\ \&\ y \in C(x)))]$$

$$(2)\ (\forall x)(x \in L_i) = \mathrm{df}\ C(x) \subset \bigcup_{k=1}^{i-1} L_k\ \text{[ibid., p. 13].}$$

The relation "$<$" is a partial order relation such that $k = \langle \Sigma, < \rangle$ is a partially ordered set.

The final concepts of interest here are those of *qualitative novelty*, *emergent properties* and *absolutely emergent properties* [ibid., p. 30]. The following definitions are then offered:

(DB_6) Let x be a concrete particular with a set of properties Px(t) at time t, and a set of properties Px*(t′) at time t′, t′ > t. Then:

(1) the set of *total qualitative novelty* occurring in x during the time interval [t, t′] is the symmetrical difference:

$$n_x(t, t') = Px(t) \triangle \bigcup_{t \leqslant r \leqslant t'} Px^*(r)$$

(2) The set of *emergent properties* of x in the time interval [t, t′] is:

$$e_x(t, t') = \bigcup_{t \leqslant r \leqslant t'} Px^*(r) - Px(t)$$

(3) the set of *absolutely emergent properties* of x during the time interval [t, t′] are those such that

$$e_x^a(t, t') = (e_x(t, t') - \bigcup_{y \in A} \bigcup_{r \in T} (Py(r)))\ \&\ (y \neq x)\ \&\ (r \leqslant t').$$

This account achieves a precise explication of some central concepts of the "Systemic Emergent Realist" world view, but the analysis if not without its peculiar metaphysical problems. The difficulty lies in asserting, via use of the identity sign "=" that concrete systems are sets: since sets are conceptual objects, this conflicts with Bunge's conceptual-concrete distinction. And to claim that one is engaging in *mathematical modelling* rather than in giving *real definitions* is to stand open to the charge of engaging in an irrelevant practice in this context. Phenomena may be mathematically modelled in any number of ways.

It is proposed that we accept systems, such as socio-systems, biological systems and so on, as concrete particulars in our ontology.[31] A concrete system is simply a particular which has a composition, an environment and a structure. Use of the "calculus of individuals" [Goodman, 1966] can be made to restructure Bunge's definition.

We take as primitive the relation "$\sqsubset$" read "is a part of". On this basis we can give an "anti-nominalistic" definition of a fusion "Fu" [Leonard and Goodman, 1940, p. 47].

$$(D_1)\ W = Fu(y, Gy) = df\,(\forall y)(Gy \rightarrow (y \sqsubset W))\ \&\ (\forall u)$$
$$[(u \sqsubset W) \rightarrow (\exists v)(\exists z)[(v \sqsubset u)\ \&\ (v \sqsubset z)\ \&\ Gz]]$$

where y, u, v, z and W are particulars and G is a property. The definition could be easily modified to take into account further properties of y by simply substituting "$G_1y\ \&\ G_2y\ \&\ \ldots$" above for "Gy".

On this basis we can define the composition C(x) of a particular x as follows:

$$(D_2)\ C(x) = df\ Fu(y, Gy)\ \&\ (y \sqsubset x)$$

The environment of a particular x is defined as follows:

$$(D_3)\ E(x) = df\ [Fu(y, Gy)]\ \&\ {\sim}\ [y \sqsubset C(x)]\ \&\ (\exists z)[(z \sqsubset x)\ \&$$
$$((y \triangleright z) \quad (z \triangleright y))]$$

The structure S(x) of a particular x is defined as follows:

$$(D_4)\ S(x) = df\ (\exists F)\ Fx\ \&\ (\exists y)(\exists z)[y \sqsubset x)\ \&$$
$$(z \sqsubset x) \rightarrow ((y \triangleright z) \vee (z \triangleright y))],$$

where the properties of x are either bonding or non-bonding properties we will also need later, the concept of *essential structure*:

$$(D_5)\ ES(x) = df\ Nature\ (x)$$

Such matters will be discussed in the last chapter.

With regard to the concept of levels, we can hold much to the ordinary English account given prior to definition (DB_5), with the qualification that levels are fusions of particulars and *do* in fact act and interact upon each other, that is, there is *downward and upward causation* in the world [Popper and Eccles, 1977, p. 19].

Does our position meet (MCA)? (MCA)(1) and (2) are met by specifying the nature of the particulars, properties and relations of the system. (MCA)(3) is met because not every particular is a system: the basic elementary particles or various bootstrap particles [Chew, 1968] lacking a composition are not systems, nor is the Universe, having no environment. Our definition is general, a virtue one would expect from metaphysics, but it is not vacuous.

Having said what systems are, and having also said what emergent properties are in a quite abstract fashion, some further more empirically orientated discussion is needed to clothe these abstract bones. Of course in the space available here, the clothing shall not be as cosy as one would wish, especially if one dwells in the ice wastelands of the PSU *Weltanschauung* – but that failing is unavoidable, and to say nothing at all, a far worse intellectual sin. Onwards with purity. . .

(1) Quantum non-separability is a consequence of the elementary principles of quantum mechanics [D'Espagnat, 1971, 1973]. Insofar as quantum mechanics is taken at "face value", it seems that in the consideration of a composite system S with components I and II, quantum non-separability does not permit us to attribute in isolation some specified states to I and II [Shimony, 1978 (2)]. This seems to indicate that on this interpretation of quantum mechanics, there are emergent, ontologically irreducible systemic properties.

(2) D'Espagnat [1976, p. 236] has argued that the thesis of macro-objectivism, that macro-scopic objects are in some given region of space and not in some other region cannot be reconciled with "ordinary quantum mechanics". Of course this statement is controversial as all such metaphysical statements in this field are. If we are forced to accept that this is the case, then we face a problem: surely macro-objectivism is as well confirmed as any theory could ever be: my desk is in the region of space near me, not in a region of space six miles from Mars. The argument which D'Espagnat employs to reach this metaphysical antinomy appeals to this assumption:

(R) Macro-objects are composite systems in principle describable by means of the algorithms of quantum mechanics, namely state vectors and density matrices.

This antinomy is readily resolved by rejecting (R), the reductionist assumption. Once (R) is rejected, contemporary physicalism is seen to be false even for commonsense objects such as tables.

(3) Quantum mechanics also gives us good reason to believe that downward causation occurs: the extended charge-current distributions of particles are dependent upon the nature of the physical system to which they are components, as well as upon the surrounding environment [Noyles, 1975, pp. 24–26] [Anderson, 1972].

(4) At a "higher level" we find further evidence of downward causation and ontologically irreducible properties. Firstly, instances of downward causation include the interaction between an X-ray beam and a crystal lattice, the photoelectric effect and the relationship between incoming photons and gravity collapsed stars. Secondly, as Platt [1961] has detailed, there are properties of molecules in the 5–50 atom range which have no counterparts in diatomic molecules, and many triatomic molecules and molecules in the 50 to 500 atom range that go beyond the properties of their respective chemical subgroups.

If one had the opportunity one could write a treatise moving upwards through biochemistry to psychology, "Sociology" and ecology, detailing how the sciences can be interpreted in the light of the metaphysical *Weltanschauung* of Systemic Emergent Realism. Physics examples are taken here because they are the "toughest cases" to deal with.

Now it will be argued at this point, as J. J. C. Smart [1963 (1), pp. 50–52] has done, that the thesis of emergence is untenable, since, for example, it is possible to "deduce the properties" of NaCl from the *purely physical properties* of Na and Cl alone, by spectroscopic techniques with reference to the quantum theory of the chemical bond. But in reply, it is pointed out, first, that we cannot "deduce properties", since the relationship of logical deducibility holds between sentences or propositions, not properties. Second, spectroscopic methods with respect to purely physical non-relational properties of elements are primarily concerned with ascertaining rotations, translational, vibrational and electronic energies [O'Dwyer et al., 1978]. Having a quantum theory of chemical bond only shows that we have a quantum theory of the chemical bond – *not* that "physics explains chemistry" [Bantz, 1977, p. 118] or that *all* the properties of a compound can be ascertained from a consideration of its constituent elements in a *non-circular fashion* [Schievella, 1973].[32] The argument here, so widely accepted, is of the same validity as the claim that because a purely biological explanation of the physiology of people's speech capacities is given (it is claimed), therefore all of socio-linguistics is reduced to physiology: it is fallacious. In both cases I prefer to speak of the primary science acting as a "foundation" for the secondary science, ending a potentially infinite regress of secondary science explanations by a primary science explanation.

Smart's argument fails to show that there are no *ontologically* emergent properties, but at best shows that emergent properties are rationally explicable [Bunge, 1979]. This in itself does not constitute grounds for ontological elimination via Occam's razor, since the Systemic Emergent Realist would maintain that neither particulars, properties nor relations are being *unjustifiably* multiplied. If one doubts this in any single case, the burden of proof is upon one to perform the reduction *and* demonstrate the unreasonableness of the proposal. True, a physicalist ontology is *ontologically simple*; but it is another question as to whether it is *ontological adequate*. The ontology of the *ontological* nihilist is even more economical: nothing exists at all. If considerations of solely ontological economy dictate our world making, then physicalists are recommended to become ontological nihilists.[33]

Preferability of the Systemic Emergent Realist Metaphysical *Weltanschauung* comes from its ontological problem solving power. It is of course not possible here to detail all the arguments necessary to provide a solution to these cited problems, rather all we shall do here is state a direction of argument. Eddington's two tables problem is essentially a problem generated by the PSU ontology. Given a levels conception of reality, and given also that particulars are not ontologically identical with their structures, we need not become involved in the paradox of ascribing mutually incompatible properties to one particular. Tables are not "clouds of electrons", nor are people expressive totalities of DNA molecules, the epiphenomenal secretions of "selfish genes".

Further, there is little reason to accept the PSUist distinction, and maintained inconsistency between the Manifest and Scientific images of human beings as a proposition in a metaphysical theory. Rather an alternative strategy is to modify *both* positions to produce a satisfactory metaphysical *Weltanschauung*, and in doing so, reconcile areas of conflict. This is quite easy to say, but the practical task of addressing the multitude of issues involved here cannot be addressed other than by entire intellectual communities; we can here, however, at a bare minimum, reject the PSU incompatibility thesis.

The Systemic Emergent Realist response to the central state and eliminative materialist positions on the mind–body problem, is to agree that there are no mental substances/souls, but to affirm the existence of irreducible mentalistic properties and their associated causal powers [Jones, 1972] [Swartz, 1975] [Dewan, 1976] [Sperry, 1976] [Bunge, 1980], thus avoiding all the recognized difficulties facing the various materialistic and idealistic identity theses [Cornman, 1971] [Swinburne, 1982] and rebutting the eliminative materialist challenge to Culturology. This is obviously not to state a philosophical psychology; it is rather to state an initial response to physicalism, to point to a solution, not to arrive at the destination, but to state physicalism's inadequacy nonetheless. All the other problems of the philosophy of psychology are still with us.[34]

My rebuttal of eliminative materialism then is that it rests upon a fundamental flawed system of metaphysics. But the eliminative materialist has a response to this. Both Paul Churchland [1981] and Richard Rorty [1982] have come to question basic assumptions accepted here. Churchland imagines a time where a new neurological "language" will be adopted, where properties of truth and falsity and relationships of entailment no longer exist. Rorty looks forward to the end of metaphysics and the coming of a post-philosophical culture, where all there is, is "ways of interpreting, ways of interpreting".

As we considered earlier, some might argue that Churchland and Rorty are outrightly inconsistent [Everitt, 1981] but such a claim, as I have argued elsewhere [Smith, 1983 (5)], cannot be sustained. Indeed the response to this argument, which I have awarded verbally against Rorty ("Pragmatism and Philosophy" Seminar, Adelaide University 12/7/82) may be summarized as:

(1) Rorty either argues with the metaphysician of he does not.

(2) If he does not, then the metaphysician has no argument to consider, and can hence ignore him.

(3) If he does argue, then he is playing the philosophical language game and is inconsistent.

Therefore
(4) Rorty's position is either inconsistent or can be ignored.
The reply by Rorty was that the "pragmatist" is not arguing but is cautioning against the Philosophical enterprise itself. But a second difficulty with this argument, I now realize is premise (2), the assumption of ignorance. This is a questionable premise because Rorty and especially Churchland, *are* making open political options about the future state of human society and our practices. It is clearly futile to attempt to *argue* against these positions, since they openly advocate an end to Philosophical argument (but insofar as they do argue (inconsistently) our criticisms in the previous chapter still hold). Clearly if one believes that these programmes will be disastrous for human society, and our own self-image, only power-politics remains.

6.0 Conclusion: The state of the argument

In this chapter I located Sociobiology within the framework of the metaphysical *Weltanschauung* of *Physicalist Scientific Unificationism* (PSU). It was not implied that this system of metaphysics is readily recognized by the working Sociobiologist. Nor is the relationship between Sociobiology and PSU one of entailment. Rather, the PSU metaphysical Weltanschauung is a plausible one to locate Sociobiology, since Sociobiology's public relations spokespersons [Wilson, 1975, 1979] have frequently advertised the unification powers of Sociobiology, and offers material which PSU will use regardless of the feelings of Sociobiologists.

We then detailed the *Weltanschauung* of PSU, including its metaphysical commitments, logical, epistemological and methodological orientations. The long arguments which followed were an attempt to attack PSU on each of these points. It was argued in section 4 that all the major accounts of the relation of theoretical reduction are conceptually, epistemologically or historically flawed. Then it was suggested that we abandon utterly reduction-talk or else if we are in a position of allegedly concluding a debate, as I am, treat this concept only as a low grade aesthetic device. Unification could be obtained more plausibly by "world-picture making", by attempting to place modes of knowledge inquiry within metaphysical *Weltanschauungs*. I argued that the popular "Australian-American" physicalist-materialist position was ontologically unsatisfactory. We should seek an alternative: Systemic Emergent Realism is one direction, at present an admittedly vague direction, but certainly less vague than our opponents' position.

Of course on this point I have not said all that should be said about Systemic Emergent Realism (granted critic, we have merely introduced a thesis) but this is not my aim, for here I intend to say the first word not the last. Indeed, if I am to follow the logical course of my own argument, some remarks must be made on an alternative if my critique is to be successful. At the least I am making explicit assumptions usually implicit in the writings of social theorists, and this seems a good thing, if intellectual honesty is an object of value.

Thus we turn now to an examination and critique of Sociobiology, having provisionally rebutted the philosophical challenge to Culturology raised in the last chapter.

CHAPTER 4

Causal-explanatory reductionism I: A philosophico-biological critique of Sociobiology

1.0 Introductory statement of the argument

In terms of a *dialectic*, the argument of this chapter is the antithesis of that of chapters 2 and 3. I will argue there that a philosophico-methodological and biological critique can be made of Sociobiology. This claim in itself is interesting and controversial, but is not exceedingly radical. The arguments developed by Richard Burian [1978], Richard Lewontin [1977] and the Sociobiology Study Group [1977], would seem to strike at the heart of any attempt to "explanatorily" reduce "Sociology" and Culturology to Sociobiology, if it was shown that Sociobiology was epistemologically and methodologically untenable. Yet I believe that one cannot argue on these grounds against Sociobiology, without accepting the implications of one's argument for neo-Darwinism and received genetics itself. Burian [1978, p. 390] puzzles over attempts to say what fitnesses actually are, recognizes that this argument will raise severe foundational problems for population genetics, then cautiously backs away. This however, leaves the Sociobiologist the ready option of arguing that since in fact this argument is too strong, it is thus an ineffectual criticism of Sociobiology. The same will of course be said about various epistemological arguments against Sociobiology, such that it is unfalsifiable: if this criticism is sound, then what prevents the same objection being used against neo-Darwinism itself?

It will be argued in section 2, that Sociobiology is a logical and methodological extension of the principles of the neo-Darwinist synthesis. Thus Sociobiology is not as Lewontin [1977] classifies the position "a caricature of Darwinism", but rather, if Sociobiology is false or methodologically untenable, neo-Darwinism cannot be taken to be generally true. Social behaviour escapes its adaptionist and selectionist explanatory nets. Here I will argue that Sociobiology fails first on internal groups, second because the neo-Darwinist synthesis itself as a cosmology is theoretically, empirically and philosophically flawed. I will maintain, even more radically, that neo-Darwinism is not even a theory of Evolution, nor is Sociobiology even in its most recent, form, a theory of social behaviour. Hence, I invert the argument of the "crisis of Sociology" tradition. In doing so, we are faced in full with the question of what the alternative is. We will consider this question in chapter 5.

Before turning to the task presenting an exposition of the prerequisite biological material, the idea of a causal explanatory reduction requires explication. Consider a sociocultural phenomena P_k. Then if there are causal theories T_B and T_S, with respective explanations E_B and E_S of P_k, then a causal explanatory reduction of E_S to E_B occurs if and only if either:

(1) E_B is true and E_S false and/or

(2) There are sound philosophical arguments against E_S, but not E_B and/or against T_S but not T_B.

What I hope to capture in such an analysis, are cases where we have some socio-cultural phenomena, such as the Eskimo sacrifice on the snow, which can be explained (E_S) through some socio-cultural theory of Eskimo cultural tradition (T_S) and a competing explanation (E_B) from Sociobiology making use of say kin selection theory (T_B). We need to capture in our account, the idea that Sociobiology and the sociocultural sciences "battle for territory". We shall say that two explanations, E_1 and E_2 of P_k compete, if and only if:

(1) It is not possible for E_1 and E_2 to both be true, or

(2) It is not possible for E_1 and E_2 both to be philosophically tenable.

The hostility which Sociobiologists have towards an *autonomous* social science which does not make anything but non-controversial and trivial recourse to biological principles, even in the recent Wilson-Lumsden gene-culture coevolutionary theory [Wilson and Lumsden, 1981] (to be documented in due course) stands as immediate evidence for my claim.

Some logical and conceptual properties of causal-explanatory reductions (E_R) can be noted. The relationship is: (1) *irreflexive* (i.e. $\sim E_R(x, x)$); (2) *asymmetric* (i.e. if $E_R(x, y) \rightarrow \sim E_R(y, x)$; (3) *antisymmetric* (i.e. $E_R(x, y)$ & $E_R(y, x) \rightarrow x = y$); (4) *transitive* (i.e. $E_R(x, y)$ & $E_R(y, x) \rightarrow E_R(x, z)$), but the relation is not connected or strongly connected. It is not the case that $\Box(E_R(x, y) \quad E_R(y, x))$ since it may well be that both the x and y explanations are either false or theoretically untenable.

The argument of this chapter is that a causal-explanatory reduction of Culturology to Sociobiology cannot in principle occur because Sociobiology and neo-Darwinism are theoretically and philosphically flawed. In the next section we shall review in detail the latter positions.

2.0 The Theoretical foundations of the neo-Darwinist synthesis and an explication of the theoretical structure of sociobiology

2.1 Staticism, creationism and evolutionary cosmologies

If we admit that the world does in fact exist, that processes are occurring in it independently of my thoughts/ideas, and indeed occurred in the past before I or any other person existed (excluding perhaps Divine persons), then the question of the origin of the world is one which must be addressed by any *Weltanschauung* which attempts to give a comprehensive view of reality. It is assumed here then, that it makes sense to speak of a "past", of the world having a history. Further it is also assumed that hypotheses such as Bertrand Russell's claim that the world might have come into existence merely five minutes ago, can be rejected on methodological grounds. A sketch of an argument is this: such a hypothesis is inadequate due to the arbitrary limits which it places upon human cosmological inquiry if accepted [Slote, 1970], and also because no plausible generative mechanism for the creation of the world, especially only five minutes ago was given by Russell. This hypothesis therefore fails the very basic necessary condition for scientific explanations [Harré, 1970] [Bhaskar, 1978 (1)].

We may distinguish between three broad cosmologies attempting to provide answers to the questions of origins: *Staticism*, *Creationism*, and *Evolutionism*. The position of Staticism proposes that the universe has existed approximately the same in its nature as it is at present, and the same kinds of organisms have always existed. Consequently there is no cosmological problem of origins at all, at least in accounting for the various types of organisms for these have always been. Individual organisms might arise either directly from the parents as an outcome of the process of reproduction, or they may be spontaneously generated by material objects themselves. This process could be directed by a Divine Personal Being or Beings, or by other spiritual or Ideal Processes. In the most radical form of Staticism, the monism of Zeno and Parmenides, the possibility of change and motion is denied, for both of these concepts, are argued by a series of ingenious paradoxes, to be incoherent [Salmon (ed.), 1970].

Creationist positions deny the assumption of Staticism, that both the world and the kinds of organisms have always existed. They have, it is maintained, been created by a Divine Being(s) or Transcendental Process(es), or else designed from some basic eternal stuff (which may be material, spiritual, mentalistic or neutral stuff) by such Being(s) or Process(es). A Creationist position would then be, except for the stated point, indistinguishable from Staticism. The basic form and order of the universe is then temporally fixed, there is neither transformation of form nor emergence of novelty without direct Divine or Transcendental intervention. In a contemporary Judeo-Christian "Scientific Creationist" cosmology, for example, it is alleged that *naturalistic* innovation and integration is contradicted by the second law of thermodynamics [Morris et al., 1974]. Creationism may thereby depart from Staticism at this point, seeing the World as essentially in a process of decay and increasing disintegration and entropy. In short there is an eschatological element in Creationism absent from Staticism.

The *Evolutionist* position, by contrast to Staticism and Creationism, holds that there has been both a transformation of form of organisms and the emergence of novelty in the universe. Central to any Evolutionist position is the idea of *transformation of Form*, with a trend of both increasing complexity and increasing differentiation of Form. From a cosmological viewpoint the universe itself has throughout time undergone Evolution from the primeval state of the Big Bang, to the formation of star systems, galaxies and solar systems. Restricting our attention to the problem of the origin of life, the Evolutionist will allege here that life on earth arose initially from certain primeval organisms, which in turn had arisen from the self-organized assembly of various bio-molecules. These primeval organisms were in turn transformed into more complex organisms, some of which were transformed into organisms of one "type", others into organisms of another "type". This resulted in a branching "tree of life", linking all organisms through the relationship of "ancestral transformation".

With regard to the primeval organisms, there are various hypotheses or origin. The most common one, cited above, is that the primeval organisms arose from non-living chemical compounds in a primeval "soup" of such chemicals [Oparin, 1953] [Miller, 1953] [Miller and Orgel, 1974]. Debate has centred around what in fact the primeval world was like, in particular whether a hydrogen-rich primordial reducing atmosphere ever existed.[1] Contrary to the biology text books, many geochemists have been suspicious of this hypothesis and there have been a number of attempts to formulate modified or alternative hypotheses [Kerr, 1980]. Alternative hypotheses take life to have arisen from another solar system [Crick and Orgel, 1973], perhaps by comets [Hoyle and Wickramasinghe, 1978] [Hoyle, 1982]. Traditional objections to the latter hypotheses are methodological in nature: they merely push the origins question one stage back, indeed to a stage which makes empirical inquiry quite difficult. The traditional rejoinder has been that such difficulties, and gross underdetermination of theory by data, infect any such model.

Thus, the most reasonable conclusion for us to draw here is that this area of science is most alive and exciting with its controversies, and there is anything but a general consensus as to the origin of life scenario, although the primeval soup scenario remains the most common one cited in the text books.

If it was the case that living things did arise from non-living bio-molecules, or arrived on this planet from outer space, then it is still the case that organisms from micro-organisms to mammals are all evolutionary related. The next step must be to outline a theory of the generative mechanisms involved here. The most widely accepted account is neo-Darwinism. Let us examine this position.

2.2 Neo-Darwinism

There are various ways in which the position of neo-Darwinism can be defined, but we would do best to look for definitional illumination from a philosopher of biology, for part of their job is to clarify such matters. In the work of neo-Darwinism's chief philosophical defendant, Michael Ruse [1973, 1979, 1981,

1981], the mechanisms of Evolution is the "natural selection" of "adaptive variants" (caused through "random genetic mutations") in an environment. In particular it would seem that natural selection is dependent upon: (1) heritable variations in individual organisms; (2) the capacity of organisms to reproduce such that there is a tendency for a Malthusian population growth rate to occur such that populations will tend to outstrip available resources; (3) available resources are fixed at any point in time. Given no migration, a struggle for survival will occur (given also no social co-operation to either limit population increases or equally distribute food) and in the "long run" variants possessing adaptations will tend to survive and reproduce. Other organisms lacking such adaptations will not in the long run contribute as many (or any) offspring to succeeding generations, so that the composition of the population changes over time. It is alleged that if this process occurs long enough, the transmutation of species will occur by the accumulation of small physiological, behavioural and morphological changes. Let us direct our discussion around this intuitively described model.

Ruse's account involves two further assumptions: (1) evolutionary changes occur slowly (the thesis of *phyletic gradualism*; (2) neo-Darwinism, by contrast to competitors such as Lamarckianism, saltationism and the theory of orthogenesis, all which have in fact "failed to explain the basic facts of heredity" [Ruse, 1973, p. 113], has achieved such an honour. Lamarckianism is thus on par with the flat earth theory [ibid., p. 109]. Presumably, if a critic suceeded in successfully criticizing the theoretical cogency of neo-Darwinism, then immediate difficulties will also be raised for "the theory of heredity" and associated sciences. Let us take up Ruse's second point first.

By contrast to Ruse's position, it is maintained that neo-Darwinism is dependent upon population and molecular genetics, and thus of course may well be consistent with them, but does not explain "the basic facts of heredity" [Rosenberg, 1980 (1)]. It may be the case that Lamarckianism conflicts with knowledge derived from population and molecular genetics, in which case a *prima facie* case is made against Lamarckianism. Even if this was so, it does not follow that neo-Darwinism explains the basic facts of heredity. A theory of Evolution must explain, and not merely be consistent with, palaeontological data. It must give some mechanism which accounts for the increase in complexity of organisms from "molecules to man", for the transformation of form, and the origin of species. This point leads me back to Ruse's first assumption.

R. B. Goldschmidt [1940] was an early critic of phyletic gradualism, proposing that large-scale evolutionary change (macroevolution), such as between say reptiles and mammals, involved macromutational changes.[2] This resulted in "hopeful monsters" which arose full-blown from progenitors. Such a radical saltationist position suffers extensive problems [Dobzhansky, 1951] [Stebbins, 1950], but contemporary "saltationists" are not so easily beaten. S. J. Gould rejected the gradualist/uniformitarian assumption that geological change involves uniformity of rates and material conditions in 1965 [Gould, 1965]. Later along with Niles Eldredge [1972, 1977] they advanced their well-known model of *punctuated equilibria.* This model, they took to solve at one stroke the well known problem of the adaptativeness of the intermediate forms and the gaps in the fossil record, by taking stasis as data. Fossil species exhibit morphological gaps, appearing without clear ancestry, and exhibit evolutionary stasis. Gradualists had usually used *ad hoc* explanations to account for such gaps: imperfections result from the geological destruction of the required fossils. In many cases, especially with "soft organisms" lacking strong skeletons this would no doubt be true. Further fossils are formed under very restrictive conditions, for most often environments consist of biological and geological mechanisms which remove such traces very quickly. Therefore it is scientifically unreasonable to hope for a "perfect/near perfect" fossil record. The issue of dispute is unlikely to be resolved by collecting more fossils, unless of course gradualists come up with a record which is unlikely to be achieved given the above-stated physical limitations. What is ultimately at issue here is a metaphysical dispute over the interpretation of data and the rate of process of evolutionary change – merely the biological equivalent of the problem of the continuity versus discontinuity of the world (seen especially clearly in the philosophy of physics in the difficulties of Zeno's paradoxes, and in "Sociology" especially in theories of revolution and piece meal social change).

To account for problems such as the gradual formation of the mammalian eye, gradualists must postulate a sequence of functional intermediatry forms. This, Gould [1977 (1), 1977 (2)] has argued, raises theoretical difficulties in that incipient stages of structures may not be functional at all, and may be adaptively disadvantageous and not explicable by well-known preadaptional strategies in any non-*ad hoc* way. An example is the splitting of the maxillary bone of the upper jaw into a front and rear section connected by a movable joint in bolyerine snakes, which Frazzetta [1970] maintained arose from the Boidae snakes (lacking this jaw structure) as non-extreme hopeful monsters. Such a division was taken to have arisen suddenly because such a structure can be either present or absent – there is no intermediate situation. True, gradualists might argue that such modifications are in actual fact "minor", and that the classification of bolyerine snakes as a new family was unreasonable. This dispute would now have to hinge upon rather difficult questions of what precisely are "small" and "large" evolutionary innovations, and this debate will be shadowed by the paradox of the sorites [Cargile, 1969].

According to neo-Darwinist gradualists, a new species arises from an ancestral population involving a large part of the ancestral species' geographical range and large numbers of the ancestral population: a daughter species arises via a gradual transformation of the ancestral species. By contrast, the model of punctuated equilibria proposes that new species arise in a sub-population occupying an area on the periphery of the geographical range of the population. Intense selection pressures exist here because new favourable mutations are not lost in the intensity of the population through environmental changes which may occur faster than fixation of the mutant genes in the gene pool. The new daughter species arises then from the sub-population, whereas the rest of the population does not change significantly throughout evolutionary time. The fossil record should then consist of fossils static throughout their range, as these will come from the population, since speciation does not occur in the local sections of the range. Evidence of speciation in the peripheral region would be slight because there would be less fossils formed.

Here is not the place to review the contemporary quantum speciation debates [Stanley, 1979]. What should be noted is that in both models of speciation, the mechanism of Evolution is the natural section of variations in turn caused by random genetic mutations.

2.3 Hereditable variation and its mutational origin

Whilst we do not yet have in this discussion anything more than a vague and intuitive account of natural selection, it can still be noted with justification at this point, that the mechanism of Evolution must lie in the process by which variations are produced. This is so because of the simple reason that any selection process requires "raw materials" to initially select from; natural selection is itself operable only if hereditable variations exist. Yet according to the Hardy-Weinberg law, perhaps the foundational law of population genetics (to be discussed in the next chapter), in the absence of mutation, migration and selection, the gene frequencies in a population remain constant. Thus Evolution, like accelerations, will not occur without a generative mechanism/force. Indeed, the more genetic variation there is in a population, the greater the opportunity for natural selection to occur. This "fact" is captured in *Fisher's Fundamental Theorem of Natural Selection*: the rate of increase in the average fitness of a population is equal to the genetic variance of fitness in that population (where "fitness" appears to be taken by Fisher as the relative reproductive efficiency of organisms in a population).[3] Consequently the question of how much genetic variation exists actually in populations is of crucial importance to any neo-Darwinist approach: if all individuals are homozygous at exactly the same alleles no change in gene frequencies of such alleles could occur purely by selection (even if the number of organisms altered).

There are two conflicting models of population structure which have been advanced: the so-called *classical and balanced models* [Dobzhansky, 1955]. The classical model, associated with geneticists such as Müller, Kimura, Crow, Ohta, King and Jukes, maintains that at nearly every locus every individual is homozygous for a "wild type" gene (i.e. a gene "causing" a phenotype characteristic of the majority of individuals of a species in a population in a given environment) and is heterozygous for rare deleterious or beneficial alleles at a few loci. Natural selection acts to eliminate deleterious mutations, consequently the fittest genotypes are those homozygotes for the wild-type alleles at all loci. Favourable mutations maintained in the population are extremely rare, but must of course occur under the assumption that Evolution occurs.

The *balanced* model is associated with Dobzhansky, Ford, Ayala, Spassky; indeed it is taken to be the prevailing "paradigm" of evolutionary genetics. This model proposes that nearly every locus of every individual is heterozygous, so that there is strictly speaking no such thing as wild type genes, and the number of alternative alleles at each locus is large, to guarantee permanent heterozygosity. Natural selection occurs in the form of *balancing selection*; polymorphisms (referring to those distinctly different forms of individuals of a species, occurring in roughly constant proportions in an inter-breeding population) can be "balanced" and maintained in a population at more or less stable equilibrium frequencies. Such polymorphisms may also be maintained in the population by non-selectional factors such as random genetic mutations and genetic drift [Dobzhansky et al., 1977, pp. 108–116].

We shall turn to some of the interesting epistemological problems facing population genetics, and in turn neo-Darwinism later in the course of reviewing material from Lewontin [1974]. Before passing on, something must be said by way of clarification of the idea of neo-Darwinism's random genetic mutations. Gene or chromosome mutations,[4] are *random* in so far as:

(1) They occur independently of conscious acts of agents;
(2) They are rare exceptions to normal DNA functioning;
(3) There exists no mechanism for prediction of when, where, and how mutations will occur *and most importantly mutations are unorientated with respect to adaptation*.

Even Dobzhansky admits that mutations are *disordering processes*, which he compares to "printers' errors" [Dobzhansky, 1974, p. 313], and he admits that the beneficial mutations which are the real generative mechanisms, leading to order, increases in complexity, adaptation and the transmutation of form, are very rare indeed [ibid., p. 313]. Of course all our talk about "adaptation", "benefits", and "harm" should be relativized to an environment. If we wish to speak of some characteristic which is "absolutely harmful", this (assuming that this idea is coherent) can readily be done by quantifying over "all environments".

It is generally conceded that a *prima facie* inconsistency exists in a neo-Darwinism between the *chance* factor of mutations and the biological order exhibited by adaptation. This inconsistency is much like the paradoxes of implication, which students of classical extensionalist and model logics first encounter; they are phenomena to explain away. In the case of neo-Darwinism, this is the *anti-chance* factor. I shall argue later, that this defence is unsatisfactory, and that neo-Darwinism is without a cogent mechanism for evolution.

2.4 Natural selection and evolutionary change

The principle of natural selection provides the basis of neo-Darwinist accounts of Evolution, but its explication is far from evident. A reasonable explication has been proposed by Brandon [1978, 1980]. Brandon gives the following generalized version which is applicable to any level (gene, individual, group, higher taxa) at which natural selection is taken to occur:

(NS) "(Probably) If *a* is better adapted than *b* to their mutual environment *E*, then *a* will have a greater reproductive success than *b* (in *E*)". [ibid., p. 428].

I shall discuss in detail the concept of adaptation later in the chapter. Here we note, to aid us in our comprehension of Sociobiology the logico-methodological status of (NS). Brandon takes the principle (NS) to be a general organizing principle, devoid of empirical biological content, but capable of structuring biological explanations of differential reproduction. Thus one such instance of (NS) is this:

(NS*) (probably) If moth M_1 has darker coloured wings than M_2 in E_1 then M_1 will have more offspring than M_2 in E.

To gain a more detailed biological grasp of selectionist explanations the idea of Darwinian fitness must be stated. The idea of Darwinian fitness (w) is most usually explicated in the text books as follows: Darwinian fitness w is a *relative* rather than an absolute measure of reproductive and survival efficiency. In both [Dobzhansky et al., 1977, p. 101] [Ayala and Valentine, 1979, p. 123], we are told that Darwinian fitness w is related to the *selective coefficient* s by: $w = 1 = s$, where s measures the reduction in fitness. How either of these values can be empirically ascertained without circularity is a mystery, and Dobzhansky et al., whilst remarking that "Darwinian fitness is a composite, a product of various advantages and disadvantages", ultimately adopts the circular and theoretically inadequate equation of the "fittest" with "the parent of the largest number of children reaching adulthood" [Dobzhansky et al., 1977, p. 101]. I will argue that the frequently cited circularity argument can be restated and defended against philosophical defenders of neo-Darwinism later in this chapter.[5]

The main categories of selection discussed in the literature are: (1) Normalizing; (2) Balancing; (3) Diversifying; (4) Directional; (5) Sexual; (6) Kin; and (7) Group Selection. The latter three are of central importance to this work and since the question of the "unit of selection" is so immediately apparent in this dispute, we will pass to this issue with only brief identification of these other categories. In *Normalizing* or stabilizing natural selection, the mean value of the phenotypic distribution is maintained, but the variance of this distribution is reduced, assuming a normal distribution. In *Diversifying* selection, individuals at either end of the phenotypic distribution are selected for, such that individuals in a distribution around the mean of the initial distribution are selected against.

In *Directional* selection, individuals with phenotypic distributions at one end of the distribution are selected for. *Balancing* selection has already been mentioned above.

2.5 Sexual, kin and group selection: The sphere of Sociobiology

These concepts move us immediately into "Sociobiological territory" insofar as the entitled concepts were devised to account for behavioural phenomena which were quite puzzling from the perspective of natural selection unsupplemented by these additional concepts.

The puzzling phenomena which Darwin found with respect to sexually reproducing organisms was the dimorphism between males and females. Darwin believed that it was implausible that this could arise by natural selection (as he saw it) and argued that males (in particular) struggled with other males for the possession of females, or females made a mating partner choice on the basis of some specific property of the males such as the beauty of displays (e.g. in the peacock). Neo-Darwinists came to view sexual selection as merely a part of natural selection [Lack, 1966], and many quite rightly, I think, came to question the anthropormorphic nature of such explanations, not merely on behaviourist grounds, but simply on epistemological grounds: what after all are the peacock's aesthetics and how do we really know what female peacocks like? Hence the concept of sexual selection was redefined to refer to any deviations from purely random mating amongst organisms in the same locality.

Sociobiologists have in part followed Darwin rather than the neo-Darwinists with regard to the issue of sexual selection, the ground preparing work in published form being presented by Robert Trivers [1978] in his theory of *parental investment*. The term is defined as follows:

> ". . . any investment by the parent in an individual offspring that increases the offspring's chances of surviving (and hence reproductive success) at the cost of the parent's ability to invest in other offspring". [ibid., p. 55].

"Investment" is taken as a primitive term (it is not explicitly defined), but is taken to include metabolic "investments" in primary sex cells [ibid., p. 55] as well as any behaviour which benefits the young in surviving. The definition is a consequential one: what is important for Sociobiologists are outcomes rather than intentions, effects rather than unmanifested tendencies.

Each offspring is thus viewed by Trivers as an investment of both energy and genes. To increase the investment in one offspring through increased cleaning and feeding, will tend to decrease the investment in the other offspring, as the parents' times and energies are limited resources. The size of the parental investment in an offspring O_1 is measured by its negative effects on the other offspring $O_2, O_3, ;.., O_n$. The *total parental investment* of an individual parent, is the sum of its investments in each of its offspring, and Trivers argues that a total parental investment which leads to maximum net reproductive success, would be favoured by natural selection against other "investment strategies". The *typical parental investment* is the total parental investment divided by the number of offspring produced by the parent during the season, which can compete for parental investment.

What governs the operation of sexual selection is, in Trivers' opinion, the *relative parental investment* of the respective sexes in their offspring. The total number of offspring produced by one sex of a sexually reproducing species is necessarily equal to the number produced by the other, as it takes one male gamete and one female gamete in sexually reproducing species, to form a (diploid) zygote. However when specific individuals are considered, it is clear that one sexually reproducing organism may well mate with a number of organisms of the opposite sex of its species throughout time, to the exclusion of other organisms of its species, of the same sex as this organism. If the *only variable* between the sexes is typical parental investment per offspring (an assumption which is false for human beings given the demonstrable existence of sociocultural forces), then the sex whose typical parental investment is greater than the other sex will become a limiting resource for the opposite sex [ibid., p. 56]. Therefore there will be competition amongst members of the sex whose parental investment is relatively less than that of members of the opposite sex, since such members whose parental investment is relatively less can increase their productive success by investing in the offspring of a number of members of the opposite sex, members who have a higher typical parental investment. It is usually the female of most mammals who is the limiting resource, since the females raise the young. If however the typical parental investment for males was greater than that of females, then males would be a limiting resource.

This model faces a general empirical difficulty: ascertaining the various required levels of investments. From a practical, experimental point of view, no human researcher can survey the activities of some relatively long living organism (say living in excess of a few years) in its natural environment in sufficient detail to ascertain the required investments. For surely, to continue the economic metaphor, investments and profits from such investments must consider *opportunity costs*, that is, costs foregone in various investment strategies which could also have been taken. Without knowledge of the biological options open to an organism in its environment we do not obtain investment values. But such knowledge cannot be accurately obtained. It is unclear, surveying a nest of birds, what the mother's metabolic investment in fact was: this would require bio-chemical knowledge of systems which existed in the past in a variable state, which no longer exist. Such operational difficulties make the model difficult to empirically support and falsify.

We should also note that a model can only be justifiably applied in explaining a certain class of phenomena, if in fact the assumptions of the model meaningfully can be applied to the phenomena in question. It was mentioned above that if the only variable between the sexes is typical parental investment per offspring, then it follows that the sex whose typical parental investment is greater than the other sex will become a limiting resource for the opposite sex. This inference can be successfully blocked by pointing out for a considered species that the conditions stated in the premise fail. This it is maintained in passing, is the case for humans where there are a multitude of sociocultural variables (in most societies) operative. There exists no theoretical model which has incorporated these relevant variables.

I turn from these initially sceptical remarks to the ideas of kin and group selection, and in turn clarify the idea of "inclusive fitness". This brings us to the heart of the Sociobiological programme, but it is not the case that Sociobiology can be understood merely from such considerations. It is itself an evolving doctrine and there are significant recent developments which must be examined. The ground we thus travel now is one well worn.

Two fundamental problems have plagued neo-Darwinism. First, how can so-called "altruistic behaviour" be accounted for, that is, how can evolutionary sense be made of behaviour which *prima facie* seems detrimental to the fitness of individual, but which benefits the species or local population? A second problem which immediately arises from this problem, is at what level does natural selection operate? This question is, as Michael Ghiselin [1981] argues, tangled up with the "species problem", but we shall not make our own attempt to sort out the latter tangle here.[6]

E. O. Wilson [1975] has given a widely cited definition of "altruistic behaviour": "when a person (or animal) increases the fitness of another at the expense of his own fitness, he can be said to have performed an act of *altruism*" [ibid., p. 117]. If the person or animal raises its own fitness by lowering that of another person or animal, it is said to be engaged in selfish behaviour, and spiteful behaviour if it reduces its own fitness in order to lower the fitness of another. Altruism is a property of behaving organisms. A gene or set of genes will be said to be altruistic if and only if they are the generative mechanisms taken as the cause of altruistic acts. We shall look more closely at the issues centred around the concept of the "selfish gene" in the next chapter.[7]

For Wynne-Edwards [1962], genetically altruistic phenomena were considered readily explicable at the level of the group or species. This thesis was rigorously attacked by Williams [1966] and by the following tradition of individual and gene selectionists, who set about to develop models which avoided appealing to group selection. In turn group selectionists have developed some interesting mathematical models which demonstrate the logical possibility and empirical infrequency of group selection [Maynard Smith, 1976]. However most Sociobiologists would violently disagree with Wilson's [1975] equation of kin selection as a special case of group selection [Dawkins, 1976, p. 101]. Sociobiologists are individual selectionist almost to a man; thus it must be an embarrassment to find such open and unashamed group selectionist heresies in what has become a discipline defining text.

There are three theories which individual and gene selectionists accept as providing a unified theoretical framework for the explanation of genetic-altruism "the central theoretical problem of sociobiology" [Wilson, 1975, p. 3]: (1) *kin selection*; (2) *parental manipulation*; and (3) *reciprocal altruism*.

W. D. Hamilton in a series of papers [Hamilton, 1963, 1964 (1), 1964 (2), 1971, 1972] stated and developed in mathematical detail a genetic model of altruism, centred around the concept of *inclusive fitness*. This, Hamilton defined as equal to 1 + individual fitness $\bar{w}$ + sum of the effects of this individual upon the fitness of individuals *weighted* with respect to the degree of relatedness of the affected individual. Hamilton argued that the condition necessary for the evolution of altruism is $k > 1/r$, where "k" is the ratio of gain to loss in fitness from the altruism, and "r" is the *coefficient of relationship*, the proportion of genes identical by descent of the respective agents. Altruism then is contingent upon the variables of: (1) respective genetic relatedness; (2) the magnitude of the benefit and (3) the magnitude of the cost. Hamilton came to modify this formulation such that $k > 1 + (F_A/2r_{AB})$, where "F_A" is the inbreeding coefficient for individual A (the inbreeding coefficient being the probability that both alleles at a locus are identical by common descent) and "r_{AB}" is the probability that a gamete of A has the same gene as a gamete of B.

West Eberhard [1975] suggested that inclusive fitness by defined as the *lifetime* sum of effects upon relative such that:

(1) Inclusive fitness a_i = df classical fitness (a) + kinship component (a_k).

Now $a_k - 2 \sum_{i=1}^{n} [(r_{By} - \bar{r}) \Delta a_B]$, where "n" is the total number of individuals whose fitness is affected; "Δa_B", the total life time effects on their fitness, and "r_{By}" the probability of the individual having a given gene in common with the young of the beneficiary. A new formula may now be given: $k > r_{Ay}/r_{By}$, where "r_{Ay}" is the probability of an individual having an identical gene by descent with the young of the donor (A) or the beneficiary (B) respectively [ibid., p. 5]. In this formula:

$$(2) \quad k = \frac{a_{B2} - a_{B1}}{a_{A1} - a_{A2}}$$

where "a_{A1}" and "a_{B1}" are the personal fitnesses of donor and beneficiary without the altruistic actor of A, and "a_{A2}" and "a_{B2}" are the respective fitnesses following the act. The frequently made objection by sociologists, that it is difficult for the recognition of kin to be made, hence invalidating the model does not hold since: (1) what is important for Sociobiology is the consequences of behaviour, and insofar as such acts are not performed in accordance to genetic self interest, selection against such individuals will occur

[Schulman, 1978]; (2) the model can also be represented without *explicit* reference to the coefficient of relationship [Orlove, 1975].

Hamilton originally advanced the model of kin selection to explain the phenomena of the evolution of the sterile worker castes in the social *Hymenoptera*. He argued that by this strategy workers maximized inclusive fitness, since given that unfertilized eggs developed into males and fertilized eggs into females, females will share 50% of their genes from both their mother and father, and consequently have 75% of their genes passed on.

Alexander [1974] has argued that sterility occurs in some types of termites where males are diploid and where worker termites are related by 50%. Trivers and Hare [1976] have also criticized this model, and proposed a reworking which has in turn been criticized by Alexander and Sherman [1977], whose hypothesis of *local mate competition* has been criticized in turn by Wilson and Oster [1978 (2)]. According to, for example, Alexander and Sherman's model of local mate-competition, the imbalance in the sex ratio is explained by proposing that it is a strategy whereby competition between males for females is reduced. Why produce 50 males if only 25 males can do the job? What a waste of investment! The assumption here is that if father M* and mother F* produce two males M_1 and M_2 and these have to compete for matings, then one male would have been more economically produced energy wise. True, it is in the parents' reproductive interests to produce only one male, but how at the cytological level are such wishes manifested, if there are no alternative strategies in the population for natural selection to select for? There has been no generative mechanism given whereby feedback between both the number of females in the population and the reproductive interests of the parents would regulate the sex ratio, without engaging in circular argument. Like much of the genetic calculi of the Sociobiologists, this is a dream of good fortune, which awaits the material means of accomplishment. Second, even granted Alexander and Sherman's premises, if M_1 and M_2 compete, this is not necessarily evolutionary irrational; we know from the theory of sexual selection that competition for females by males can be selectionally advantageous, even though M_1 and M_2 are kin. Indeed via kin selection theory, if M_2 is much "fitter" than M_1 it may well pay M_1 to let M_2 fertilize the females, for a conflict may reduce *both* fitnesses (if for example M_1 dies and M_2 is wounded allowing some other male to fertilize the females) thus lowering the inclusive fitness of M_1.

Ruse [1979, p. 47], who has given a similar summary to that presented here, concludes that the notion of kin selection is quite powerful; on the contrary these repeated failures and deep controversies over even altruistic behaviour in social *Hymenoptera* can only reasonably make one a black pessimist about what must then be, the near fantasy use of this model with reference to human sociality.

Alexander [1974] has proposed another model of altruism: *parental manipulation*. Here parents manipulate an individual into becoming an altruist, and altruistic behaviour evolves because of benefits to the parents' fitness. Parents are in a unique position to manipulate offspring, due to the general dependency of the offspring upon the parents, lowering the probability of rebellion in comparison with a situation where perhaps a relative is manipulated. As West Eberhard [1975, p. 18] points out, it is practically difficult to distinguish cases of parental manipulation from cases of kin selection, especially in situations where the parent may have died, or is absent, but had initiated some process which results in altruistic behaviour by an offspring at some future point. Yet this does not in any way constitute a defect in the theory itself. Some illustrative examples of parental manipulation are these: trophic eggs in insects [Wilson, 1971], cannibalism in hawks [Alexander, 1974], homosexuality in humans [Trivers, 1974] and even, believe it or not, family squabbles over farmlands [Ruse, 1979, p. 68].

The final theory of altruism to be reviewed is Trivers' theory of *reciprocal altruism* [Trivers, 1971]. This model is taken to explain altruistic behaviour between members of different species. Kin selection theory was, we recall, devised to deal with the problem of altruism. As Sociobiologists accept that for different species, the coefficient of relationship is 0 or undefined, it follows that either $k = 1/0 = \infty$ or the model is inapplicable. If the former, then it is theoretically impossible for an altruistic act to be performed in this situation. If so, then either the kin selection model faces counter-examples, or else the kin selection model is satisfactory but is inconsistent with the theory of reciprocal altruism, since then the behaviour in question would be both theoretically possible and theoretically impossible. It is best to regard the model as inapplicable, and to recognize the need for a supplementary theory (unless a model not involving coefficients of relationships is accepted). This Trivers claims to have supplied.

Trivers' argument is that an act by a donor A, benefiting another organism B, will be selected by natural selection if the benefit to A from beneficent behaviour to B (Ba) is greater than the cost to A of the original beneficent act (Ca) measured with respect to long-term effects on "inclusive fitness" [ibid., pp. 35–36] (disregarding the effects on kinship), that is, $Ba > Ca$. If on the other hand, a model not involving coefficients of relationships is considered, for reciprocal altruism between kin: $Ba > Ca - 2r_{By}(Bb - Cb)$, where "Ba" is the benefit to A from beneficent behaviour to B; "Ca" is the cost to A of the original beneficent act; "Bb" is the profit to B of aid; "Ca" is the cost to B of reciprocation and "$2r_{BY}$" is the increase in inclusive fitness of A for each offspring added to B's reproduction. We should note however that Trivers [1971, p. 35] defines "altruism" as that behaviour benefitting organisms *not closely related*: but if this is so, then the theory of kin selection is rendered useless, or the extension of the model of reciprocal altruism to kin illegitimate – or else such *prima facie* "altruistic acts" between kin are not altruistic in Triver's sense.

The theory has received extensive use, but perhaps the best known and most frequently cited use is that made by Trivers himself to account for warning calls in birds and cleaning symbioses. Unfortunately for Trivers, it is not the case that symbioses are cases of reciprocal altruism, since the required loss of fitness of one organism does not occur and repayment is immediate, not in the future. There is "cooperation", but not genetic altruism or reciprocal altruism. Nor is the case of alarm calling an example of reciprocal altruism. Calling birds perform altruistic acts to be sure, but the point of the example is that there are non-calling birds which by definition do not reciprocate even in the future, hence logically invalidating Trivers' model.

At this point in the discussion, we take note of the state of the argument. Sociobiology is a position which studies the biological basis of social behaviour, especially altruism, with reference to the framework of the evolutionary biology of the neo-Darwinist synthesis. On this basis, no qualitative difference is drawn between human agents and non-human organisms, for it remains an assumption of this orientation that all the properties of an organism can be given adequate explanations within such a theoretical framework, supplemented by bio-chemistry and physics. Insofar as it is shown that this is not so, a case is made for the explanatory limitation or incompletedness of this position. Sociobiology, however, remains a legitimate extension of the neo-Darwinist research programme: as we have seen, the problem of altruism was a general problem facing neo-Darwinists, who were for theoretical and *a priori* reasons sceptical of group selectionism. Consequently I believe Lewontin's claim that Sociobiology is a caricature of neo-Darwinism [Lewontin, 1977] is to be false, since we cannot differentiate these theories as distinct conceptual entities.

2.6 Optimization foundations of Sociobiology

The basic problem of optimization theory is to minimize some scalar E, which is the value of a function of n parameters, where n is usually finite, i.e.

$$(OP_1)\ \text{minimize } E = f(x_1, x_2, \ldots, x_n).$$

Given the formulation (OP_1), maximization problems can be readily formulated:

$$(OP_2)\ \text{maximize } f(x) = -\text{minimum } \{-f(x)\}$$

There are various mathematical methods which may be used to solve various optimization problems such as the use of the calculus, linear programming and games theory [Adby and Dempster, 1974]. Indeed it can be proved that for linear programming problems capable of solution, there exists a two person zero sum game, the solution of which is a solution to the linear programming problem. Further the principle theorem of linear programming can be proved by means of the minimax theorem of games theory [Luce and Raiffa, 1958, pp. 419–423].

The theory of games has received application in Sociobiology, especially with regard to the definition of the concept of *Evolutionary Stable Strategy*. Most importantly the "central theorem" (or correctly axiom, since there is no proof of the following proposition from other axioms) of Sociobiology, that individuals *maximize their inclusive fitness* [Barash, 1977, p. 63], despite difficulties raised by changing environments [Slobodkin and Rapoport, 1974] [Levins, 1968] receives near universal acceptance in this discipline. This optimality assumption receives near universal acceptance in the sisterfield of Ecological Sociobiology which will be reviewed in chapter 7.

Biologists are usually quick off the mark to reject the claim that they are committing the Panglossian fallacy of assuming that organisms do those things which maximize the chance of survival of their species. Maynard Smith [1978 (1)], in a paper attempting to defend biology from this charge, unfortunately presents excellent evidence earlier in his paper [ibid., pp. 33–35] to believe that Pangloss is alive and well. For example, in most theories of the sex ratio, he tells us [ibid., p. 34], it is the number of copies of a gene acting in a parent which is maximized; in theories of the evolution of life history strategy it is the expected number of offspring produced in a lifetime which is maximized, and finally in foraging theory, the basic assumption is that the animal is maximizing its energy intake per unit time spent foraging. In Sociobiology the optimization assumption is utterly explicit, elevated to the grand honour of being a "central axiom".

This is hardly surprising given the structure of the basic optimization problem. Indeed, even situations such as an animal being faced with a problem of taking in a required amount of food without an excess of toxins [ibid., p. 34] can be readily viewed as a minimization problem for toxins under food restraints. Nor is it at all clear what an "optimal level" might be apart form maxima and minima considerations. Consider an example from economics. A firm may choose a profit level which is not a maximum, but is sufficient to allow the firm to contribute in business and expand. To say that the "optimal level" is the lowest level at which this can occur, is to fall back into stating a minimization problem. Alternatively one is faced with a range of values (possibly infinite) from this lowest level to the maximum profit level under constraints. For the lowest level plus 1 cent or 1.1 cents or 1.11 cents etc. are also solutions. Thus the idea of an optimal level distinct from the maxima and minima is unclear. Thus if the "Panglossian assumption" is false in biological contexts, then we do not have an optimization problem as defined by (OP_1) and (OP_2) on our hands, and alternative mathematical models must be sought.

The relevant mathematical models for Sociobiology come from the theory of games, and it would perhaps be useful to give a non-technical account immediately. Games theory is that branch of mathematics concerned with the study of formal decision problems or "games". The central concept of the theory is that a *game tree*, which is "implicitly defined" for a game by a postulate set specifying: (1) the set of "players" (null, one, two or n, where n is a positive integer); (2) the set of "strategies" for each "player" specifying those alternative choices which can be made when it is the "player's" turn to move;

(3) sets of statements specifying what choices have been made by other players on previous moves; (4) sets of statements defining "the termination of the game" and (5) pay offs to each "player" associated with the "outcome" of the game (i.e. that state in which the game terminates) [Rapoport, 1970, p. 53]. Usually games are represented by a matrix array in which the elements are strategies: it is thus said to be represented in *normal form*. An N-person game would be represented by an N-dimensional matrix with an n-tuple of pay offs in each element space. A two-person game is a rectangular array where the horizontal rows and vertical columns represent the strategies of the players. In two-person zero sum games, where the sum of the payoffs to the two players is always zero, the largest payoff is the strategy which contains *saddle points*: the maximin principle. A saddle point is an element space in the matrix where the raw pay off is minimal and column pay off maximal. For two-person zero sum games of *perfect information* (i.e. players always know the actual choices made previously by other players) each such game contains at least one saddle point, and consequently at least one best strategy for each player. In *mixed strategies* each available strategy is assigned a set of probabilities. It has been proved that for all two-person games, where the sum of the pay offs of the two players are the same, there exists a pair of strategies which are in equilibrium with each other. If one player chooses such a strategy, then it is the best strategy for the other player to adopt as well. For non-constant sum games, where the sum of the pay offs differ, it can be proved that each such game has an equilibrium point, such that if one player adopts an alternative strategy to the other who is at this point, the former cannot improve the pay off. It is too difficult to give any gloss here on N-person games. I hope, however, that we should be sufficiently clear about these basic concepts to clarify those concepts of Sociobiology which are of immediate concern here.

A concise statement of the application of games theory to evolutionary theory is given by Maynard Smith [1978 (2)] and Dawkins [1980]. Here we are interested in organisms (plants and animals) adopting pure or mixed strategies and pay offs in terms of inclusive fitness or Darwinian fitness. Let a population P consist of organisms adopting strategies I or J with frequencies p and q respectively, where $p + q = 1$ and the pay off in fitness to those adopting strategy I is $E_J(I)$ is the opponent adopts strategy J. Then the fitnesses of these strategies are:

$$(1) \quad f(I) = p.\, E_I(I) + q.\, E_J(I)$$

$$(2) \quad f(J) = p.\, E_I(J) + q.\, E_J(J).$$

I is said to be an *evolutionary stable strategy* (ESS) if and only if:

$$(3) \quad (\forall I)(\forall J)\{[(I \neq J) \,\&\, (E_I(I) > E_I(J))] \vee [(E_I(I) = E_I(J)) \,\&\, (E_J(I) > E_J(J))]\}$$

or more simply, if the majority of individuals adopt I, then a mutant strategy such as J will not invade the population. It is alleged by individual selectionists, that except in special rarely achieved circumstances, a population of group selection playing organisms would be invaded by a "selfish player". But we should note that this only occurs if the necessary mutation which "causes" the selfish player occurs (or the necessary migration) – otherwise selection simply cannot operate.

An example from Parker [1970] will illustrate what Sociobiologists can do with ESS's. The female dung flies of *Scatophaga stercoraria* visit cow pats to mate and lay eggs, and the frequency of visits decreases as the pat dries. Males cannot maximize their fitness by going to fresh pats, since other males will be doing this, and competition will be great. So the strategy adopted will depend upon what other males are doing: if they leave a pat quickly he should stay, but if other males stay on, he should move. The solution, is they believe, an ESS which is a mixed strategy having a negative exponential distribution, such that the expected number of matings for early leavers is the same as that of late leavers. Consequently this is the strategy which *would* be established by natural selection over alternative strategies which could be adopted.[8] An array of other examples are discussed in non-mathematical detail by Dawkins [1980].

This completes our exposition of early Sociobiological theory. The position has undergone an extension recently and in the next section I will chart some of the most important coast line navigated.

2.7 Extensions and developments of Sociobiological theory

The most significant recent development in Sociobiological theory is the Wilson–Lumsden gene-culture co-evolutionary theory [Wilson and Lumsden, 1981]. The position has been taken by many to overcome the inherent theoretical reductionism of mainstream Sociobiology [Trigg, 1982]. I, however, interpret Wilson and Lumsden as having developed the most extensive critique of the "pure culturalist" position in the Sociobiological literature. The book is of course of importance, since the encultured nature of human beings is given very explicit recognition, but this point, it will be maintained, confirms rather than falsifies my initial statement.

The thesis of their book is that there exist sets of epigenetic rules, these being rules which are "encoded in the DNA" [Wilson and Lumsden, 1981, p. 370], which determine "Culturgen" choice. Culturgens are a new attempt to define the old untenable idea of a unit of culture: Culturgens are sets of relatively homogeneous artifacts, behaviours or mentifacts sharing either a common essential nature, or at least bearing a significant family resemblance insofar as they constitute a polythetic group. So the class of bronze swords (to use their own example [ibid., pp. 28–29]) and bronze flange-tanged Erbenheim swords, are different Culturgens with respect to the variable of "swordness", but swords are different Culturgens from axes. We are told the Culturgen choice is determined by epigenetic rules, which is an outrightly absurd thesis, given that what we actually regard as a polythetic group is admitted to be subject to a very high bias factor by the theoretical interests of the classifier. So there is probably an infinite (denumerable) number of these Culturgens, and there is most likely not an infinite number of epigenetic rules. There is possibly even an infinite number of sword Culturgens, since we can arbitrarily draw very fine distinctions in our classifications (e.g. introducing time factors, area of manufacture, historical factors such as past ownership of the sword and so on). Further, since theoretical entities, including mathematical entities are also Culturgens, there is good reason to believe that the total number of Culturgens is non-denumerable. It is debatable as to whether there is even a denumerable number of physical particles in the world, let alone epigenetic rules. If you wish to escape this conclusion by saying that only very broad polythetic groups are epigenetically encoded, then this is a substantial departure from the Wilson–Lumsden position, necessarily weakening the position further. We note in passing that their book contains no argument that sword choice, to use their own example, is in anyway under genetic control.

Primary epigentic rules operate at the level of the sensory filtering of perceptual information, insofar as relatively automatic physiological processes are involved. *Secondary epigenetic rules* operate upon all other information displayed in the perceptual fields including "the evaluation of perception through the processes of memory, emotional response, and decision making" [ibid., p. 36]. Together these rules determine *usage bias curves* which affect the probability of using one Culturgen, like a red biro to write a note, over another one, such as a blue biro to write a note. On this basis, they define three positions: (1) *pure genetic transmission*; (2) *pure cultural transmission*; and (3) *gene-culture transmission*. In pure genetic transmission the probability of transmission of Culturgens will be limited to a single choice; in pure cultural transmission all choices are equiprobable, and in gene-culture transmission "more than one culturgen is accessible and at least two culturgens differ in the likelihood of adoption because of the innate epigenetic rules" [ibid., p. 11]. We should note that for the pure cultural transmission position, not all choices are equiprobable, since there are inevitable cultural biases existing (if and only due to the availability of Culturgens). Wilson–Lumsden are presumably referring here to the epigenetic biases which exist. Their argument of interest to this work, stripped of its mathematical detail, is an attempt to eliminate the pure cultural transmission position. We shall subject that argument to a critical examination in due course.

There are a number of other developments which should also be cited. Underlying all of these developments is an attempt to produce "balanced bio-social theories" [Baldwin and Baldwin, 1980]. Baldwin and Baldwin list three criteria which such balanced bio-social theories should meet:

(1) they must not be biased towards extreme "nature" or "nurture" positions but devised "to contain an empirically defensible mixture and interaction of biological and environmental factors" [ibid., p. 5];
(2) they must present multi-causal models of behaviour integrating phylogenetic and ontogenetic causes into a harmonious theoretical framework [ibid., p. 5];
(3) they must "also adjust the weighting of phylogenetic, ontogenetic, and interaction effects according to the phylogenetic level of the species being studied" [ibid., p. 6].

Baldwin and Baldwin argue that neither ethology or Sociobiology (in its pre-1981 form) are balanced bio-social sciences. They do not in their paper present an answer to the grand problem of presenting a general model of the nature of such theories, beyond listing some useful methodological tenets for bio-social research: (1) integrate existing data from the social and biological sciences; (2) examine, criticize, test and modify work from the biological sciences; (3) conduct original sociological research on biological mediators of behaviour; and (4) conduct original sociological research on animal behaviour and social organization. Such methodolgoical tenets may be accepted even by those would attribute significant causal efficacy to social processes and cultural phenomena in their own right: here it would be maintained that the results of such studies are of little significance in understanding in a detailed way, much of human social behaviour.

Needless to say, we are not told what really balanced biosocial theories are. This would involve saying what precisely an "empirically defensible mixture and interaction of biological and cultural factors" actually is. Baldwin and Baldwin seem to believe that a Culturological position must "refuse to acknowledge the biological nature of *Homo sapiens*, the animal" [ibid., p. 4]. This is hardly more than a repetition of the stereotypes which Baldwin and Baldwin initially told us they had abandoned. Further, the search for bio-social theories is not as methodologically plausible as it might seem. No one has yet produced any plausible non-trivial bio-social account of social phenomena such as the development of paraconsistent logics, the quantum theory or the composition of a poem. Balanced bio-social theories may thus lead us on practically fruitless methodological wild goose chases.

A further interesting development has come from Plotkin and Odling-Smee [1981]. The authors reject what they call the "Monolithic theory" of Sociobiology, the position that evolution occurs through the single process of the natural selection of alternative alleles. Their rejection of the Monolithic theory of Sociobiology is right in my opinion, but not for the reasons which they give. They maintain that the Monolithic theory fails to explain sufficiently the evolution of complex social behaviour, since genes have a high degree of permanence, whilst many social behaviours may change during the lifetime of the organism [ibid., p. 226]. The objection is not as strong as one would wish, since the Monolithic theorist may argue (ignoring the question of the *ad hoc* nature of this reply) that such permanent genes determine multiple behavioural sets or make use of the idea of preadaptations. Nevertheless, what is important here is Plotkin and Odling-Smee's own model.

Levels of evolution, on Plotkin and Odling-Smee's model, have their own specific site of information storage, known as a "referent", where it is periodically drawn under environmental challenges and translated into phenotypic adaptations [ibid., p. 228]. The first level of evolution is the genetic system, whose referent is the gene pool of a breeding population and the primary process is the change in gene frequencies in that gene pool. The translatory process of the first level is known as *epigenesis*. At the second evolutionary level, *variable epigenesis* is the process of information gained by the interaction of alternative developmental pathways with the environment, and the resultant developmental "cannalization" or "trajectory". Information gained by the second referent is constrained by the instructions of the first evolutionary level, and thus cannot be wholly autonomous, although presumably this is because genes evolved before other means of storing information evolved [ibid., p. 228]. Further, the second level is "primed" by the first level gaining information from the "minimum baseline" supplied by the first level, rather than gaining information *de novo*. Such acquired information by the second level, remains confined to the individual's phenotype, and does not itself stand as anything more than a solution to a specific problem existing at the primary level. If the second level does not solve such problems it will be changed or eliminated [ibid., p. 229].

The third level of evolution is the process of learning, and the referent is the central nervous system. The third level may or may not be "nested" under the second level, but it is under the first level, which is to say that is may be an autonomous information gaining level vis-á-vis the second level, but it is constrainted by the first. All information gained by the learner is confined to the phenotype of the individual organism, although if adaptatively successful, may ultimately be fed back into the gene pool.

The fourth level of evolution is socioculture. For such an occurrence there must be (1) a group of organisms capable of learning; (2) this group must have available a non-genetic channel of communication for the transmission of learned information. It is necessary for both sender and receiver to gain such information through third level learning processes. The referent of the fourth level is named a "cultural pool" [ibid., p. 231]. The fourth level is a nested and constrainted relation to the first level. This seems to mean that (1) it is not possible to transmit a capacity for engaging in sociocultural evolution nongenetically; (2) noncultural adaptations cannot be transmitted nongenetically.

This completes my summary of Plotkin and Odling-Smee's position. Needless to say I do not accept this modified multilevel account of Sociobiology either. Criticisms will follow later in the chapter. I turn immediately to the critique of "Monolithic" Sociobiology.

3.0 Towards a critique of Sociobiology

I will not survey to any great extent, the various epistemological criticisms which have been made against Sociobiology, as this has already been done by Ruse [1979]. Past critics have usually accepted the cogency of the neo-Darwinist theory of evolution: such a position will be questioned in this chapter and the next. Without making this move, however, the best thing which a critic can attempt to do is to criticize various parts of the theory, and certain formulations and implications of the position. This is what I shall attempt to do in this section.

The discussion will proceed as follows. In section 3.1 I will discuss empirical and theoretical limits to "Monolithic" Sociobiology; in section 3.2 some programme resistant areas for human Sociobiology will be outlined and in section 3.3 an Impossibility result for a general Sociobiology accepting classical optimization assumptions will be established. The section will be concluded by a critique of sophisticated Sociobiology.

3.1 Limitations of Sociobiological theory

The central "theorem" of Sociobiology (we refer throughout to "Monolithic Sociobiology" [Barash, 1977, p. 63]) takes the whole of the organismic world, if not the biochemical world, to be a grand Panglossian optimization machine – and perhaps it is. If so, organisms are taken to be machines for the maximization of inclusive fitness. If they do not maximize their inclusive fitness, they are thereby exhibiting evolutionary unstable strategies and will be selected against. in the long run. But why accept this *axiom*? It is not, as we noted earlier, a theorem since no proof has yet been given of this proposition from the other axioms of Sociobiology – indeed it is unclear formally what precisely are the "axioms of Sociobiology". Thus as it is an axiom, and self-acknowledged to be an axiom of central importance to Sociobiology, its collapse will do great damage to the programme.

Bell [1982, p. 79] has given a defence of the axiom of "perfection", which Sociobiologists could possibly cite in defence of their own central axiom. It is methodologically sound, Bell maintains, to assume that adaptation and selection are precision optimal instruments, and that selection alone is a sufficient explanation of organic diversity, for:

> "If its truth is denied, then whenever we are confronted with a problem whose solution is not self-evident we shall be tempted to refer it to the operation of unique causes in an unknowable past. We shall become mythologists". [ibid., p. 79].

Further, he claims, to deny this axiom points to "the prevalence of miracles", where genetic systems are makeshift and several of them may perform the same function equally well. Contrary to Bell nonoptimality is quite common in biological systems [Clarke, 1975] so Bell's implication may well be grasped without contradiction.

However, as a defence of the axiom of precision, Bell's argument is logically flawed: taken for an argument for the position, it is circular. But of course, Bell is explicit in claiming later in his book that he does not accept the truth of this axiom (even though he says: "I shall base all my reasoning on an axiom of perfection" [ibid., p. 79], making the central arguments of his book up to section 5.1 unsound). But does the stated consequence follow? Hardly. Palaeontologists certainly would not agree with Bell that they are mythologists although most would openly admit that the fossil record is less than "perfect". The claim, however, that the past is "unknowable", unless maintained as a *general* metaphysical claim, is far too exaggerated to be seriously considered, and talk of those who deny this axiom as accepting "miracles" is also nonsense.

Whilst it seems impossible to give any non-circular justification of the axiom of precision and the central axiom of Sociobiology within the structure of the respective positions themselves, these Panglossian assumptions can be falsified. Indeed, as cited above, the axiom of perfection is in fact false. In the case of Sociobiology, if the methodological defence of Sociobiology's central axiom fails through circularity, and if a plausible case is made for the falsity of this axiom, then in any advanced Sociobiological explanation, independent argument must be given for accepting that the behaviour in question is maximizing inclusive fitness. This, as we shall see later, is virtually impossible to empirically document with sufficient accuracy. Thus Monolithic Sociobiological explanations would be vitiated, preventing immediately any causal-explanatory reduction of Culturology. Note that if the central axiom of Sociobiology is rejected, then the entire theoretical structure of Monolithic optimality Sociobiology is vitiated. It does not follow from this that Sociobiology is thereby redundant, for the central axiom might well be severely qualified. Nevertheless, it is informative to refute the extremist position.

I was first led to my present interest in the relationship between *Pepsis marginata*, the Tarantula-Hawk wasp and *Cytopholis portoricae*, the Tarantula, through Watson's [1980 (2)] popular book. After reading this far in the book I was immediately suspicious of Watson's data and sources, and attempted to obtain more information. A Biosis computer search came up with no independent references for the past ten years. I have been unable to obtain Eiseley [1975] and so was led back to Petrunkevitch [1926]. The latter's experiments have been accurately described by Watson, and my initial suspicions shown to be unfounded.

Pepsis marginata lays its eggs only on paralysed *Cyrtopholis portoricae*. *Cyropholis portoricae* has a poor sense of sight and hearing, but a very acute sense of touch, attacking insects which come into range, such as beetles, quite swiftly. *Cyrtopholis portoricae* females are so savage that they will attack even the *Cyrtopholis portoricae* male outside of the breeding season [ibid., p. 368].

Cyrtopholis portoricae is extremely fast in attack; Petrunkevitch reports that it is difficult to follow what happens with the naked eye [ibid., p. 370]. Yet *Cyrtopholis portoricae* allows *Pepsis marginata* to climb all over her body, whilst searching for the articulation membrane between the maxilla, first leg, sternum and lip. Sometimes *Cyrtopholis portoricae* may rise on its hind legs, but it does not strike, or at least was not observed by Petrunkevitch to do so. Since *Cyrtopholis portoricae* can *immediately* attack other insects after probing [ibid., p. 377], it is therefore not paralysed during these encounters – indeed *Pepsis marginata* is searching for just the right spot to paralyse *Cyrtopholis portoricae*. So the most immediate counterargument against this example must be rejected. Thus, given that *Cyrtopholis portoricae* females will attack even the males of the species outside of the mating season, this phenomena is quite interesting. From Petrunkevitch's description *Cyrtopholis portoricae* makes the supreme sacrifice time and time again, even to a *Pepsis marginata* specimen with a damaged leg. There is no doubt that *Cyrtopholis portoricae* could destroy this wasp, so the disturbing question is: why not?

Sociobiologists should also be disturbed. This is by defintion a case of genetic altruism since *Pepsis marginata* does not lay her eggs on any other species of Tarantula, and surely *Cyrtopholis portoricae* makes the supreme sacrifice in fitness. After all, as we have seen, Sociobiologists are willing to classify cases of "trophic eggs" as "genetically altruistic", so it would be arbitrary to rule out this case. Now if this is a case of genetic altruism, how do they explain it? Since spiders belong to the class *Arachnida* and wasps to the class *Insecta,* kin selection theory involving coefficients of relationship cannot be used as the costs are now "infinite" or undefined. Nor, fairly obviously, can the theory of parental manipulation be used. This leaves Trivers' theory of reciprocal altruism. But the conditions for the applicability of this model cannot be met because *there can be no reciprocal altruism from a dead spider*! There is no conceptual machinery left to explain this as a strategy which maximizes inclusive fitness. Thus the central axiom of Sociobiology is not only false, but the Sociobiological theory of genetic altruism is radically incomplete. Worse, if Alexander [1979, p. 19] hopes to escape the claims of the unfalsifiability of Sociobiology (and in this instance neo-Darwinism) by maintaining that just one example of some behaviour or structure, which is for the exclusive good of another species, was known, would falsify Sociobiology and neo-Darwinism, then both are thereby false.

It will not escape this conclusion to merely label this phenomena as a predator-prey relationship. Surely a qualitative difference exists here in that *Cyrtopholis portoricae* could strike but does not. Now independently of what biological label one wishes to pin upon this phenomena, from a logical point of view this is irrelevant as long as: (1) the behaviour is a case of genetic altruism; and (2) the only model of genetic-altruism capable of dealing with this fails because its applicability conditions fail.

One might argue that *Cyrtopholis portoricae* is simply maladapted to its environment. This might well be true. If so, then by good neo-Darwinist reasoning one would expect its elimination by natural selection over evolutionary time. This is not observed. Thus the Panglossian assumption is false because *Cyrtopholis portoricae* can play an "Evolutionary Stupid Strategy" in not eliminating its enemy. The reason why *Cyrtopholis portoricae* is not itself eliminated is that *Pepsis marginata* does not lay as many eggs, and *Cyrtopholis portoricae* has no other natural enemies. Moreover, *Cyrtopholis portoricae* is safe when wedged in its burrow with the articulation membrane protected.

This is but one example, but we find upon closer biological inspection more general phenomena which raise problems for Sociobiology. Consider the phenomena of sex. Now it cannot be argued at all plausibly that sex is not within the domain of phenomena to be explained by Sociobiology (as Michael Ghiselin has suggested to me in correspondence 26/5/82) since "sex" simply is, for animals at least, a form of social behaviour (not to deny that there are obvious genetic and biochemical components here). More precisely, I take "sex" to refer to acts of sexual reproduction and associated capacities for such acts, made possible by the possession of sexual organs for producing male or female gametes. It is not important to break the circularity of this definition at present (c.f. chapter 6). What is important is that meiosis according to the central axiom of Sociobiology is genetically costly: why adopt the strategy of sexual reproduction when asexual reproduction allows one to pass on 50% more of one's genes?

Dawkins [1976, p. 47] toys for a few lines with the idea of "genes for sex" and asserts that a sufficient explanation for the existence of sexual reproduction is if sexual reproduction "benefits a gene for sexual reproduction" [ibid., p. 47]. This is circular and he knows it, and he soon loses interest in the question, moving onto some other topics in fast succession.

Another explanation of sex is that "sex is a means of mobilizing genetic variability" [Ghiselin, 1974, p. 57]. Ghiselin continues:

> "Sex acts so as to obtain variability, and, most important, *to release it at the appropriate time*. We may infer, if this is so, that its activity in giving unity to the population is a side effect of the circumstance that the component individuals of the population are mutually benefitted through bringing together and combining their separate genetical resources". [ibid., p. 57].

In short it is claimed that sexual reproduction is an advantage in the short run to the individual because the offspring of this individual, through such mobilized genetic variability, will have a high "fitness". This explanation contrasts with that of Fisher [1930] who takes sexually reproducing populations to evolve more rapidly than asexual populations, and thus survive environmental challenges which cause extinction in the latter. Maynard Smith [1971] has suggested that the most important advantage of sex is in cases of migration of two populations into a new environment, where the best adapted genotype is a combination of genes from both populations.

These explanations and variants upon this general form, despite their mathematical sophistication, do not succeed on logical and methodological grounds. The variability given by sexual recombination never results in a transformation of form necessary for evolution to occur, as one may only reshuffle the genes of an already existing gene pool. If some species need a radically new adaptation to deal with a new environment (e.g. lungs instead of gills) one does not get this by sexual reproduction. The sole mechanism for evolutionary innovation of a species is random genetic mutation, and these are indiscriminate between sexual and asexual clones. That many asexual clones of higher organisms flourish in nature, indicates that asexuality can hardly be a maladaptive trait. Further Thompson [1976] has argued that sexual recombination only has effects upon the rate of evolution in extremely small populations, and sex is certainly a widespread phenomena, present in large, as well as small populations. Finally, even if sexuality did lead to increased genetic variability this does not as a matter of logic explain the evolution of sexuality itself, as such a variability is an effect rather than a cause of sexuality.

Stanley's [1979] explanation would be totally unpalatable to Sociobiologists and neo-Darwinists who have rejected group selection for conceptual and methodological reasons [Williams, 1966] as his explanation involves selection at the level of higher taxa than that of the species. Sexual reproduction enables, throughout fairly long stretches of geological time, the establishment of a broad adaptive zone by divergent speciation to cope with the extensive extinction rate witnessed throughout such time. This mode of speciation is not possible for asexual taxa [ibid., p. 222]. But if individual selectionists are so disturbed about group selection, the thoughts of selection at higher taxa than the species are surely poisonous thoughts. Such an explanation would be rejected *a priori*.

There are a number of rigorous papers and books which cannot be considered here in the detail which they deserve [Treisman, 1976] [Maynard Smith, 1978 (3)] [Bell, 1982]. Treisman's model has been criticized by Williams [1980, p. 382] on the grounds that this model requires "intensities of selection, and changes in selection coefficients, which would be unsupportable in the low-fecundity populations for which he claims applicability" [ibid., p. 382]. Further, both Maynard Smith and Bell in their respective books [Maynard Smith, 1978 (3), p. 10] [Bell, 1982, p. 389] recognize the restrictions of their own models. Whilst it can hardly be claimed that no impressive progress has been made, the full story has yet to be satisfactorily told about the evolution of sex. Most importantly, Bell ultimately rejects the axiom of perfection in formulating his verison of the tangled bank theory [ibid., p. 390].

Williams [1980] after his own survey of the literature concludes:

"The near universality of sexual reproduction remains a major unsolved

> mystery. It is explicitly a mystery in relation to kin selection, on which so much of sociobiological thought depends". [ibid., p. 382].

Sex may thus be readily taken as another counterexample to the central axiom of Sociobiology. Indeed surely part of the difficulty must remain with the very formulation of the problem by Monolithic Sociobiology: what would be gained by an individual who eliminated meiosis *but remained otherwise the same* – given the genetic cost of meiosis, why not cheat? One answer to Williams' formulation of the problem must be that it is an incoherent question as one cannot make such a radical change in a higher organism without enormous physiological, behavioural, and in the case of humans, social changes. Having a sex (although not necessarily being of one particular sex) is a *characterizing* or *essential* property of organisms. It is therefore logically unclear what it means to "cheat" whilst remaining the same in all other respects. No generative mechanism was given by Williams for the transference of genetic wishes into biological reality.

3.2 Further programme resistant areas for human Sociobiology

For Sociobiological modes of explanation to be applicable the phenomena to be explained must be demonstrated to effect either the Darwinian or inclusive fitness of the organism. As we have seen, Sociobiologists have asserted rather than demonstrated that all properties of an organism are sensitive to evolutionary forces. It is widely recognized, even by Sociobiologists that both the *sense* and *reference* of many of their technical biological terms, and some commonsense terms differs. This is most clearly seen in the sets of concepts related to that of "commonsense altruism". In a wide sense "commonsense altruism" refers to attitudes and acts of love and care for others, which in some way benefits them, not necessarily without benefit to the giver. The concept of "genetic altruism" differs in both sense and reference to the former concept: for one thing it must assume that altruistic acts must benefit only the receiver, also in terms only of fitness. Therefore there exist innumerable counterexamples to the alleged completeness of the Sociobiological theory of "altruism", if they are really serious in attempting to explain the same phenomena which social scientists are concerned with.[9] One cannot explain what one cannot even conceptualize, and without some demonstration of the universal truth of the Panglossian assumption, much of the dealings of Monolithic Sociobiologists with human altruism remains the meaningless confusion of concepts.

I will outline in this section that I take to be some "programme resistant areas" for human Sociobiology, supplementing these above considerations.

3.2.1 Deliberate inclusive fitness limitation and the central axiom of Sociobiology

Many human societies lately have opted for *both* restraints on fertility via contraception and restraints upon reproductive success, although large increases, or at least some increase in "child-production" could occur without putting too much strain on the vast reserves of wealth of such societies. To this Colinvaux [1976] must make the question-begging denial that inclusive fitness limitations occur: contraceptives increase Darwinian fitness and inclusive fitness. Weinrich [1978] proposes that an evolved mechanism is operating "maladaptively" in a new environment, and Barkow [1978, p. 9] argues that population limitations may maximize inclusive fitness in some environments.

The Colinvaux–Barkow and Weinrich proposals are logical contraries, so both cannot be true, although both can be false. Now let us take the Colinvaux–Barkow account first. It is alleged that cases of population restraint, and deliberate fitness limitations are actually ways of maximizing fitness. This is a *prima facie* contradiction in the usual way in which "maximizing fitness" is used. No doubt, some population restraints can maximize long-term fitnesses where marginal population increases are outweighed by marginal population costs of increase. However our initial assumptions carefully rule out this possible rejoinder because we can assume at least some, however small, "child-production" expansion possibilities without long-term detrimental effects on overall inclusive fitness on members of the population, in the above societies. To say in a situation where some bourgeois family, who could easily have another child and adequately support it, are *maximizing* their inclusive fitness by not having it is straightforwardly false. In neither cases are any detailed arguments for this *ad hoc* claim offered by Colinvaux–Barkow.

Weinrich gives no argument as to why we should believe that contraceptive measures are an "evolved mechanism" which is "maladaptively" operating in a new environment, resting content in merely advancing an *ad hoc* hypothesis. What precisely is the evolved mechanism, and why take this to be a maladaptation in any case? If it is argued that any behaviour which does not maximize inclusive fitness in an environment is maladaptive, as an argument for both Weinrich's thesis, and the presupposed central axiom of Sociobiology, this is straightforwardly circular. Further, a society widely and consistently practicing reproductive restraint can hardly be *selected against*; selection operates for Sociobiology or organisms or genes. Finally, it is asserted rather than demonstrated that a "new environment" exists, unless Weinrich tautologically takes the new environment to simply be, a society with some effective contraception [Livingstone, 1980, p. 324].

The case of deliberate fitness limitations shows that human beings are not mere mechanisms for the maximization of inclusive fitness. If Sociobiologists intend to describe such behaviour as maximizing inclusive fitness, then they need an account of fitness not synonymous with fecundity of either the individual or the gene [Wilson, 1975, p. 67], and this they have failed to supply [Burian, 1978].

3.2.2 Sociobiology and incest

The phenomena of incest taboos is probably the all time favourite of

Sociobiologists, and there is hardly an exposition which fails to present the argument summarized below [Wilson, 1979, pp. 37–40] [Wilson and Lumsden, 1981, pp. 147–158] [Crook, 1980, p. 161] [Alexander, 1979, p. 79]. So popular is this argument, that even the rigorous, yet careful John Maynard Smith [1978 (4), pp. 32–33] accepts the following account. I shall argue by contrast that Sociobiologists fail once more, even on their homeground.

The universality of the incest taboo is explained with reference to the alleged facts that inbreeding between close relatives leads to reduced fertility, increased prevalence of genetic diseases, and so on [Ruse, 1981–82]. This, they tell us, is also seen in the reluctance of the kibbutz adolescences to engage in sexual activity and marriage [Shepher, 1971, 1979] [Tiger and Shepher, 1975] and the frequent breakdown of Taiwanese "minor marriages", where husband and wife-to-be are brought up together in the same household as children [Wolf, 1970]. Incest is in short a genetic disaster.

This argument hinges on the idea that inbreeding will lead to an increase in deleterious homozygous genes, and therefore to an increase in the death rate from homozygosity. But as Livingstone [1969, p. 46] points out, inbreeding would then tend to *decrease* the frequencies of such deleterious genes, as we find have occurred in small groups of primitive people (the Yanomamo Indians of South America have an inbreeding coefficient as high as 0.5, and resultant genetic damage is minimal [Livingstone, 1980, p. 318]).

More importantly we can argue that the question of homozygosity is a red herring: heterozygous genes may be defective, or even lethal, and at least as open to alleged mutational damage, so the question of the maladaptivity of inbreeding simply depends upon the nature of the gene pool. Now inbreeding is a good way to maximize one's inclusive fitness, so a *prima facie* tension exists between the respective costs and benefits of outbreeding, which is thought to balance out at the level of the first cousin [Crook, 1980, p. 162]. Since many societies prohibit first cousin marriages, we deduce as an added bonus that they are not in general maximizing their inclusive fitness (and mutant ESS's are dealt with by law).

What precisely is an incest taboo? In response to this question Rodney Needham [1971, p. 25] points out that incest prohibitions do not form a definite class; the reason for this is the often mutually inconsistent uses which have been made of this term. The Sociobiologist would define incest taboos as a prohibition of reproduction between immediate genetic kin. This does conform to some degree with the social conception of kin, at least for the nuclear family, where kin are most usually quite closely genetically related. But even for a monogamous nuclear family, an incest relationship can exist where the children may well be of a childbearing age (and hence not subjected to sexual adversion imprinting like so many of Konrad Lorenz's geese!) and not genetically related very closely to the parent at all (e.g. in adopted children or with step-parents). We should also note that homosexual incest relationships are also a socially recognized phenomena. However, here dangers of inbreeding do not exist. The Sociobiologists have confused sex with reproduction, a mistake even an experimenting adolescent would not (if it could be helped) make. Therefore the outbreeding hypothesis fails to provide a unified account of incest.

Let us turn to the kibbutz example which Sociobiologists take as a natural experiment, complete with beloved and theoretically vague, negative imprinting mechanisms. What is bizarre about the kibbutz example is that the children may not be closely related genetically at all – so Sociobiologists must claim that in general the negative sexual impriting mechanisms are indiscriminate, and that is almost certainly false as the example of the Yanomamo Indians cited above demonstrates. What is also strange, is that Tiger himself recognized in an earlier work with Fox, that this example violates "the sociological law of propinquity": nearby marries nearby [Tiger and Fox, 1972, pp. 115–116]. So if the kibbutz is the exception rather than the rule, then we need to look closely at the social relationships within these forms of life, rather than jumping with glee at a possible biological conquest and utterly vague, *ad hoc* and unexplicated talk about negative impriting mechanisms.

Kaffman's [1977] study, ironically cited by Wilson and Lumsden [1981, p. 149] in support of the outbreeding hypothesis shows that the Sociobiologists have been giving us misleading data. The kibbutz preached abstention from sexual intercourse until after high school up until the mid-sixties [Kaffman, 1977, p. 208], which the majority of adolescents upheld. But with the various social changes which occurred at the end of the sixties [ibid., p. 213] the attitude has been slowly eroded. True, there is self-imposed exogamic pairing, but this is more parsimoniously explained through the peculiar nature of the kibbutz life: the smallness of groups and the extreme closeness of living are hardly conducive to the illusions and fantasies of romantic love [ibid., p. 216]. Cohen [1978] has pointed out that compulsory military training occurs before the kibbutz member marries, which inevitably exposes the member to a greater range of potential partners. This increases the probability of not marrying a kibbutz member. This probability is further increased in that after military service these members are sent out to colonize a new kibbutz. Add to this Kaffman's testimony that there is no sign of an "adversion feeling" among these adolescents, and it seems reasonable to conclude that the kibbutz example does not support the Sociobiological theory of incest. Much the same argument can be given about the Taiwanese "minor marriages", only here we have an added cultural variable, that such marriages are culturally recognized as being inferior [Harris, 1980, p. 266].

That there exists a universal incest taboo is part of the mythology of the social sciences. There are several societies which do not seem to have any such taboo [Schneider, 1976] including the strongly documented case of brother–sister marriage amongst *commoners* during the period of the Roman rule in Egypt [Middleton, 1962, pp. 606–608].[10] Therefore alternative Sociobiological theories such as that of Parker's [1976] which presuppose the universality of the incest taboo are also untenable.

We thus conclude our presentation of some programme resistant areas to Monolithic Sociobiology in our attempt to intellectually discredit the position as a cognitively satisfactory position. The final section dealing with Monolithic Sociobiology, to immediately follow, will attempt to establish the impossibility of a *General* Monolithic Sociobiology as previously characterized.

3.3 The impossibility of a General Monolithic Sociobiology

The next chapter will be devoted to a consideration of the metaphysics of the selfish gene. Here however we endorse an argument found in Washburn [1978]. The second argument will mathematically attack the optimization structure of Monolithic Sociobiology.

3.3.1 The Washburn argument

The genetic calculus of kin selection theory is committed to the proposition that the coefficient of relationship r, is a measure of the "sharedness of genes". After all, what other sense can be made of the commonplace reasoning that is makes better genetic sense to save all of my family – mother, brother, sister, aunt and cousins than to save my father – but that the grand token of my genes as allelic-tokens-of-a-type is maximized under the specified constraints.

There are, however, qualifications which need to be made. It is debatable as to whether it can be truly claimed that a human XY contains precisely one half of its mother's genes and one half of its father's genes, even if there is a chromosome matching and no chromosome anomales. The Y chromosome in humans is quite small, and contains less "genes" than the X chromosome. Further, excluding the case of MZ twins all other relationships are statistical in nature. If n is the number of chromosomes, and full sibs have δ of them, where $\delta = 0, 1, 2, \ldots, n$, then the number of shared chromosomes has a binomial distribution $\delta \sim B(n, p)$, where p is the probability of receiving a particular chromosome. For $p = 1/2$, $\mu_\delta = n/2$ and $\sigma_\delta^2 = n/4$, where μ_δ is the mean number of shared chromosomes and σ_δ^2 the variance in the number of shared chromosomes [Barash et al., 1978, p. 356]. Thus it is logically possible that my non-MZ sister may in fact have zero of my "genes" as allelic tokens of a type.

It follows that kin selection explanations for diploid organisms outside of the relationship between parents and offspring and MZ twins are statistical in nature, (much like the structure of mainstream non-hidden variable quantum mechanics) are concerned with the behaviour of systems of organisms rather than individual organisms. Thus we may predict that there will be greater variability in sibling–sibling interactions than parent–offspring interactions, but not that these need be the case in any arbitrarily chosen individuals interactions. It *may* well be that I am as closely related to my brother as my mother. As such most popular Sociobiology [Barash, 1981] which attempts to explain the actions of particular individuals and tribes, must be statistically untenable.

These qualifications do not as such challenge the theory of kin selection with coefficients of relationships, rather they sharpen its focus. An argument culled from Washburn I think will supply such a challenge. Whilst an individual receives half of her genes from each parent, such an individual also shares both half the genes in which the parents differ, and all the genes in which the parents do not differ. Washburn appeals to the molecular research of King and Wilson [1975] which he believes demonstrates that individuals who would be considered unrelated (i.e. $r = 0$) share almost all of their "genetic material". King and Wilson compared human and chimpanzee polypeptides by means of amino acid sequencing, immunological and electrophoretic methods of protein comparison concluding: "the average human polypeptide is more than 99 per cent identical to its chimpanzee counterpart" [ibid., pp. 114–115]. Such results require very careful interpretation.

There are basically four methods of investigating genetic differences between species: (1) immunological comparisons of proteins; (2) amino acid sequencing; (3) electrophoretic comparisons; and (4) DNA hydrization. In immunological comparisons, one popular test for similarities is the strong tendency of lumping reactions to occur when rabbit serum which contains anti-chimpanzee antibodies from a prior mixing of chimpanzee serum injected into rabbits, is mixed with human blood. Those anti-chimpanzee antibodies will "clump" with human proteins similar to those initial chimpanzee proteins. This method was com bined with amino acid sequencing by King and Wilson. If the sequence in which amino acids are placed is determined by the genes, then an index of similarity can be made of what genes are in common with respect to the sequence arrangements. In electrophoresis the overall charges of amino acids in an electric field are compared. This technique, described in detail by Lewontin [1974], involves first homogenizing tissue samples to release the proteins and enzymes. These are placed in a gel of polyacrylamide, agar or some other jelly-like substance and subjected to an electric current for a few hours. The proteins in the gel are observed to migrate at a direction and rate dependent upon the molecular size and net electric charge. Then various chemical stains are added which react with the proteins to form a coloured bond at the spots of protein migration [Dobzhansky et al., 1977, p. 49] and it is both the number and position of these bonds which enables the assessment of protein differences. The final technique, DNA hybridization involves the comparison of the respective temperatures at which hybridized DNA or combined DNA separate, to the temperatures at which recombined DNA from the respective species, separate.

So much then for the biology, now for the philosophy. Plomin and Kuse [1979] deny that King and Wilson's results show that "more than 99% of human polypeptides (or the genes that code for them) are identical to those of chimpanzees" [ibid., p. 189]. They also deny that Washburn's claims of a substantially greater DNA similarity *within* the human species shows the irrelevance of individual genetic differences to the social sciences [ibid., p. 190].

Now in comment upon this, we should recognize that King and Wilson [1975, p. 114] themselves recognize that the arrangement of genes on chimpanzee chromosomes differs from that of human chromosomes. This is not even to mention that even if every structural gene was identical, biological differences could arise from regulatory gene action. Plomin and King are quite correct in pointing out that the mere alteration of a single amino acid can result in quite radical changes in protein formation as the case of sickle cell anemia testifies. But let us not lose sight of what is debated here. Whilst it remains true that for some purposes we can consider individuals as having a coefficient of relationship $r = 0$, it does not follow that they have no genes in "common". What is wrong here is our interpretation of r as somehow measuring the percentage of shared gene-tokens-of-a-type, and this interpretation is refuted by the above argument from molecular biology.

We are not concluding as Washburn did, that individual genetic differences are irrelevant to the social sciences. It is concluded that the current interpretation made of r by Sociobiologists is wrong. Therefore all Sociobiological reasoning based upon such misplaced metaphors as "shared-gene-tokens-of-a-type" in kinship analysis, therefore must be fallacious.

3.3.2 Routley's theorem and the impossibility of a General Monolithic Optimality Sociobiology

One of the most startling discoveries of mathematical logic in the past century, next to the antinomies, is the discovery that many problems are demonstratable unsolvable. For example, the problem of determining whether or not for an arbitrary Turing Machine T and initial tape inscription i, that T(i) is defined, is known as the halting problem for Turing machines. It can be very simply demonstrated that there is no such *general* Turing machine and hence that the halting problem is unsolvable [Boolas and Jeffrey, 1980].

Routley [1980 (3)] has demonstrated that an orthodox economic solution of environmental problems and that a general orthodox assessment procedure in education and "Sociology" is impossible. This startling claim seems at first glance irrelevant to the Sociobiology debate. It is the thesis of this section, that Routley's theorem demonstrates the impossibility of a General Sociobiology.

Most of the standard proofs given of Arrow's General Impossibility theorem (e.g. [Sen, 1970, pp. 40–46]) including Arrow's original proof [Arrow, 1963, pp. 51–59, 97–100] are from a logical point of view invalid. The errors involve the standard ones of quantificational use and abuse [Routley, 1979, p. 878]. More importantly from a decision-theoretical perspective, proof failure in the standard proofs of Arrow's theorem arise from an unclarity in the use of the principle of the independence of irrelevant alternatives. Routley has attempted both a repair and extension of Arrow's theorem, and I hope here to extend the scope of the theorem further. Arrow's theorem, as stated by Sen [1970, p. 42] is as follows:

(AT) There is no social welfare function satisfying conditions U (unrestricted domain), I (independence of irrelevant alternatives) and D (non-dictatorship)

Routley's theorem is as follows:

(RT) There is no rational General Ranking Method (GRM) on any usual decision structure (d.s.)

The (RT) proof is similar in meta-logical structure to the mainstream proofs of (AT) (minus the errors!), involving a *Dominance Lemma* (D.L.) and a *Voting Paradox Lemma* (V.P.L.). The impossibility result holds for any field of study which attempts to give rational GRM's, and where the assumptions of the theorem are applicable. Arrow's social welfare function, constrained by individual preference rankings, is replaced by an overall ranking constrained by factor rankings and where other dubious assumptions such as: (1) *the strong reduction assumption* (social rankings depend only upon the rankings of individuals); (b) *the assumption of methodological individualism* (social wholes are analyzable solely in terms of their individual components), can be abandoned.

It remains yet to say, at the level of informal and general explication, how in fact Routley can (justifiably) maintain what must be an utterly stunning conclusion to be reached solely on *a priori* logical grounds, namely that an orthodox economic and social theory is impossible. The argument is simply that an orthodox economic theory (for example) is but a branch of optimization theory, where various desired phenomena such as utility, well-being welfare and so on, are maximized subject to various relevant constraints such as ethical considerations, resource limitations, limitations on knowledge and technology, and so on. So the very point of orthodox economics is to produce optimization of an overall ranking of alternatives on economic systems. On the assumption that it can be shown that any reasonable overall ranking procedure must satisfy the various conditions of the Routley theorem, the impossibility of a general orthodox economic theory is established. (This is of course, not to claim, that descriptive components of such a theory are thereby false.) Routley's theorem is applicable to orthodox optimization theory, which accepts the assumptions upon which the impossibility result is premised upon. We will need then to show that the Sociobiological programme is committed to these assumptions and the required result follows immediately: a General Sociobiology is impossible.

We define from the onset various binary preference relations, taking as primitive the binary relation of "weak preference" R ("at least as good as") for individuals x, y (i.e. xRy). The relations of "strict preference" P, and "indifference" I, are defined:

(D_1) xPy $=$ df xRy & $\sim$ yRx

(D_2) xIy $=$ df xRy & yRx

Two further definitions concerning preference relations are added (a subscript, such as xP_Jy, is read "individual J prefers x against y"):

(D_3) $xD_Jy \;=\;$ df $x\,P_Jy\ \&\ (i \in F)(i \neq J \rightarrow yP_ix) \rightarrow .\ xPy$

where F is a class of individuals (or factors) and "xD_Jy" reads "j is *almost decisive* for x against y"

(D_4) $x\bar{D}_Jy \;=\;$ df $xP_Jz \rightarrow xPz$

where "$x\bar{D}_Jy$" reads "J is decisive for x against y".

Further definitions

(D_5) $C = \langle K, F\rangle$ is a decision structure (d.s.) = df K and F are non-null sets, K being a set of alternatives and F being a set of individuals or factors.

(D_6) C is usual = df K has more than two elements.

(D_7) p is a ranking on C = df p is a function which assigns for each factor i in F, a binary relation $P_i = p(i)$ (a factor ranking) on $K \times K$.

p is subject to the following requirements:

(B) (Admissibility of Factor Orderings)

$(\forall m)\ [(m \in F)(xP_my\ \&\ yP_mz \rightarrow xP_mz)]$

for x, y, z in K and m being those factors of F.

$(\forall m)\ [(m \in F)(xP_my \rightarrow \sim yP_mx)]$

for x, y, z in K and m being those factors of F.

(D) (Strong Transitivity of Overall Ranking)

$xPy\ \&\ \sim zPy \rightarrow xPz$ for x, y, z in K.

(D8) GRM* is a General Ranking Method on $C = \langle K, F\rangle$ = df GRM is a function assigning for each ranking p on C a strongly transitive and asymmetric overall ranking P on $K \times K$

(D9) GRM* is a rational GRM on a d.s. = df GRM* satisfies these conditions:

P (Pareto): $(\forall p)(x, y \in K) \cdot (i \in F)\ xP_iy \rightarrow xPy$

IP (Independence): $(\forall p)(\forall p')(S \subseteq K) \cdot (x, y \in S)(i \in F)(xP_iy \equiv xP'_iy) \rightarrow$

$(x, y \in S)(xPy \equiv xP'y)$

M (Non-Dictatorship): $(i \in F) \sim (\forall p)(x, y \in K) \cdot xP_iy \rightarrow xPy$

For $x, y \in K$ and $V \subseteq F$:

(D10) $xD\ \ y \;=\;$ df $(\forall p)\,[(i \in V)\,xP_iy\ \&\ (i \notin V)\,yP_ix)] \rightarrow xPy$

(D11) $x\bar{D}\ y\ =\ df(\forall p)(i \in V)\ xP_i y \rightarrow xPy$

Definitions (D3) and (D4) may be restated as follows:

(D3*) $xD_J y\ =\ df(\forall p)(xP_J y\ \&\ (i \in F - \{J\})\ yP_i x \rightarrow \cdot\ xPy$

for every set of admissible rankings.

(D4*) $xD_J y\ =\ df(\forall p)(xP_J y \rightarrow xPy)$

for every set of admissible rankings.

Routley's theorem is now stated:

(RI) There is no rational GRM on any usual decision structure.

Proof: The proof consists of proving a *Dominance Lemma* and a *Voting Paradox Lemma* and deducing their consequences. (For logico-masochists cf. [Routley, 1979]).

The extension of the impossibility result to Sociobiology is readily made. We interpret the relation R ("at least as good as") as the relation of being "at least as fit as", for individuals x, y. The relations of "strictly fitter than", will correspond to the relation of "strict preference" (P) above; "fitness indifference" will correspond to the "indifference relation" I, defined above, and the term "fitness" shall be used here as a primitive concept. The relations of being "almost decisive" and "decisive" may be readily given a Sociobiological interpretation: instead of speaking of x being almost decisive in preference for individual J against individual y, we may speak of individual x being almost decisive in fitness against individual y for time period T in environment E. The further interpretation of Routley's definitions is mechanical.

The theorem will only have significance if it can be argued that Sociobiology is committed to accepting those assumptions upon which the impossibility theorem is premised.

It is evident that (D5) and (D6) cannot be rejected unless Sociobiology is to avoid reference to any decision model structure at all. The impossibility result could be avoided if the decision structure was taken to be non-usual (as defined), but to do this commits one to a life of impotency, for K has then two or less elements, and this without doubt severely limits the scope of Sociobiological theory.

One may perhaps wish to deny that Sociobiology need accept any GRM at all, thus denying the applicability of (D8) to Sociobiological theory. But to do this is nothing short of denying that Sociobiological theory involves the determination of "best choice" between alternative evolutionary strategies in biological systems. This requires an overall ranking satisfying condition (D) of strong transitivity. If this was denied, then one could argue that even though x had a higher inclusive fitness than y, and it was not the case that z had a higher fitness than y it does not necessarily follow that x has a higher inclusive fitness than z. This, one might say, is a reasonable denial because the condition of strong transitivity is absurdly strong. But since it is *a priori* true that a binary relation such as "has a higher inclusive fitness than" is asymmetrical, this criticism cannot succeed as long as transitivity is accepted. Suppose for x, y, z in xPy and yPz and ~ xPz. By asymmetry ~ yPx, but by strong transitivity yPx, which is impossible by asymmetry. To deny transitivity is a move which any Sociobiologist is unlikely to make.

This leaves only principles P (Pareto), IP (Independence) and M (Non-Dictatorship) to deny, and this is the area where a counter-attack is most likely to be made.

Can the Pareto or Factor Unanimity principle be reasonably denied? To do so amounts to denying that if every factor ranks alternative x above y, then so does the overall ranking. This ranking may or may not be a necessary truth. Sen [1970, p. 22] points out that an economy could be Pareto-optimal even given massive social inequalities. But Sen's claim seems to ignore completely the rejoinder that the poor are not completely indifferent to their suffering; Pareto-optimality involves, it is recalled, these principles: (a) if everyone in a society is indifferent to alternatives x and y, then the society as a whole should be indifferent as well; (b) if at least one individual strictly prefers x to y, and every individual regards x to be at least as good as y, then the society as a whole should prefer x to y. Many of the alleged counter-examples to the Pareto condition hinge upon the presentation of irrelevant counter-models.

For the Sociobiologist, to deny the Pareto condition will amount to accepting that given an overall inclusive fitness ranking, y may rank over x, even though every factor ranks alternative x and y. To accept this is nothing short of abandoning any possibility of empirically supporting overall assessment rankings. For the empirical support can only come from the individual factor ranks – what else is there? But to deny Factor Unanimity, means that even if every factor ranks x above y, the best possible ranking, the overall ranking need not necessarily do so. What then would count as evidence for such an overall ranking?

The next principle which could be denied is the Multiple Value or Non-Dictatorship condition; the claim that there is no factor which determines overall inclusive fitness rankings in *all cases*, in *all environments*. That is to say, there are no general biological powers possessed by organisms of any species, which thereby determines overall inclusive fitness ratings in any environment. This principle is quite rightly accepted in all Sociobiological work: individuals adopt various strategies to maximize inclusive fitness which may differ throughout time and changing environments – the very point of Sociobiology is to outline these strategies.

This leaves us with only IP (the Independence Principle) to reject. It is the most controversial of all these principles, and it is here that the argument threatens to break down. Tal Scriven [1981] is the latest to allegedly present a counterexample to the independence condition; I have argued elsewhere that he does not succeed [Smith, 1983 (4)]. Independent of this, condition (IP) may be enforced by an argument outlined by Routley [1980 (3), p. 163]. If (IP) is denied then P_i' coincides with P_i for every factor i but where $xP'y$ and $\sim xPy$ for $x, y \in S$. Thus there must be a $h \notin F$ such that $uP_h'v$ and $\sim uP_hv$ for $u, v \in S$. But (IP) should then be once more guaranteed if this relevant factor is added to F. More generally, Routley has proved [ibid., p. 164] this thesis:

(RT*) For any ranking p on a usual $C = \langle K, F \rangle$ satisfying M, there is a set of factors F^+ including F and a ranking p^+ on $\langle K, F^+ \rangle$ satisfying the principles P, IP and M.

Consequently Routley's Impossibility Theorem is not blocked even by rejecting (IP).

The Impossibility of a General Sociobiology thereby follows.

3.4 The case against sophisticated Sociobiology

We thus complete our case against Monolithic Sociobiology. With no fanfare we immediately advance in our critique of Sophisticated Sociobiology. It will be argued in the subsections that follow that neither the accounts of Wilson and Lumsden [1981] nor Plomin and Odling [1981] are theoretically satisfactory and rather than clear the waters muddied and bloodied by reductionism, further cloud it.

3.4.1 Against Wilson and Lumsden

Any Culturologist must respond to the Wilson–Lumsden [1981] model (reviewed in section 2.7) by at least rejecting their claims that pure cultural transmission is an evolutionary unstable strategy. It is an argument well known to physical anthropology, that human beings have been selected for extreme behavioural plasticity and cultural adaptivity, so we must be immediately suspicious of any argument leading to the conclusions Wilson and Lumsden advance. So let us consider their arguments. They argue that: (1) pure cultural transmission requires "a genotype that encodes multiple complex fine tuning mechanisms" and "can be sustained only with precise controls that depend upon continuous genetic evolution" [ibid., p. 11], an argument which begs the question at issue, since no reason is given for the truth of the consequence, nor deductively does it follow from the antecedent. (2) They point out that the CNS is unprepared to classify equally, primary stimuli from Culturgens; here human senses are limited in that some frequencies of sound cannot be heard. True, but this is irrelevant to the choice between Culturgens which can be sensed by human beings, and this we should recall was the point of introducing the idea of usage bias curves. (3) Human beings do not exhibit undifferentiated behavioural development they tell us, in particular they exhibit a biased preference for some stimuli. True once again, but note the level of generality which this claim is pitched at: *human beings in general*. But the question asked was about possible Culturgen choice, and it is evident that the stated general facts are irrelevant to this question. Thus the argument is fallacious. Therefore neither of these points tell against the Culturology position.

The final argument is that pure cultural transmission is not an ESS, because members of a population when subject to both adaptive and deleterious Culturgens are "unable to distinguish [between] them" [ibid., p. 13], so the population is open to invasion by gene-culture transmission and culture is kept on a leash [ibid., p. 13]. But is it? There is no argument given for the premise that the members of the population cannot distinguish between adaptive and deleterious Culturgens. The claim seems in general false: bombs which destroy the universe are deleterious, as may well be ideologically soaked Sociobiology books. Adaptive Culturgens might range from philosophies of love, truth and passion, to a maternity hospital equipped with the latest baby saving devices. Of course these examples are based upon an intuitive guess about what adaptive Culturgens are with respect to pure cultural transmission (assuming the question-begging assumption has not been made, that they are those Culturgens which increase inclusive fitness) since no formal account is given of this by Wilson and Lumsden. No matter, the assumption upon which their counter-argument rests is false.

We should note at this point the absurdity and utter strawman nature of Wilson and Lumsden's pure cultural transmission position. Are all choices of Culturgens equiprobable? If this claim is understood literally it is trivially false, for my choice of using a Culturgen such as a toothbrush to clean my teeth, rather than a Culturgen such as an oxy welding torch, is guided by the nature of the task at hand. But the fact that choices are not equiprobable, if one is not suicidal, does not support the claim that there are non-trivial epigenetic biases involved in such Culturgen choice. The reader should bear in mind, when reflecting upon this, Wilson–Lumsden's own example of a Culturgen: bronze swords. Unfortunately they never give us in the promised biochemical and neurological detail epigenetic rules involving their own paradigm cases of Culturgens. The reader would be justified in inferring that their position is bankrupt.

Yet for their position to be rationally justified, they must show that: (1) the required epigenetic rules exist; (2) genetic variance in epigenetic rules exist within human populations with respect to Culturgen choice; (3) that a relationship between cultural practices and genetic fitness is established, and (4) "the existence of molecular and cellular mechanisms that directly connect genes to cognitive development" [ibid., p. 18] must be shown. These conditons are their own stated criteria of adequacy.

Wilson–Lumsden's arguments shorn of frequently unjustified and vaguely applied mathematical modelling, are very poor. With regard to point (4), they seem to feel that citing evidence of single gene mutation effects on colour division and the mutational effects on neurotransmitters demonstrates the existence of epigenetic rules. This argument, which has also been used by Wilson elsewhere [Wilson, 1979, pp.44–46], as well as by countless geneticists, I christen the "loose bolt in the computer argument". It may be summarized as follows: (1) there are undeniable cases of single gene effects, genetic mutations and diseases such as Turner's syndrome and the Lesch-Nydam syndrome, which affect social behaviour, *therefore* (2) a substantial proportion of all behavioural variation between individuals is genetically based. Despite the psychological plausibility given to these arguments by the many illustrative examples, graphs and coloured pictures of the genetic textbook, the argument is *logically unsound*. It is formally *invalid* insofar as one could argue on the same methodology from the situation of some social behaviour SB_1 being caused by a blow to the brain, to the absurd conclusion that a substantial proportion of all human behavioural variation between individuals is caused by the blows to the brain. The arguments involve an elementary quantificational fallacy, committing a some-to-(nearly)-all inference. A loose bolt may cause a computer to very significantly misbehave, but the operation of computers is not primarily determined by loose bolts at all. Genetic defects are like bolts in computers: they are only of causal relevance (in a nontrivial sense) if they are "loose".

Not only is the above argument fallacious, but we are not given an example of any epigenetic rule "torn down" at the molecular and cellular level. Much of their discussion, it was noted, is taken up with outlining species-specific traits, the relevance of which to the central thesis of Culturgen bias is not demonstrated. It would seem that if we cannot hear certain sounds, then a "musical piece" composed of such sounds is not strictly speaking a musical piece Culturgen at all (even if it is a Culturgen in some other classification). It hardly seems informative or true to speak of usage bias here: better to say that we cannot hear these sounds at all. There is then no *usage* bias at all.

The secondary epigenetic rules discussion also rests upon fallacious arguments. A strong case is argued that mouth movements in smiling are species specific, present even in idomide-deformed children [ibid., p. 77]. But the genetic variability of this characteristic then is precisely zero, as its *variance* in populations is zero, so if it is taken as a Culturgen (which by definition it is not) it violates condition (2); if it is not, then the argument is irrelevant to the question of Culturgen bias. Other cases where epigenetic rules might exist are: (a) incest avoidance; (b) sex differences in book carrying behaviour; (c) differential attachments of mothers to new born infants and (d) valuation and decision making. Condition (4) as I have repeatedly said throughout this discussion is not satisfied for any example in the book. Whilst (a) can be plausibly argued to meet (3), the kibbutz argument is as I have already argued is fallacious (cf. section 3.2.2). The book carrying behaviour example is nothing but an attempt to clutch at straws: that females more frequently carry books against their chests and males, books at their sides on many U.S.A. campuses, may or may not be true [Jenni and Jenni, 1976], but this cited paper develops no arguments, presents no evidence for reasoned belief in genetic dispositions for this behaviour. Jenni and Jenni speculate that this is theoretically possible, but theoretical possibilities are not necessarily theoretical actualities. Wilson–Lumsden's use of this secondary source is therefore highly suspect.

The same style of criticism can be made of Wilson–Lumsden's use of O'Conner's et al. [1977] material on differential attachments of mothers to new born infants. Here it is alleged, mothers who had more frequent contact over a two-day period (six hours) had a much lower rate of parental mistreatment relative to a control group which did not when observed twelve to twenty months later. This they explain, by postulating a sensitive period of interaction which enhances the development of attachment. But no theory is given about this, just the *ad hoc* assertion that this occurs is made, thus violating conditions (1) and (4). The possibility that this six-hour magical inprinting period is causally irrelevant and that a whole range of unobserved and uncontrolled social causal mechanisms intervening within this twelve to twenty month period is not considered, as the mothers are not observed during the intervening time. Their claim is therefore methodologically invalid.

It is concluded that the Wilson–Lumsden position if non-trivial, is untenable. No case has been made against Culturology, and the supporting arguments for the central theses of their book are dust.

3.4.2 Against Plotkin and Odling-Smee

The theoretical system proposed by Plotkin and Odling-Smee is in many ways correct, and in many respects incorrect. To begin with, the levels view of biological reality, whilst not original needs reemphasis in these days of genetic atomism, raving physicalism and mechanism. It is also true that "adaptations" are the results of gaining and storing "information" about the world [Plotkin and Odling-Smee, 1981, p. 227]. However, I would also take all of these various levels to stand in causal relations, but Plotkin and Odling-Smee's entire position is based upon such information-theoretical definitions. There are difficulties with such approaches, which will be discussed in the next chapter.

What remains problematic is precisely the relationship between the fourth level of evolution, socioculture and the first level of evolution, the genetic system. The sociocultural level is taken to be constrained by the instructions of the first evolutionary level, and is not wholly autonomous. True, but the genetic system does not exist independently and autonomous of other higher level biosystems such as the cellular environment either. Nor, contrary to Plotkin and Odling-Smee, is the third level of evolution the process of learning autonomous from variable epigenesis, as they assert. This is quite incredible, one would have thought that developmental biological mechanisms if anything "constrain" or "nest" learning levels, and if not, then it is unclear why something as plastic and abstract as socioculture must be respectively constrained in any non-controversial way. The comments in the latter parts of Plotkin and

Odling-Smee's paper [ibid., p. 261] do not in my opinion significantly aid in clarification. This leaves enormous gaps in places where such gaps should not exist if the argument is to be convincing. For example, they state:

> "This notion of level also implies that at each separate level of evolution, the mechanisms of selection and the units that they select will likewise be different. This point was not demonstrated in our earlier paper, however nor do we have the scope to do this now" [ibid., p. 228].

Nevertheless, there are a number of hints about the nature of the genetic constraints on socioculture. They hold for example that "at least some aspect (one channel) of communication must be the product of the first and possibly the second level" [ibid., p. 232]. This is no doubt true, since speaking involves muscular movements and nerve transmissions of impulses, etc. It is of more interest when they assert that:

(1) it is not possible to transmit a capacity for partaking in Sociocultural evolution nongentically;

(2) noncultural adaptations cannot be transferred nongenetically.

Both these principles would be greeted with universal acceptance from the biological sciences; I think from the perspective of artificial intelligence and the technological sciences both claims can be questioned. It is no longer unintelligible to contemplate the creation of artificial intellects [Boden, 1977] [Sloman, 1978] which might transmit the information for the creation of other intellects or androids nongenetically. It is theoretically possible that such androids, mentally functioning isomorphic to us, might represent a way for humanity to escape the limits of biological bodies. Under this scenario, parts of the human body, such as excellent eyes, might be transferred from body to body as might copies of one's brain. Of course this scenario is a technical science fiction. Yet it does question the intuitive plausibility of (1) and (2). This is especially evident by even current day organ transplants which refute (2).

I am much less critical of Plotkin and Odling-Smee's sophisticated Sociobiology than I am of the Wilson–Lumsden position, but it is maintained that neither position is satisfactory.

3.5 The state of the argument

Let us review the state of the previous long and difficult argument. Our argumentative strategy has been to demonstrate that Sociobiology is a theoretically untenable cognitive enterprise. The counter argument here consisted first of a critique of the central axiom of Sociobiology and the axiom of precision. Further I attempted to show that even the Sociobiological theory of altruism is incomplete, as there are counter-examples to its principles, that is genetically altruistic behaviour whose very description violates the applicability conditions of the stated models. Then various programme resistant areas to Monolithic Sociobiology were stated and our critique concluded by arguing for the impossibility of a General Monolithic Sociobiology. In section 3.4 a case was developed against even "Sophisticated Sociobiology".

The reader may well be anxious with the sceptical blood baths witnessed within these pages; indeed as Mates [1981, p. x] observes people usually want positive contributions, not arguments to tell them that inherent limitations exist in some field of cognitive inquiry. As should be evident from my remarks throughout this work, I maintain that the establishments of theoretical defects and limitations is a positive contribution. Further, I believe that in debates such as the Sociobiology debate we should take the opportunity to examine basic assumptions, and if necessary reject such assumptions. The arguments of many critics of Sociobiology seem to me to hit further than their stated targets and to penetrate with poisonous accuracy to the heart of neo-Darwinism. Therefore in this final section I will follow my sceptical inquiries to their logical conclusion.

3.6 The critique of neo-Darwinism

The account given of neo-Darwinism earlier is one which assumes Panglossian selectionism and adaptationism, and whilst these assumptions are quite common [Gould and Lewontin, 1979] critique of such assumptions will not be sufficient to establish the radical conclusions which I require. I hope however before this discussion is concluded to meet and beat alternative variants.

3.6.1 Neo-Darwinism is not a satisfactory theory of Evolution

Let us accept for the moment the account of Panglossian neo-Darwinism and ask whether we have a satisfactory cosmological theory of *Evolution*. Following von Bertalanffy [1969], Ho and Saunders [1976] and Gruner [1977], Evolution is taken as a process of increasing differentiation of form and increasing complexity of organisms throughout geological time. The transition from quite primitive to higher organization, as I said at the onset of this chapter, is central to any satisfactory theory of Evolution as deserving explanation rather than ignorance. Can neo-Darwinism account for this increase in complexity and differentiation? I believe that it cannot.

First, let us ignore in this chapter questions about the origin of the first cell, of spontaneous emergence [Holland, 1976] and the circularity problem involved in the origin of the genetic code [Wigner, 1961] [Popper, 1974, p. 270] [Harris, 1965, pp. 191–192] [Mercer, 1981] [Hoyle, 1982]. The following argument will not be effected by the answers given to the question of origins, so we may assume that we have a world populated by the primeval organisms. The neo-Darwinist mode of explanation, whilst recognizing phenomena such as pleiotrophy and allometry will in general seek to give adaptationist stories about the selectionist virtues of increased complexity. Yet if the claim is made that selection will favour more complex organisms, we must recognize that this claim is false. A species (or organisms of a species) may well have a better chance of survival by becoming less rather than more complex [Gruner, 1977, p. 712]. If on the other hand it is alleged that heredity, variation, adaptation, multiplication, competition and selection are necessary conditions to account for the Evolution of all life on this planet, then the allegation must also be rejected. Bacteria and other simple organisms are very hardy, many strains being capable of surviving great environment contingencies, and in reproducing through mitosis they leave no "parent" as such to die, and thus stand as far better candidates for immortality than Dawkins "selfish genes" [Dawkins, 1976]. One wonders why Evolution occurs at all given adaptationist reasoning.

If we look more closely at the way "Evolution" is viewed by population geneticists we obtain even more startling conclusions. Let "n_x" be the total number of copies of a particular allele x in a population; "N" be the total number of all alleles at this locus, and the frequency of gene x, $F(x) = n_x/N$. Differentiating by the "chain rule":

$$(1) \qquad \frac{d(F(x))}{dt} = F(x) \left[\frac{\frac{dn_x}{dt}}{n_x} - \frac{\frac{dN}{dt}}{N} \right]$$

where "(dN/dt)/N" is the average rate of increase of all the alleles at the same locus, and $dF(x)/dt = 0$ for $F(x) = 0$ or $\frac{dn_x}{dt}\Big/n_x = \frac{dN}{dt}\Big/N$. Now $dF(x)/dt$ increases for $\frac{dn_x}{dt}\Big/n_x > \frac{dN}{dt}\Big/N$ and decreases for $\frac{dn_x}{dt}\Big/n_x < \frac{dN}{dt}\Big/N$. But note, that this so called "Evolutionary Change" in frequency at a locus, is consistent with the population as a whole going extinct [Wilson, 1975, p. 80] or migrations of the suitable genotypes occurring, and hardly explicates Evolution as understood here.

An objection to the above argument, which both von Bertalanffy [1969, p. 67] and Gruner [1977, pp. 713–714] have in my opinion adequately dealt with, is that increasingly diversified descendents enable a ready exploitation of quite diversified environments, so enabling a more widespread increase in numbers. But there is no necessary connection between complexity and diversification, as quite primitive organisms may be highly specialized, and more complex organisms may become more specialized by becoming more primitive. Second, maximum life support may also be efficiently obtain by flexibility rather than diversity; highly specialized organisms are always in danger of extinction due to radical environmental changes. Third, the argument is question-begging since it does not explain *how* in fact such organisms arose for selection to get under way.

Implicit in this chapter is the thesis that natural selection is not an Evolutionary mechanism. This is not to deny that selection occurs; on the contrary the evidence that selective processes occur is good [Ruse, 1982] but even here it is another matter, as we shall see below, whether neo-Darwinism correctly describes it. Rather, what is being asserted is that selection does not create, it only eliminates. Does this fly in the face of thousands of years of success by plant and animal breeders and one hundred years of laboratory work? I think not. This argument presupposes the validity of traditional ideas of the effects of sexual recombination on the rate of Evolution, which are open to question [Thompson, 1976]. Further, from a logical point of view the argument not only accepts implicitly questionable gradualist assumptions [Stanley, 1979], but one can argue, according to the logic of this position, that breeding failures must thereby count against Evolution. To argue in this fashion leaves Evolutionism readily open to demolition by citing the well known fact that artificial breeding results in "a reduction of fitness for life under natural conditions" [Falconer, 1960, p. 186]. Therefore this extremely popular rejoinder is logically invalid.

Again, it might be argued that what really is at fault is our own naive characterization of neo-Darwinism. Indeed, one must recognize multiple selective peaks [Wright, 1931] random noise factors [Lewontin, 1957] pleiotropic gene action and allometry [Gould, 1977 (3)]. Excluding random noise factors which will be considered shortly, of these phenomena only allometry goes some way towards explaining increased complexity, such as for example antler increases in deers. It goes no way towards supplying a general explanation for the increase in complexity in Evolution, and must at least presuppose the existence of structures such as antlers to be subject to allometric forces. The *Bäuplan* [Gould and Lewontin, 1979] is not itself explained.

Would the problem be solved by pan-neutralist theories? Pan-neutralist theories [King and Jukes, 1969] [Crow and Kimura, 1970] maintain contrary to neo-Darwinism, that most evolutionary change is due to neutral alleles, which are neither harmful nor useful to the organisms in question. Evolutionary change occurs through *random genetic drift*. According to the texts, drift is taken to be a case of the general phenomenon of sampling errors [Dobzhansky et al., 1977, p. 157] and consequently is a property only of small samples/populations.[11] The argument is however fallacious. A sampling error or the standard error of the mean is a property of a mathematical construct and is itself incapable of causal interaction in biological systems.[12] Regardless of these pieces of epistemological finery, pan-neutralism gains great plausibility from its neat solution to *Haldane's dilemma* [Haldane, 1957] which plagues pan-selectionism.[13]

It has only recently been argued that from the perspective of pan-neutralism most "genetic variations are of no physiological [i.e. phenotypic] significance" [Ho and Saunders, 1979, p. 577]. This conclusion seems inescapable given the pan-neutralists own premises, if it is admitted that mutational changes are neutral. (Note, we accept implicit in all such talk that we are positing neutrality in an environment). There thus would seem to be nothing for either natural selection or random processes to "work on" to produce major phenotypic changes. So why "attach such importance to *genes* in evolution? [ibid., p. 577]. The answer is "no reason" if pan-neutralism is accepted. This seemingly leaves the neo-Darwinist without an Evolutionary mechanism at all.

But not quite. The remaining strategy would be to reject pan-neutralism, and yet to avoid Haldane's dilemma, viz. that there are some rare beneficial mutations under an environment. Such mutational changes will be unorientated with respect to adaptation, and will in a specified environment remain very rare exceptions to the usual workings of the biosystem and if they are not generally neutral are deleterious, perhaps even lethal [Dobzhansky et al., 1977, p. 68]. However, very rarely a newly risen mutation may increase the "adaptation" of organisms when they move into a new "niche". So such mutations are like the random replacement of single English letters in English words. The passage from a single word such as "WORD", to another such as "GENE" illustrates this process: WORK, WORE, GORE, GONE, GENE, where we change one letter at a time and still produce at each stage a meaningful word [Maynard Smith, 1970, pp. 266–267]. Thus, the old Creationist argument about the evolution of a complex structure such as the mammalian eye is met by pointing out that there may well be a set of funtional, or even neutral intermediate stages, much like our words. And indeed there are a variety of light receptors from simple single photosensitive cells to the mammalian eye.

The argument does not succeed since it rests upon, as I shall argue in the next chapter, false assumptions about the metaphysics of ontogenesis: development occurs in an integrated rather than atomistic fashion [Gould and Lewontin, 1979, p. 594] and it is such an assumption which is presupposed by Maynard Smith's counter-argument. Regardless of this, the question will then centre around the probability of obtaining the intermediate stages by random mutations [Moorhead and Kaplan (eds.), 1967] [Salisbury, 1971] [Spetner, 1970]. In particular one needs a biosystem capable of processing information from the particular type of eye if it is adaptive, or if it is neutral, one needs a biosystem which "cushions" the phenotypic effects of the neutral useless eye and does not make this neutral structure ultimately maladaptive. So it is hardly just a matter of changing a few "genetic letters" as Maynard Smith implies.

Let us localize the problem. If it is the case that the principle generative mechanism, adaptation-building-random-mutations (RM) are quite "improbable" and the probability of either deleterious (DM) or neutral mutations (NM) is in general for all environments much greater than is: $\Pr(DM \vee NM) \gg \Pr(RM)$ and $\Pr(DM) > \Pr(RM)) \& (\Pr(NM) > \Pr(RM))$, neo-Darwinism is certainly incoherent regardless of what interpretation of the probability operator we accept. Since it is also accepted that RM mutations are more likely to be adaptive when small than when large, there will exist two counteracting probabilities to RM: (a) the probability that small advantageous changes may be simply eliminated through rapid environmental changes and new competitive challenges; (b) the probability of a build up over time of semi-lethal (recessive) genes and the occurrence of the occasional lethal mutation. Thus if one can argue from the mere possibility of beneficial mutations to Evolutionary order, one could just as easily argue from the existence of the countervailing probabilities to the conclusion that Evolution does not occur at all. Since we have rejected Creationism as methodologically absurd, the initial neo-Darwinist assumptions must be false.

The mutation claim itself is premised upon certain implicit biological assumptions, which are open to criticism. N. D. Cook [1977, p. 128] points out that mutational frequency has been observed to decline when post-irradiation conditions are unfavourable to RNA and protein synthesis and have increased when such conditions are favourable. He suggests that the establishment of mutations is dependent upon the cell environment. It is suggested that mutability is itself a phenotypic character subject to genetic control, rather than *merely* the outcome of external factors such as background radiation. What is satisfying for my Metaphysics in these remarks, is the recognition of the cell itself as a *system* rather than a collection of biological atoms, or the mere epiphenomenal shell surrounding the all-important genes. This is the first step in the rejection of the dogma of the "central directing agency" of the gene [Webster and Goodwin, 1982] [Ho and Saunders, 1980] and the movement towards a trans-Mendelian genetics.[14]

What empirical evidence exists which indicates that random genetic mutations can do what the neo-Darwinist says that they do? Certainly not the claims of the immunity of micro-organisms to antibiotics or insects to insecticides. What we must suppose here is that in a subpopulation of insects or micro-organisms, the spray acted as a mutagen causing an adaptational immunity to occur. Or that it just so happened that a mutation occurred which produced enzymes which had a chemical structure such that the chemicals could be detoxified. But this explanation is incoherent. Either the organism was immuned or it was not. If it was not, then given a lethal dose of the chemical spray, it dies. There is no chance of an immunity to build up from scratch instantaneously. If it was immuned, and this is most likely [Dobzhansky, 1974, p. 325], then we simply have a case of preadaptation. What therefore seems likely in such cases is that the various subpopulations of insects already had powers of resistance, most likely in the form of a very flexible range of enzymes. The inflexible simply died off.

Nor is the famous case of industrial melanism in *Biston betularia* evidence of the occurrence and manifestation of the creative powers of random genetic mutations. The example illustrates clearly the operation of natural selection, but with regard to our central issue of concern, it is invalidated by the basic formulation of the argument, for it is admitted that *both* speckled and melanic forms have existed, although preindustrially the melanic forms were extremely rare. It is amazing to believe that it might just so happen that the right mutation out of the infinity which could occur, does occur causing just the right colour changes, and might even be considered a "materialist miracle". Surely it is scientifically reasonable for one to search for some additional mechanism.

Dobzhansky [et al., 1977, p. 66] cites the work of Ayala [1966, 1969] as providing evidence that random mutations play the causal role required of them by neo-Darwinists. In one experiment, males of *Drosophila birchii* of a population, received doses of X-rays, by contrast to the control group which did not. The population numbers fell relative to the control as deleterious mutants were eliminated. But afterwards the population numbers rose at a faster rate than the controls. Does this indicate that some mutations induced by radiation may be favourable to their carriers? Not necessarily. Another hypothesis is simply that all but the strongest *Drosophila birchii* were eliminated since these did not have to compete with moderately strong *Dropophilia birchii* which lacked preadaptive immunities, and the population thrived. All of these examples strongly support my own view of natural selection as primarily an eliminative rather than a creative force.

It follows from these arguments that neo-Darwinism is without a plausible generative mechanism at all.[15] The argument presented is in itself hardly original, but it is one which I believe is sound. Further objections of a specific biological nature will be given in the next chapter, so my critique will be taken to fall into two parts. In the next chapter I will consider alternatives to the neo-Darwinist position. Before passing on towards further considerations of objections to neo-Darwinism a point of some philosophical interest should be noted. If neo-Darwinism is incoherent as a theoretical framework, the thesis of convergent realism [Newton-Smith, 1981] must be, as Laudan [1980, 1980] notes is wrong. Neo-Darwinism has not produced an "advance in approximation to the truth". In fact, as we shall see in the next chapter, many reputable biologists believe that it has been a step in the wrong direction.

3.6.2 Problems with fitnesses

Without heritable variations there would be no natural selection. To justify selectionist arguments, it is necessary to support both the claims that heritable variation exists and that natural selection has occurred. This raises problems about actually measuring the fitness and amount of genetic variation in the population; in particular if natural selection operates on very small random variations, it would seem, and indeed is, difficult to detect such variations.

Let us note from the outset, that even granted the thesis of gene selectionism, common to most mathematical theories of population genetics and evolutionary dynamics, is an assumed selective value for the period of time of their studies. Kojima [1971] has argued cogently against this assumption. It seems that the assumption has been made more as a matter of mathematical convenience, than on the basis of factual evidence [ibid., p. 284] since a mathematical model involving time-dependent, frequency-dependent or density-dependent fitness values is mathematically messy: frequently most environment factor-dependent fitness functions are simply unknown. Thus if we adopt Kojima's methodological tenet that non-constant fitness is a rule in population studies, we can deduce an interesting epistemological conclusion: that in most cases fitness values even over a period of time are unknown. Methodologically we must begin with the attempt to find environment factor-dependent fitness functions. But if these are frequently unknown our study simply does not get off the ground. Moreover, there are even more cogent reasons for believing that such studies will not get off the ground. The reasons will now be detailed.

There are, as was pointed out earlier, two positions on the question of how much genetic variation there is which can be the basis of neo-Darwinism evolution: the classical theory (CT) and the balance theory (BT). Lewontin's remarkable and misunderstood book [Lewontin, 1974] explores this issue up until the early 1970's and I will draw heavily upon this work here. What precisely are CT and BT debating and what is its significance for us? The variance debate is over the question of what proportion of the locus of a typical diploid genotype is heterozygous [ibid., p. 23]. It has proved virtually impossible to decide between CT or BT at the population level and studies at the molecular level, similar to those detailed in section 3.3.1 involving especially gel-electrophoresis techniques have seemed to produce results contrary to the predictions of CT.

In response to this CT has undergone a transformation: the new classical theory (NCT) has replied to this evidence by arguing that the substitution of a single amino acid, although molecularly detectable, is in most cases of no phenotypical significance, and remains a *neutral mutation* from the perspective of natural selection. Note that NCT cannot be criticized on the grounds that there are cases where single amino acid substitutions do have large consequences, or that there are balanced polymorphisms such as sickle cell anemia, since the point at issue is whether a large proportion of natural selection is explicable in these terms, not whether there is none at all [Lewontin, 1974, pp. 198–199].

Yet both NCT and BT suffer defects; BT because it predicts tremendous inbreeding depressions which are not observed, among other defects [ibid., pp. 266–267]. NCT is unsatisfactory because it requires population sizes for all species to be effectively the same and because there is too much variation from locus to locus in the quantity of divergence between populations (among other defects [ibid., p. 267]). Consequently the meaning of the genetic variation observed in natural populations is unclear.

Lewontin himself locates, I think correctly, the problem in the theoretical structure of population genetics itself. First is the problem of *empirical sufficiency*: there are too many parameters in the theory which are not measurable to a satisfactory degree of accuracy to enable the discrimination between alternative hypotheses. Second, population genetics can only refer to equilibrium states and steady states, whereas in fact it should be considering historical phenomena. Yet it is the nature of a system in equilibrium, that at least locally, no trace of historical information remains. This is problematic insofar as historical explanations are often required to discriminate between alternative, mutually inconsistent hypotheses, leaving the discipline with an epistemological difficulty [ibid., p. 269]. Third and most importantly, population genetics involves a general tendency towards *genetic atomism*, where in fact the genome is not a collection of "biological atoms" [ibid., p. 318]. This criticism is one which shall be developed in more detail in the next chapter and some of our socio-theoretical remarks in Appendix 1 depend upon the cogency of this. Yet the criticisms are reasonable, and the conclusion of the argument needs more forceful statement than has been given by Lewontin: the theoretical structure of population genetics, despite its technical mathematical beauty is utterly theoretically and epistemologically flawed. Logically, a position such as Sociobiology must therefore disintegrate.

Of equal relevance, Lewontin [ibid., p. 236] also points out that:

> "To the present moment no one has succeeded in measuring with any accuracy the net fitnesses of any genotypes for any locus for any species in any environment in nature".

The practical difficulties of measurement are insuperable. For species with synchronized populations such as annual plants, fitnesses are ascertained by measuring the number of fertilized zygotes produced per fertilized zygote in the previous generation, but in populations with overlapping generations "the age-specific mortality and fecundity schedules for each genotype need to be measured" [ibid., p. 235]. To estimate the fecundity of genotypes requires having the complete age-specific mortality and fecundity schedules, and this is extremely difficult to obtain except possibly for human beings [ibid., p. 237]. If we however require, given the games theoretical structure of Sociobiology, estimates of fecundity-opportunity-costs, the problem arises again, as the difficulty is to say how some hypothetic event might affect fecundity hypothetically, or contrary to fact.

The strategy of measuring partial fitness through the estimation of individual components of fitness could in principle show the existence of some differential fitness. Here however there are problems of statistical power, and in particular the need for an adequately large sample size presents itself. Yet this method and the strategy of correlating frequencies of alternative alleles with spatio-temporal differences in the environment, does not answer the question of how much selection occurs in nature. Attempts to demonstrate that in laboratory conditions there was selection for some allele at a polymorphic locus are not in Lewontin's opinion at all impressive: the carefully designed and controlled experiments of Yamazaki [1971] failed to detect any selection.

These difficulties are paralleled by a set of well-known philosophical difficulties facing the concept of fitness. Some critics would allege that this concept itself renders neo-Darwinism theoretically untenable since the statement of the theory of natural selection is a "tautology", or else in a strict sense is "metaphysical and unfalsifiable". A brief listing of some of the sources of this argument are as follows: [Himmelfarb, 1962] [Smart, 1962] [Manser, 1965] [Barker, 1969] [Macbeth, 1971] [Grene, 1974] [Peters, 1976] [Løvtrup, 1976] [Quadagno, 1979]. There has been a strong tendency amongst such critics to want to "have it both ways", to claim that the neo-Darwinist theory of natural selection is a classical tautology *and* that neo-Darwinism is a successful metaphysical research programme. But classical tautologies have zero empirical content, so it is difficult to see how this could be, if in fact this is the case.

The tautology question need not be of much bother outside of the framework of classical logic, where in certain relevant systems, tautologies do not have zero content [Routley, 1980 (1), pp. 936–937]. Yet I believe the tautology claim is not in itself the real problem facing neo-Darwinism; rather it is the claim that the selectionist argument is itself *circular*. Consequently to point out that there are falsifiable consequences of neo-Darwinism [Flew, 1966] [Williams, 1973] [Ruse, 1980, p. 72] is an irrelevant argument. The circularity argument takes, as Dobzhansky et al. [1977, p. 505] note, the following form:

(1) All alternative genetic variants which increase the Darwinian fitness of the organisms which possess this variant, will increase in frequency overtime at the expense of those genetic variants with lower fitnesses, in a specific environment.

(2) In environment E, the genetic variant G_1 confers higher fitness than G_2 in E.

Therefore,

(3) G_1 will increase in frequency relative to G_2 in E.

It is then argued that if Darwinian fitness is defined in terms of actual survival and reproductive success, as many text books do [Ayala and Valentine, 1979, p. 123] and Sociobiologists do [Wilson, 1975, p. 585], the explanations involving fitness differences are circular and therefore, whilst not strictly speaking false, are devoid of explanatory power. This is so because if type G_1 is "fitter" than type G_2, then G_1 is leaving a higher average number of offspring than type G_2, but the difference in fitness of G_1 and G_2 does not explain the difference in the average offspring contributions of G_1 and G_2. No description of a generative mechanism is given here [Mills and Beatty, 1979, p. 265]; at best natural selection could only be, as E. O. Wilson [1975, p. 67] admits, a restatement of history. If this is the case the view of natural selection as a creative mechanism, cannot be sustained.

Further, the explication of "fitness" in terms of actual survival and reproductive success is highly counterintuitive. Survival and reproductive success may be a result merely of cosmic coincidence, not of any superior characteristics of an organism. Thus is the case of two MZ twins MZ_1 and MZ_2, where MZ_1 is accidently killed before reaching a reproductive age, and MZ_2 successfully reproduces, then the *individual fitness* of MZ_1 is less than that of MZ_2. For the case of the fitness of *genotypes* rather than that of individuals, consider the single autosomal dominant gene resulting in Huntington's chorea, an invariably fatal destroyer of the CNS. However, people frequently reproduce before knowing that they are afflicted. Barash [1981, p. 26] suggests that such people tend to have much higher reproduction rates than unafflicted people. Thus such a gene may if uncontrolled increase in relative frequency in the gene pool, even though the health of the population in general will decrease.

There have been some very interesting attempts to avoid this problem [Maynard Smith, 1969], but critics have in my opinion succeeded in demolishing even the most interesting accounts [Grene, 1969] [Waddington, 1969]. Perhaps under pains of vicious circularity or infinite regression, this concept should be taken as primitive. This is the position taken by Mary Williams [1970] in her frequently cited axiomatization of neo-Darwinism. Let us therefore consider her position.

Williams gives first an intuitive outline of the meaning of the term "fitness", and then attempts to formulate axioms which explicates the meaning of the term in question. She states that fitness is:

> ". . . a measure of the quality of the relationship between an organism and its environment, where the environment of an organism is the set of all external factors which have influenced it during its life. (Thus, every organism has one and only one environment, so fitness is not a function of the two variables *organism* and *environment in its environment*) [ibid., p. 358].

The fitness if a biological organism in an environment in which it lives, represented by "$\phi(b)$", is however taken by Williams in *direct contradiction* to these remarks to be a function of both organism and environment: "Axiomatization of the complete theory will require that fitness be defined as a function of both organism and environment" [ibid., p. 359]. This contradiction is not removed by saying that $\phi(b)$ is equivalent to $\phi(b, e_b)$, where e_b is just the actual environment of b, so the dyadic function $\phi(b, e_b)$ becomes the monadic function $\phi(b)$. To begin criticism, "b" is a constant designating an individual, not a variable, and "e_b" is a constant as well, designating the individuals environment, not a set of environments.

Here, $\phi(b)$ is taken to be a positive real number, not equal to zero, "since any organism which has lived at all has some fitness" [ibid., pp. 358–359]. Let "B" be the set of all biological entities, and "b" be a variable which has as its domain individual biological organisms, and let R be the set of real numbers. Then:

Axiom D3:

$$(\forall b)(b \in B \rightarrow \phi(b) \in R)$$

The concept of a k-ancestor requires definition. Let "b_1", "b_2", "b_3", . . . , "b_n" designate biological organisms: Let "$\gtrdot$" be the dyadic relation of "is a parent of" such that "$b_1 \gtrdot b_2$" reads "b_1 is a parent of b_2". Then:

Axiom B1:

$$(\forall b)(b \in B) \rightarrow \sim (b \not\triangleright b))$$

Let the symbol '$\triangleright$' designate the relation "is an ancestor of":

Definition B1

$$b_1 \triangleright b_2 = \text{df}\,(b_1 \not\triangleright b_2)(\exists S)[(S = \{b_3, b_4, b_5, \ldots, b_{k-1}, b_k\} \;\&\; (b_1 \not\triangleright b_3 \not\triangleright b_4 \not\triangleright \ldots \not\triangleright b_{k-1} \not\triangleright b_k \not\triangleright b_2)]$$

Also for $b_1 \triangleright b_2$, then $\sim (b_2 \triangleright b_1)$.

To define the concept of "k-ancestor", we need further definitions. First is the definition of the concept of the *descent from b′ to b″* ($R(b', b'')$):

Definition B2

$$R(b', b'') = \text{df}\,(\exists S^*)(S = \langle b_1, b_2, b_3, \ldots, b_k, b_{k+1}\rangle) \;\&\; (b' = b_1) \;\&\; (b'' = b_{k+1}) \;\&\; (k \in \{\text{Natural Numbers}\}) \;\&\; (k \geqslant 1) \;\&\; (\forall C)((1 \leqslant i \leqslant k) \;\&\; (b_i \not\triangleright b_{i+1})))$$

Second is the length k of the descent from b' to b'':

Definition B3

$R(b', b'')$ is of length k = df. There are $k + 1$ biological entities in $R(b', b'')$.

On this basis we can define the concept of a k-descendant ("k-des (b'', b'):

Definition B4

$$\text{k-des}(b'', b') = \text{df}\,(\exists S^{**})(S^{**} = \{R(b', b'') \text{ of length } k \text{ from } b' \text{ to } b''\})$$

And finally, k-ancestor ("k-ances(b', b'')"):

Definition B5

$$\text{k-ances}(b', b'') = \text{df k-des}\,(b'', b').$$

Let "$v_1(b, k)$" represent the sum over all k-ancestors of b of their respective reproducing offspring, and let "$v_2(b, k)$" be the number of k-ancestors of b. The estimate which Williams gives of the "average fitness" of the k-ancestors of b is as follows:

$$(3) \quad \text{Ave. Fitness (k-anc(b))} \approx \frac{v_3(b, k)}{v_2(b, k)}$$

The operational definition of $\phi(b)$ is as follows:

$$(4) \quad \phi_0(b) = \sum_{k=1}^{n} (0.5)^k \, v_3(b, k),$$

where "n" is the number of generations for which data are available.

Williams gives two other axioms D4, D5 which she takes to expand the concept of the "fitter". Both of these axioms make reference to the concept of Darwinian subclans, which is also a primitive term. Subclans are not individual organisms so it follows that Axiom D3 is effectively Williams' only account of fitness for an individual organism.

There are a number of severe conceptual difficulties facing Williams' position. The first is that we do not have any clear idea what the fitness of an organism $\phi(b)$ is. Since by Definition D3 the fitness of Darwinian subclans is explicated by means of $\phi(b)$ and latter axioms D4 and D5 are irrelevant, Williams fails to tell us more than the fitness of b may be represented by a real number.

Here operational definition (4) is of no help to us. It is unclear why (4) is chosen as an operational definition rather than some other of a non-denumerable infinity of mutually inconsistent fitness functions. As Williams admits "there are drastically different functions which satisfy the axioms and hence could be used as operational definition" [ibid., p. 359]. The point remains that even if Axioms D4 and D5 were relevant, there would still be a number of mutually inconsistent fitness functions. Just two of these is enough to warrant an epistemological crisis, for given Axiom D3, it is not necessarily the case that $\phi^*(b_1) \geqslant \phi^{**}(b_1) \vee \phi^*(b_1) < \phi^{**}(b_1)$ if one assigns simultaneously mutually inconsistent fitness values. Consequently Williams can only recommend to the working biologist to "choose whichever of these functions is most suitable for his particular type of work" [ibid., p. 359]. The formalism plus the operational definition, despite the "nice properties" of the latter, do not aid us here. Williams clearly needs further axioms about the fitness of an organism if we are to have a meaningful analysis. Axioms D4, D5 and the remaining definitions are of no help to her here.

There also seems to be some counter-intuitive properties which (4) possesses. Most population genetics accounts of fitness are concerned with expected numbers of progeny, not particularly with numbers of ancestors [Crow and Kimura, 1970, p. 178] [Mills and Beatty, 1979] [Kimbrough, 1980]. The definiens of (4) involves essentially the ancestor concept. This in itself is in considerable tension with the intuitive exposition of fitness: there fitness is a property of an organism–environment totality. Williams accepts that fitness is "somewhat hereditary" meaning "[that] there is a correlation between how much fitness an organism has in its environment and how much its offspring have in their environments" [Williams, 1970, p. 358]. Here she is appealing to concepts of descendency. It seems any definition which is to be consistent with her intuitive remarks must make use of the concept of descent, not ancestry. A very young pre-reproductive age organism, or one which by accident does not mate would have a fitness of zero if we rework (4) by using the concept of k-descent, thus conflicting with Axiom D3 and a previous statement that $\phi(b)$ is a positive *non-zero* real number. Thus Williams' intuitions conflict with the very formalism devised to explicate them. Consequently I cannot agree with Rosenberg [1982] who maintains that the circularity problem can be broken by taking "fitness" as a primitive and accepting Williams' formalism.

Mills and Beatty [1979] reject the accounts of fitness discussed here so far. They also reject attempts to explicate the concept of fitness with respect to some concept defined independently of survival and reproduction, on the grounds that it is difficult to say what the "fittest" bacteria and the "fittest" human being have in common given that such properties would be highly context-dependent and environment-relative. They themselves offer a "propensity interpretation of fitness".

The fitness of an organism is its *propensity* to survive and reproduce in a specific population and environment [ibid., p. 270]. They introduce two definitions of "fitness": "fitness_1" of individual organisms, and "fitness_2" of types. Propensities, we are told, are properties of individual objects and are ontologically real [ibid., p. 272]. They reject (rightly I think) the claim that there is a unique number of offspring which an organism has a propensity to produce, so fitness_1 cannot be in general the mode of the distribution of propensities, since there may well be no mode at all. It is supposed that the fitness_1 of an organism o_1 in an environment E_1 is the expected number (e_n) of descendants which o_1 will leave in E_1. The expected number e_n, is the weighted sum of the values of the possible outcomes, where the possible outcomes $o_1, o_2, \ldots, o_n$ are the different number of offspring contributed; the values $1, 2, \ldots, n$ are the outcomes corresponding to the number of offspring, and the weighting probability $\Pr(o_i)$ is the propensity of o_i to contribute i offspring.

The fitness_2 of a *type* of organisms, is the average fitness of the tokens (i.e. the organisms) of the type. The highest average fitness_1 is said to be the maximum fitness_2 in E. Then the relative fitness_2 of type X in E is the fitness_2 of X in E divided by the maximum fitness_2 in E. On this basis the concept of natural selection may be defined as follows:

> "Natural selection is occurring in population P in environment E with regard to organisms x, y, z (members of P) = df x, y, z differ in their descendent contribution dispositions in E and these differences are manifested in E in P. Natural selection is occurring in population P in environment E with regard to types X, Y, Z (included in P) = df members of X, Y, Z types differ in their average descendant contribution dispositions in E, and these differences are manifested in E in P" [ibid., p. 283].

Whilst this analysis avoids the first circularity problem it has its own difficulties. For one thing, the circularity problem is not resolved but only replaced by another problem: the propensity is the cause of the differences in actual rates, but we only obtain our knowledge about the properties of this propensity through knowledge of "causes and effects reflected in actual retrospective and prospective rates of reproduction" [Rosenberg, 1982, p. 272] – which merely reopens the circularity once more. I hardly agree with Rosenberg's proposal to take the term as primitive, regardless of the question of the cogency of Williams' formalism. We would need a clear intuitive grasp of this concept, which is totally lacking given that there are disagreements about what fitnesses are properties of, and how in fact they are to be ascertained.

The position of Mills and Beatty is also flawed insofar as no theory of the determination of propensity values is given, so we cannot ascertain even in prinicple the fitness_1 of an organism. Indeed this is a general problem facing also propensity theories of probability [Kyburg, 1974]. They also maintain that so long as measured *actual* reproductive differences are statistically significant, *actual* reproductive differences may be regarded as measures of reproductive capacity. But this proposal is open to the Huntington's chorea counter-example where reproductive capacity hardly indicates "health/fitness/adaptation".

It may be possible to rework Mills and Beatty's position if we can satisfactorily understand the idea of "adaptation". This seems the neo-Darwinist's remaining hope.

3.6.3 Problems with adaptation

Before concluding that we are unlikely to obtain any satisfactory explicit definition of "adaptation", let us consider the recent account by Bunge [1979, p. 104], followed by that of Brandon [1980]. The concept of adaptation is so difficult to analyse Bunge maintains, because in actual fact, there are three concepts which are not distinguished here which should be: (1) the *suitability* of a bio-system to a function of high biovalue for the organism; (2) the adjustment of an organism to its environment; (3) the fertility of a bio-population. I remain sceptical as to whether concept (2) is any improvement over the explicandum; indeed even concept (1) is problematic. "Biovalue" is explicated by means of the health function:

$$\text{(H)}\ V(a, b, e, t) = H_A(b, e, t) - H_{\bar{A}}(b, e, t)$$

such that $V(a, b, e, t)$ is the value of a subsystem a for organism b in environment e at time t; $H_A(b, e, t)$ is the degree of health of b in e at t with a, and "$H_{\bar{A}}(b, e, t)$" is the degree of health of b in e at t without a. But an organism b is healthy if it possesses all the properties described by Bunge in postulate 3.1 (he means postulate 3.2 or nonsense results) [Bunge, 1979, p. 86]. This however tells us that healthy organisms are just those organisms which are living things (by definition 3.1 [ibid., p. 80]) and this is hardly informative, and most likely false, as the case of Huntington's chorea testifies.

Brandon [1980] has offered the following account of the *principle of natural selection*:

> "(Probably) If *a* is better adopted than *b* to their mutual environment *E*, then *a* will have greater reproductive success than *b* (in *E*). [ibid., p. 428].

Brandon follows to a remarkable degree the structure of the argument which we have seen Mills and Beatty [1979] develop. He rejects attempts to define adaptedness independently of reproductive success, for the reason that there is no such independent property that is invariably selected for. Adaptedness he treats as a dispositional property and relative adaptedness as a dispositional relation:

> "More formally adaptedness can be fined as an expected value, i.e., as a weighted sum of probabilities. For each reproducing entity *O* and environment *E* we have a range of possible numbers *Q* of offspring, Q_1^{OE}, $Q_2^{OE}, \ldots, Q_n^{OE}$ and for each Q_i^{OE} our ecological theories associate a number represented by $P(Q_i^{OE})$, which is the probability (or chance) of *O* leaving Q_i offspring in E. With this we can define the adaptedness of *O* in *E* (symbolized as $A[O,E]$ as follows:
>
> $A[O,E] = \Sigma P(Q_i^{OE}) Q_i^{OE}$" [ibid., p. 431].

Thus on this account the adaptedness of *O* in *E* equals *O*'s expected fitness value. We may say *a* is better adapted than *b* in *E* if and only if $A[a,E] > A[b,E]$.

Brandon's position parallels, as I stated, the Mills and Beatty [1979] account of fitness. The previous criticisms which both Rosenberg [1982] and myself made may be used against the above position.

From such defeats we may no doubt wish to take the concept of adaptation as primitive. The image of the properties of xerophytic plants in a harsh desert environment has rightly strongly forged intuitive images of adaptation. We know that trophic vegetation will not grow in such an environment and this insight is something *close* to being a *synthetic apropri* truth. Given the nature of the water loss it seems necessary that we would not, excluding miracles caused by God, find such plants in the desert. But what would a maladaptive organism in its environment look like? The argument of this chapter has established that it is certainly *not* one merely not maximizing its inclusive or Darwinian fitness. For Manser [1965] a maladapted organism is one which is either dead or dying: what lives must be adapted to life. Second, we only recognize and individuate adaptations in so far as they promote survival: the darker *Biston betularia* are only taken to be adapted to a dirty industrial environment insofar as selection has actually been observed (indication colour is a significant property), and if it was not, then these characteristics would not be classified as an adaptation in the first place. We should note that this argument is concerned precisely with an adaptation of an organism to an environment and not with some concept of absolute adapation, so one frequently made rejoinder immediately fails [Dobzhansky, 1968, p. 6, 1977, p. 42]. Hence it is concluded that no explanation is given by the concept of adaptation, only a redescription of the fact that the organism exists in its environment. Circular and vacuous arguments are frequently psychologically seductive, and adaptationist arguments are quite voluptuous.

One might argue against this by simply saying: so what? The problem with this is that adaptationist arguments because infected with the antinomies which Kant believed infected the Rationalistic metaphysics of his own day. One can readily give theoretically plausible adaptationist arguments for too many phenomena observed to persist in the biological world. True, there are examples of phenomena such as the appendix in humans which are vestigal organs (although well-known debates exist about the possible immunologic functions of the appendix), but it is clear that one can explain mutually inconsistent phenomena. This underlies Lewontin's [1977] criticism about the unfalsifiability of many adaptationist arguments. To give but one example, we might argue that the courting of males in *Drosophila* [Lewontin, 1979, pp. 12–13] could be explained by the fact that the females are resistant to multiple inseminations (which is true), so that a maximization of fitness would occur if males court indiscriminately, increasing the probability of meeting unmated females. But suppose this behaviour was not observed: given the same empirically true statements, one can argued that such a behaviour would waste energy since with males courting the probability of striking a receptive female will decrease. Note that it is not adequate to argue that one merely decides from experiment which explanation to use: this is to rob the position utterly of any predictive power at all, and to fail to provide a satisfactory explanation of why the phenomenon in question ultimately occurs which would seem to be the very point of such so-called "ultimate explanations".

The problem with my counter-arguments might be taken to lie in viewing the concept of adaptation to be, as Stern [1970, pp. 41–42] calls it, an "all-or-none" concept. It is a problem which may well be resolved by the introduction of a new concept, that of "relative adaptation". Here, "adaptation" as Stern views it refers to the power of the organism to produce on average more offspring than would be produced in the absence of those properties giving such powers [ibid., p. 44]. But how in fact do we obtain the estimates of the average offspring production which the organism O_i would have, contrary to fact, if it did not have its adaptations? One faces here an epistemological paradox. What exists in the world is organism O_i producing its lifetime of offspring and then dying. To determine that some alleged properties are adaptations we need a control, but the control can be only O_i minus its adaptations by Stern's definition for an individual organism. Yet even if one could surgically remove O_i's adaptations one could then no longer empirically ascertain the offspring production of organism O_i with its adaptations. It is no good arguing that one merely compares the offspring production of O_i with that of another organism of the same species O_i^*, which has its adaptations surgically removed, since we are interested in the individual fitness of O_i not that of O_i^*, and even to use O_i^* as an estimate of O_i's fitness (or average offspring production), we need some idea of O_i^*'s unmutilated fitness/average offspring production. But now we have landed ourselves in a vicious infinite regression.

There are further epistemological problems facing adaptationism. Lewontin [1978, 1981] has denied that Evolution can be a process of adaptation of organisms to their environment. Organisms create and define the niches which they occupy; this is most obvious in the case of human beings (where no adequate account of the human environment can ignore the realm of culture and abstract entities cf. ch. 8), but is also true even for non-human animals and plants. If Evolution is described as a process of adaptation of organisms to their niches, then niches must pre-exist organisms; but empty niches are easy to generate, so easy that an infinity of empty niches can be described simply by altering the specifications of some present actual niche. This robs the concept of all explanatory and predictive power [Lewontin, 1978, p. 159]. Consequently Evolution is not, and cannot be a process by which organisms adapt to their environments because organisms are either already adapted, extinct or going extinct. As Lewontin [1981, p. 245] puts it: "It may be more illuminating to see organisms as *changing* and, in the process, as reconstructing the elements of the outer world into a new environment that is sufficient for their survival". This interpretation fits well with the views expressed by myself, and seems to be a confirming confession of the explanatory bankruptcy of the concept of adaptation.

In his 1978 paper, Lewontin attempts to escape the paradox of adaptation in arguing that natural selection operates to maintain the State of adaptation of species in constantly changing environments. But as Steele [1979, p. 3] points out, this seems at best, little more than a redescription of the evolutionary process, not a solution to the adaptation paradox. Worse, it conflicts with Lewontin's other statements in the same paper which stress that all organisms "create and define the environment in which they live" [Lewontin, 1978, p. 159]. He admits that the adaptationist programme is quite problematic [ibid., p. 168]. Yet he does not recommend its abandonment since, he reminds us, adaptation is a real phenomena, demonstrated for example by the similarity of responses of organisms to environmental challenges, such as the differing mechanisms of water locomotion in the whales, seals, penguins, fish and sea snakes [ibid., p. 168]. The argument is fallacious. What we have here is simply different – *not similar* mechanisms consistent with morphological constraints imposed by the nature of the organism in its environment. The form of locomotion of many other sea creatures is physically impossible for large masses such as whales. There is no similarity of response here to the problem of moving through the water because the mechanisms in question are utterly different.

4.0 Conclusion: The state of the argument

The concept of adaptation is a useless theoretical concept and we have failed to explicate either this concept or that of the concept of fitness after a comprehensive survey of the best accounts in the literature. The problems and paradoxes associated with these concepts destroy any intuitive plausibility which they might have, rebutting attempts to treat these concepts as primitive as a satisfactory move. Thus the fault must lie in the structure of neo-Darwinism; if it is vitiated, so must be Sociobiology. Therefore, according to the idea of a causal-explanatory reductionism developed here, no causal-expanatory reduction of Culturology to Sociobiology as defined, is even in principle possible. Thus our principal thesis is established.

In the next chapter we shall attempt to further our scepticism by examining some theoretical and metaphysical issues arising from an analysis of the "selfish gene" concept, and more generally, gene action in cell systems. It will be my thesis that neo-Darwinism lacks utterly a satisfactory theory or even metaphysical image of epigenesis. *Even if* the idea of an adaptation was non-problematic, since adaptations are not created by natural selection (in my argued view), an adequate theory of evolution must presuppose an adequate theory of development. Some alternative to neo-Darwinism must be sought, and we shall consider some alternatives in the next chapter. The present author does not believe that any sufficiently detailed alternative yet exists. Yet there are many promising theoretical developments which might never be explored if the present ecclesiastical worship of neo-Darwinism continues.

CHAPTER 5

Causal-explanatory reductionism II: The metaphysics of the selfish gene

> But each thinking geneticist, in interpreting his factual data and in trying to fit his results into the total theoretical structure of his science, does it under the conscious or subconscious influence of his basic philosophy or *Weltanschauung* in regard to genetical thought. [Goldschmidt, 1954, p. 703].

1.0 Introductory statement of the argument

Before an alternative(s) to neo-Darwinism can be considered, the critique developed in chapter four must be extended one stage further. It was stated there that the faults with Sociobiology illustrated basic defects in the neo-Darwinist position. My previous argument was concerned quite extensively with the semantical, logical and methodological defects of neo-Darwinism. The concerns here are primarily metaphysical and empirical. In particular I hope to refute the neo-Darwinist view of the organism, quite correctly described by Brian Goodwin [1979] in the following words:

> There is an intrinsic dualism in this tradition [i.e. neo-Darwinism] which arises from its two-fold conceptualization of organisms as adapted to the external environment on the one hand, and as vehicles of internal hereditary factors on the other. Since the internal factors, a collection of genes which for any individual is called its genotype, do not themselves interact with the external environment, a third mediating structure enters the theory: the phenotype, which is what common-sense recognizes as the organism with its morphology and behaviour. It is then assumed that the phenotype is reducible to or determined by the genotype, so that the latter remains primary in the definition of the organism. Living things are then the product of two forces, those of heredity and those of natural

> selection, giving continuity and adaptation to biological process. The theory thus has an explicitly historical and utilitarian basis. All problems in contemporary biology tend to be reduced to these terms.
>
> Since the collection of genes which defines an organism has no constraints imposed upon it as a whole, the only requirement being that the characters it specifies should be adapted to the environment in which the organism lives, the phenotype has no defining characteristic as a total organized entity: it is an atomic collection of mechanically-interacting parts, each of which can be independently varied by the process of natural selection [ibid., pp. 109–110].

It will be argued that such a view of the organism, including its more sophisticated variants is wrong and should reasonably be abandoned and that biology, especially population genetics, needs a substantial metaphysical and theoretical facelift. Indeed the idea of the gene as a "central directing agency", excellently described by Goodwin in the previous citation is wrong; the idea of the gene-chromosome complex as the units of inheritance must also be recognized as a gross oversimplification as must positions such as "the central dogma of molecular biology" and the genotype–phenotype distinction. The admittance of non-Mendelian mechanisms makes the genes only one quite important part of a complex mutually interacting bio-system.

A science of genetics should be concerned with two closely interrelated problems: organisms seem to possess some properties which their parents possess and other properties which their parents and other organisms in the popultion do not possess; that is to say, within populations of organisms *hereditary* and *variability* of these characteristics cited above exist. What are the generative mechanisms responsible for this? Further, organisms have a form or pattern, and the relation of this to the behaviour and mode of life of such organisms, is as Robert Rosen [1972] observes "a dominant component of physiology on the one hand and an understanding of evolutionary processes on the other" [ibid., pp. 1–2]. An adequate genetic theory must consequently contain a theory of *ontogenesis* and *morphogenesis*. It must explain the developmental processes of the organism from conception to adulthood, and especially if form or phenotype is caused by the genome, and developmental processes directed by it, then some theory of how this is done must be given.

There are two main metaphysical positions which can be taken as answers to the question of the relationship between the genome of the organism and its subsequent characteristics which it comes to possess through development: *preformationism* and *epigenesis*. Stated briefly the preformationist position maintains that ontogenesis is primarily a process of the unfolding of preformed properties of an organism, without increasing the complexity of the organism and without the emergence of new properties. The epigenetic position maintains by contrast that the organism starts from a lowly organized state and that both increasing complexity and new properties emerge during ontogenesis.

Classical preformationism claimed that microscopic studies had even revealed that the head of a sperm contained an entire person, which merely enlarged during ontogenesis. This position was criticized by Buffon, writing in the eighteenth century in his *Histoire naturel des Animaux* [Jacob, 1973]. Classical preformationism conflicted with atomic theory insofar as the infinite regress of people inside sperms would imply that some such preformed people must be smaller than atoms, leading to a logical contradiction if such people are also taken to be composed of atoms. Buffon was also impressed by the regenerative powers of certain organisms such as the acquatic worm and the hydra, and especially with the capacity of the crayfish to regrow amputated limbs. Such occurrences were taken to present problems for classical preformationism, since it must be supposed that the organism must constantly borrow missing parts for such regenerations from potential descendants.

During the nineteenth century preformationists such as Wilhelm Roux and August Weismann accepted that a *prima facie* increase in complexity occurred during ontogenesis [Jacob, 1973]. However, this complexity arose through an unfolding of a pattern which was already possessed by the germ cells. Germ cells differed from the soma cells in that they possessed a special "substance", the "germ plasm", which had the capacity to produce an individual of the same species. Such germline cells are protected from environmental influences, and can acquire no properties which they are not already predisposed to possess by the germline cells. In direct contradiction to Lamarck, the environment can have no direct creative impact upon ontogenesis. The thesis that there is no ontogenetic interaction between the soma and the germline cells is known as Weismann's dogma, and must be distinguished from the Watson–Crick "central dogma of molecular biology" which shall be discussed later. Both positions are aptly named.

Jacques Monod [1972] gives a concise statement of the modern preformationist position in these words:

> No supplementary input of information other than the genetic is necessary; nor it seems, even possible, as the mechanism as we know it leaves no room for any. And to the extent that all the structures and performances of organisms result from the structures and activities of the proteins composing them, one must regard the total organism as the ultimate epigenetic expression of the genetic message itself. [ibid., p. 107].

The term "epigenetic" is not used in the above defined sense here, and the term "preformationist" should be substituted. This is not to say that all neo-Darwinists are preformationists; some such as Dobzhansky would seem to be most difficult to classify precisely in this respect. Yet, on the other hand there is little reason for such moderates not to accept Monod's position: the defects of their alternatives shall be reviewed shortly.

It is on the question of form, in the field of *morphogenesis*, (and especially ontogenesis) that neo-Darwinism and the received neo-Mendelian view of genetics faces its greatest difficulties. As we shall also see shortly, the genetic material of the cell in a cellular system primarily functions to produce proteins. How, for example, does one explain the morphogenesis of physiological phenomena of an organism, such as its limbs, digestive and nervous systems, and if we consider the claims of Sociobiology, how does one *explain* any social behaviours' morphogenesis? To this, it might be said that the behavioural and physiological blueprints/programmes are possessed/contained within the genome. I hope to show in the course of section 2 that this response cannot succeed.

2.0 The nature of the gene: Mendelian genetics, quantitative genetics and molecular biology

I wish to argue in the course of this section that there is at present no satisfactory account of the nature of the gene; more precisely that the generative mechanisms of evolutionary genetics (including Sociobiology) on the one hand, and molecular and developmental biology on the other, are mutually contradictory. Consequently attempts to give classical positivist theoretical reductions of "genetics" (presumably quantitative genetics) to molecular biology [Goosen, 1978] [Ruse, 1976] [Schaffner, 1976], including so-called token-token reductions [Kimbrough, 1979], fail.[1] Further the central presupposed dichotomies of received genetics are far from clear, reasonable and empirically supported. We cull much biological material from the following highly respected texts: [Ayala and Valentine, 1979] [Dobzhansky et al., 1977] [Ehrman and Parsons, 1976] [Goodenough, 1978] [Mays, 1981] [Plomin et al., 1980] [Watson, 1977] and the two "best books" on the philosophy of biology [Hull, 1974] and [Ruse, 1973].

2.1 The genotype–phenotype distinction

Neither of the terms "genotype" or "phenotype" were used by Mendel in his classic paper [Mendel, 1866 in Davern, 1981]. Indeed, they were first used by Johannsen [1909] [Dobzhansky et al., 1977, p. 29]. Johannsen's experiments were with French beans: he produced allegedly pureline beans and claimed on this basis that further selection for characteristics would not occur as the offspring beans bore no greater resemblance to their parents than to other pureline individuals so that there was no variation in the population. However, Johannsen's data did not support his claims since the inferred parent offspring correlation was not zero but 0.348 [Pearson and Weldon, 1903]. Ho and Saunders [1980, p. 12] have concluded from this that the belief that there is an absolute distinction between hereditary and non-hereditary characteristics is one lacking in empirical support. This is probably an overstatement and would be met by received genetics by denying that Johannsen had produced purelines at all. We need to search elsewhere for reasons for rejecting the genotype–phenotype distinction.

Despite the fact that the genotype–phenotype distinction receives a prominent place in contemporary genetics books, the distinction itself is utterly unclear. In Dobzhansky [et al., 1977], the phenotype is taken to be that which is empirically observable ("what we can see" [ibid., p. 29]) and the genotype, the sum total of the hereditary materials of the organism. Ehrman and Parsons [1976, p. 6] take the genotype to be "the sum of the genes" possessed by the organism, and the phenotype to be the "appearance" of the organism which is causally dependent upon the genotype. Ayala and Valentine [1979, p. 50] identify the genotype as the "genetic information" which the organism has inherited, whilst accepting the Dobzhansky et al. definition of the phenotype. Mays [1981, p. 6] by contrast, defines the genotype as "the genetic information itself", whilst the phenotype is the "organism's expression of the gene". Goodenough [1978] identifies the "genome" as the genetic information possessed by the organism [ibid., p. 27], the genotype as "the genetic makeup" of the organism [ibid., p. 87] and the phenotype as the sum of the *traits* of an organism, a trait being "the particular expression of a gene" [ibid., p. 87]. Ruse identifies the genotype as "the whole collection of an organism's genes" [Ruse, 1973, p. 15] and the phenotype as the "organism's *physical* characteristics" [ibid., p. 16, emphasis added]. Hull follows Goodenough in taking the genotype to be the genetic makeup of the organism, and is content to define the phenotype as "the rest of the organism" [Hull, 1974, p. 15].

There is no detailed defence of any of these definitions in the cited texts; the authors I assume must therefore suppose that the rationality of these definitions is self-evident. But since these definitions conflict we can hardly agree. Ruse [1973] gives an interesting discussion of the relationship between the genotype–phenotype distinction and the observation-theoretical distinction (which incidently raises grave problems for the Dobzhansky et al. definition of the phenotype) but presents no defence of the former distinction. Many of the definitions employ quite problematic information-theoretical and cybernetic concepts, the arguments to support this claim being given later. Further many of these definitions are problematic in themselves: on Ruse's account for example non-physical properties of organisms are not part of the phenotype, and as a hard line neo-Darwinist he therefore inherits the problem of explaining the ontogenesis of these properties. Of course Ruse could defend physicalism on an independent basis, but to do so is only to further plunge genetics into the philosophically boiling waters of controversy. On Hull's definition such non-physical properties are part of the phenotype. On Goodenough's definition of the phenotype it is arguable that the "particular expression of a gene" is nothing more than those produced polypeptides, if we consider structural genes. It is arguable that the particular expression of regulator genes might well be the switching on or off of structural genes – in which case the genotype–phenotype distinction begins to blur. Finally on Mays' account, what is the "organism's expression of a gene"? Do organisms "express" genes? But Goodenough has also told us that genes are themselves "expressed". What is this expression? The cited texts do not enlighten us on such matters.[2]

Our best policy might now be to attempt to understand what these "genes" are. The early parts of our texts generally take the genes to be the hypothetical units of inheritance, which are taken to obey the laws of *segregation* and *independent assortment*. A common statement of these laws is as follows: the law of segregation states that for sexually reproducing organisms each parent contributes only one particulate unit at every locus and the probability of any gene of a parent being transmitted is the same as the probability of the corresponding gene at the same parental locus. The law of independent assortment states that if $G_1, G_2, G_3, \ldots, G_n$ are genes in a sexually reproducing organism and $G_1 \neq G_2 \neq G_3 \neq \ldots \neq G_n$, then the probability of receiving a particular gene is independent of the probability of receiving any other gene at a different locus, i.e. the conditional probability $P_r(G_k/G_{k-1}) = P_r(G_k)$ and $P_r(G_{k-1}/G_k) = P_r(G_{k-1})$. There is independent segregation of genes at respective loci.

Now if scientific laws must be universal truths, then Mendel's laws must be rejected as being scientific laws as there are clear counter-examples to both laws [Kimbrough, 1979] and a rigorous deductive derivation of the Hardy–Weinberg law is threatened.[3] Further, one cannot then implicitly define the gene as that entity truly described by the postulate set of Mendel's laws. The phenomena of sexual reproduction in humans gives a counter-example to the first law. Even if we accept that some structural genes may be "redundant", it is generally accepted that the Y chromosome contains "less" genes (or more correctly less possible genes) than the X chromosome, as it is smaller. If one restates the law to be applicable rather to chromosomes, chromosome aberration studies, especially of non-disjunction (the failure of two homologous chromosomes of a pair to go to opposite poles at the first meiotic division so that one daughter cell has *both* chromosomes, and the other neither [Bridges, 1913, 1914]) refute it. The second law faces the counter-example of *linkage* which has been defined now precisely by virtue of being such a counter-example: linkage is the relationship between two or more non-allelomorphic genes such that they tend to be passed on from generation to generation without independent assortment, being split up only by crossing over processes during chromosome pairing at meiosis. As Kimbrough [1979, p. 396] points out, this strategy has the effect of turning Mendel's second law into a tautology, especially when qualifications are added for other counter-examples such as meiotic drive (any process which results in alleles being represented in unequal proportions in the gametes) and segregation distortion [Crow, 1979]. In neither case do we individuate these "genes", the atoms of inheritance.

2.2 Alternative explications of the concept of the gene

It has just been argued that an attempt to characterize the genetic atoms of inheritance with respect to Mendel's laws fails, as these laws are either strictly false or straightout tautologies. We must therefore search for some other true set of statements which might be taken to explicate the concept of the gene. We can reject from the outset the postulate that one gene "causes" one character, as most geneticists did in the early part of this century. The phenomena of *multiple genes*, *pleiotropy* and *epistasis* clearly refute this position, although it seems to have been received in some of the unclear metaphorical writings of some Sociobiologists. The relationship in general between the genetic atoms of inheritance is therefore not 1:1, but many:many.

Perhaps then the answer to our question might be given by molecular genetics. After all, the reductionists tell us that genetics now is essentially a chapter of molecular biology so an excursion into such a field should not disappoint and frustrate us, if their claims are to be accepted as true. We must begin our trek.

The principal chemical molecules of molecular genetics are the *proteins* and the *nucleic acids*, both of which are polymers. Polymers are chemical aggregates or unions of covalently bonded monomers. Monomers are chemical compounds consisting of a finite series of molecules of a single molecular type. Nucleic acids are polymers of nucleotides. Nucleotides are chemical compounds formed from sugar with 5 carbon atoms, phospheric acid and a nitrogen containing base. There are two kinds of nucleotides: *ribonucleotides* and *deoxyribonucleotides*. Ribonucleotides are chemical compounds consisting of a *purine* or *pyrimidine* base kind, bonded to ribose which is itself esterified with a phosphate group. Deoxyribonucleotides consist of the same purine or pyrimidine base kind only bonded to the sugar 2-deoxyribose, which is itself bonded to a phosphate group. The purine base kind consists of the bases adenine (A) and guanine (G); the pyrimidine base kind consists of the bases thymine (T), cytosine (C) and uracil (U).

DNA (deoxyribonucleic acid) is a polymer consisting of two complementary chains of nucleotides. Since base pairing can only occur between A and T, and C and G, if one slice of DNA consists of the nucleotide sequence TGGATCTAT, the complementary chain must be ACCTAGATA. RNA (ribonucleic acid) is a polymer of ribonucleotides. The molecule is usually single stranded and contains a ribose sugar instead of DNA's deoxyribose, and as well as the pyrimidine uracil (U) instead of DNA's thymine (T). Double helical RNA molecules have been described for certain viruses, where RNA serves as the "genetic material" [Goodenough, 1978, p. 13].

The second important class of biomolecules are the *proteins*. Proteins consist of polypeptide chains crossed linked together in a variety of ways. Polypeptide chains are molecular chains of three or more *amino acids* bonded by a peptide linkage. Amino acids are chemical compounds containing an amino group

(NH_2) and an acidic carboxyl group (COOH), with the amino group always attached to the carbon atom next to the carboxyl group, as well as a variable grouping of atoms R, usually a carbon chain or ring (general formula: $R-CH-NH_2-COOH$). There are twenty common amino acids which occur in nature as well as a few rare ones; hundreds and perhaps thousands of amino acids combine to make each protein. The term "peptide linkage", refers to a covalent bond between two amino acids, where the amino group of one amino acid is bonded to the carboxyl group of another with the elimination of water (H_2O).

DNA has two functions. First is self-duplication. The double helix structure and base pairing restrictions give the DNA molecule the power of self-duplication, which occurs during cell division. The DNA helices unwind, the base pairs separate, one on each strand of the helices and act as templates for the replication of a complementary strand. The second major function of DNA is translation from the "genetic code" to protein (in particular enzyme) production.

Following the Jacob–Monod operon model [Jacob and Monod, 1961] we may distinguish between *structural*, *operator* and *regulator* genes. Structural genes specify the sequence of amino acids which are *translated* ultimately into polypeptides. Operator genes are short segments of DNA which are next to the structural gene, and act as an activating or repressor mechanism for the transcription activities of structural genes: both of these as a unit are known as an *operon*. Operator genes are in turn activated or repressed by *regulator genes*, which produce a repressor which binds to the DNA site of the operator gene, which in turn suppresses the structural gene. The repressor however is activated if it binds with a *regulatory metabolite* produced either by other operons or by substances such as hormones [Korner, 1970], and the uninterrupted transcriptional activities of structural genes is permitted. Structural genes may in turn produce enzymes which metabolize the regulatory metabolite, so that the repressor will shut down the operon. In turn, the regulatory metabolite may build up again, starting the process over again. It is known that local anaesthetics such as procaine can also perturb the translation of proteins [Granett and Villarejo, 1982].

Descriptions of protein synthesis are heavily laden with information-theoretical terminology, the use of which is quite questionable, and will in the course of this chapter be questioned. Nevertheless since my texts use such terms, so shall I for the point of exposition. The so-called "genetic information" is taken to be "encoded" in the sequence of nitrogen bases of DNA, in particular the *codons*. These are sequences of three adjacent nucleotides in messenger RNA or the corresponding DNA that code for a particular amino acid. The genetic code itself is the correspondence relationship between nucleotide triplets in messenger RNA and the amino acids in the resultant proteins.

In *transcription*, "information" in the base sequences of DNA is transmitted to messenger RNA, and by translation, such "information" is converted into specific amino acid sequences which are ultimately synthesized into proteins.

Very schematically the process of protein synthesis proceeds as follows [Watson, 1977, ch. 11]. *Promotor sites* adjacent to the potentially transcribable sequences contain "information" as to which strand of DNA shall serve as the template. A process similar to the duplication of DNA occurs resulting in the production of a complementary messenger RNA sequence. The process is "primed" by a RNA polymerase enzyme, which links together ribonucleotides, which the DNA has "lined up". The translation process consists of the messenger RNA molecule moving into the cytoplasm and connecting with the ribosomes, the site of protein synthesis. Transfer RNA molecules attach to amino acids and pair up with messenger RNA. On transfer RNA are various anticodon sites, complementary to the codon. Transfer and messenger RNA's engage in codon–anticodon pairing, bringing the amino acids into position, and from which, to skip much biochemical detail, polypeptides are formed.

So much then for a familiar story. The central dogma of molecular biology is that "genetic information" "flows" from DNA to messenger RNA to proteins, i.e.: DNA → mRNA → protein and never: protein → mRNA → DNA [Watson and Crick, 1953]. This doctrine is logically distinct from Weismann's doctrine that genetic information cannot penetrate the barrier protecting the germ line cells from the soma [Steele, 1979, p. 36]. The discovery of reverse transcription, where RNA serves as a template for the synthesis of complementary DNA chains, through the enzyme *reverse transcriptase* has occurred and seems to be widely accepted [Baltimore, 1970] [Temin and Mizutuni, 1970]. More controversially, models of *reverse translation* have been proposed [Mekler, 1967] [Cook, 1977], which one would reasonably interpret to challenge the central dogma. However Cook believes that if the central dogma is interpreted to mean that when macromolecular information is built into a protein it does not escape, the central dogma stands. Maynard Smith [1982, p. 50] also maintains that it is information "not material or energy" which flows from DNA → mRNA → proteins, and from germline to soma, but in neither of the reverse directions. What is problematic with such defence is that we must make an unjustified reification of "information" ("not material or energy") to save the central dogma, since as the description above establishes, proteins and nucleic acids are in a web of interactions which is evident from the above description of the Jacob–Monod operon model. But more shortly.

Armed with enough molecular biology to carry our argument through, let us return now to the problem of the explication of the concept of the "Gene". Many biologists have accepted the distinction drawn by Benzer [1962] between the *cistron* (the unit of function), the *muton* (the unit of mutation) and the *recon* (the smallest unit of crossing over) as a good replacement for the concept of the Mendelian gene. Michael Ruse [1973, p. 206] has maintained that an identification of the cistron with the "molecular gene" and the muton and recon with various nucleotide pairs of DNA would enable a solution to the consistency problem of Mendelian genetics and molecular biology.[4] This problem is not solved by this strategy for a number of reasons.

To begin, if the muton is taken as *the* unit of mutation, then our explication is inadequate since the use of the descriptor, marked by the article "the" is a misuse: there is no unique unit of mutation as anything from a single nucleotide to a whole chromosome or set of chromosomes can mutate. Mutations may occur at any point on the DNA. One could just as well eliminate the concepts of both the muton and recon and speak of point segments of DNA. These hardly seem to be the central directing agencies which neo-Darwinism requires.

Deeper problems occur in isolating the reference of the term "cistron". One may take the cistron to be the smallest unit of "function" as identified by the *cis-trans test* [Goodenough, 1978, p. 376]. Here the individuation of functional units is made by the dual considerations of mutation and crossing over. Two mutations, M_1 and M_2 are in the *cis-position* if and only if the mutations are on the same homologous chromosomes (i.e. those chromosomes which pair during meiosis, the genes of which "influence" the same characteristics). They are in the *trans-position* if they are on opposite homologous chromosomes. Now via crossing over, either two mutations in the cis-position can come to occupy the trans-position or two mutations in the trans-position, the cis-position. Let C_1 and C_2 be two functional units on the chromosome. It is argued then that if M_1 and M_2 are in either the cis or trans positions, no functional disturbance is observed if M_1 and M_2 are not both in C_1. If M_1 and M_2 are both in C_1, a functional disturbance may exist, as the trans-position may function significantly differently from the cis-position. The reasoning here is based upon the *complementation principle*: if one homologous chromosome is a mutant in gene G_1 and the other in gene G_2, then a mutant G_1 product is synthesized by one gene but a normal G_2 product is synthesized by the other, and a normal G_1 product by one gene, but a mutant G_2 product by the other, so that normal functioning occurs as no products are missing. Therefore, it is concluded, the two mutations are on two different genes. Mutants in gene G_1 carried by both chromosomes, would not enable normal G_1 products to be produced, hence the mutations lie on different genes.

There are however difficulties with the above argument. It would not and is not intended to establish the existence of all theoretically possible functional units. On the one hand, even though no detectable gene products may be missing, the gene may in fact through epistasis have a substantial effect on the function of other genes so that we individuate "too much", taking the functional unit to be larger than it should be. On the other hand, the mutations may be so insignificant, as pan-neutralists propose, that there is hardly any measurable effect on functioning at all. In such a case, it might well be argued that the cistron is not defined, and the proposer could rightly point that that none of these considerations show that we cannot individuate cistrons in *some* experimental circumstances. Against this my master-argument is that the cis–trans test cannot do any more than establish that for two mutations M_1 and M_2, they are either both on a functional unit, or both or not. But this does not precisely tell us for any mutations, what the cistron is, that is, it does not individuate it. If for example M_1 and M_2 are both on the same chromosome, perhaps the cistron is the actual chromosome itself, or the length of the chromosome $L - \epsilon$, where $L - \epsilon > d(M_1, M_2)$, where d is the relation of "distance between". If we wish to establish that the mutations M_1 and M_2 lie on different genes through non-complementarity, we still cannot infer that the respective cistrons are of any particular length. Indeed, the cis–trans test is logically consistent with the conjecture that instead of one cistron being individuate, what we observe at a molecular level is the product through epistasis of a number of cistrons. Therefore Benzer's categories do not resolve our initial difficulties in stating the sense and reference of the term "gene".

Another strategy consists in equating "the gene" with those segments of DNA which respectively produce one molecular product. This however is strictly inadequate since neither regulator nor operator genes produce polypeptides. Indeed, and we should note immediately the relevance of the following statement to Benzer's cistrons, that about 50% of the total 3×10^9 nucleotide pairs in the human haploid genome of highly repetitive DNA sequences, which seem to do nothing more than simply be reproduced, serving as Bodmer [1981, p. 672] remarks, "no obvious function" for the organisms that harbour them. Hence there could only be a very modest number of genes in the total genome that code for molecular products, the rest of the DNA segments being composed of intervening sequences and flanking regions between "genes".

Matters are much worse for the gene-molecular product model. As Walter Bodmer dramatically puts it, this model was "blown to pieces in 1977" [ibid., p. 668]. The entity coding for a single polypeptide is a collection of smaller units, separated by non-coding sequences, and a one to one relationship between DNA sequences and products does not exist:

> . . . a protein product may be made up of many different coding regions corresponding, perhaps to its different domains, and that a given coding region may contribute to more than one gene product. [ibid., p. 668].

The so-called "split gene" phenomena [Crick, 1979] has been observed in both viruses and higher organisms. Along the DNA segment, there are stretches of base sequences which are not "encoded" in mRNA. These sequences are known as intervening sequences or "introns", whilst those sequences which are expressed in mRNA are known as "exons". In immunoglobulin production in the mouse, cellular DNA is rearranged to eliminate unnecessary DNA sequences [ibid., p. 265]. But by far the most common mechanism is splicing. Here RNA polymerase makes a transcript of both introns and exons, and then removes the unnecessary sequences before the RNA moves to the cytoplasm. Splicing may hence be defined as "the mechanism by which a single function RNA molecule is produced by the removal of one or more *internal* stretches of RNA during the processing of the primary transcript [ibid., p. 265].

What is the philosophical relevance of these molecular biological considerations and how precisely do they support my argument? It is noted with interest that Crick himself states that "any precise definition [of the term "gene"] would be premature" [ibid., p. 270] at the present time. Therefore it follows at the present time we have no precise definition of this term. Nor do we have any very clear ideas of what the atom of inheritance is. Clearly much is known about the structure and function of DNA and RNA, but our descriptions of such structures and their functioning can proceed without use of gene talk at all. Therefore I am compelled by my own argument to carry the mitigated scepticism of Lewontin [1974] further: molecular biology may be taken to show that the generative mechanism of population genetics does not exist. Thus much of population genetics and Sociobiology mut be rejected as false because they are speaking about entities which do not exist. Let me grind my spurs in the wound: they are like theories of combustion which appeal to the mechanism of phlogiston. Whilst some of their claims may be true, the positions are strictly speaking false. Thus of course the principal thesis of this work is established immediately.

2.3 The metaphysics of heredity: Beyond neo-Mendelianism

Since I am not a sceptic, matters cannot be left as section 2.2 left them. Further I have many critical promises yet to be fulfilled. The argument may be best progressed by charting an alternative metaphysical image of the process of heredity which transcends the limitations of the neo-Mendelian position. Further, I would wish to transcend the PSU view of the step ladder model of the relations of the sciences and suggest that genetics can no longer ignore social theory.

A theory of inheritance is a theory about the transmission of matter and characteristics from one generation to the next. The view of Weismann that there exist one particular biological atom which is the sole mechanism of heredity and which stands as the sole "central controlling" agency of the cell is one which should be rejected. As Wahlsten [1979] observes "When it [i.e. the stated position] collapses, the familiar models of quantitative genetics will also collapse" [ibid., p. 458]. Such familiar models must assume the genotype–phenotype distinction as well as the *additivity* of the effects of genotype and environmental mechanisms upon the phenotype [Falconer, 1960, p. 112]. The additivity assumption is presupposed by all linear and higher polynomial models of variance. Once however the neo-Mendelian theory of heredity is rejected, including the genotype–phenotype distinction, the received model of the analysis of variance also collapses. It follows then that the numbers game played by those still dealing with IQ testing for example and with the separation of its hereditary and environmental components *in any conceivable way*, even if interaction factors are recognized, is scientifically meaningless.

We must first ask whether it is true to say that the genes are the "central controlling agents" of the organism as much of popular "selfish gene" Sociobiology seems to presuppose. This view is refuted by developmental neurobiology where the received position seems to be that the genome could not specify even which hemisphere each of the 10^{11} human neurons could occupy, let alone "preform" an entire human brain [Clarke, 1981]. Further the claim that the genome is the central directing agency of the cell operating as binary on-off devices, and organisms merely "randomly aggregated nets of chemical reactions", "molecular automata" [Kauffman, 1969, p. 437] is empirically questionable. Very simple nuclear transplant experiments establish the vital relevance of the non-nuclear parts of the cell for ontogenesis. Transplanting the nucleus of an egg from one species into the enucleated egg of another does not yield ontogenesis past the blastular stage [Webster and Goodwin, 1982, p. 33]. Therefore the non-nuclear parts of the cell are not randomly aggregated epiphenomena. Further the green algae *Acetabularia mediterranean* regeneration of apical structures can occur in the absence of a nucleus [Goodwin, 1979, pp. 111–112].

Further evidence against Kauffman's position comes from research on the ciliated protozoan *Paramecium aurelia*. This has given strong evidence for non-Mendelian hereditary mechanisms. As Sonneborn [1970] has stated:

> There seems to be no present reasonable escape from the conclusion that the determinative role of cortical microgeography [of *Paramecium aurelia*] cannot be accounted for by the action of DNA regardless of whether the DNA is nuclear or mitochondrial, in basal bodies or elsewhere in the cell cortex. [ibid., p. 354].

Nanney [1977] has been led from a study of such *Paramecium* literature to the formulation of three generalizations:

> (A) Cells with identical genes and molecular composition may have different hereditary patterns, (B) Genes, through their prescribed molecular derivatives, contain the modes of pattern permutation and define the states of greatest stability. (C) Cells with identical hereditary patterns may have entirely different genes and molecular compositions. [ibid., p. 27].

A further interesting phenomena also observed in *Paramecium*, first observed by Jollos [1921] is that of *dauermodifications* where acquired characters may persist for many succeeding generations after the environmental agent has been removed. The existence of dauermodifications has been confirmed in both insects [Harrison, 1928] and mammals [Fuji, 1978] [Martin, 1956] and plays an important role in the evolutionary theory of Ho and Saunders [1979, 1980] to be reviewed below.

Despite this evidence neo-Darwinists/neo-Mendelians can always argue, as Maynard Smith [1982, p. 51] has done in reply to Webster and Goodwin [1982], that such considerations are only "special cases". In reply it can be said that it only took two dark clouds on the horizon of classical physics to motivate the development of an alternative theoretical framework. Given the philosophical, logical and methodological problems facing the neo-Darwinist/neo-Mendelian position, the search for an alternative is in order. There are many well recognized non-Mendelian mechanisms of inheritance – enough to warrant not only a chapter in D. I. Patt and G. R. Patt's *Introduction to Modern Genetics* [1975] but also the call for a new point of view.

Such a call is in order once the genetic atomism of neo-Mendelianism is rejected. A more appropriate image for a science of heredity/genetics is that of a field [Webster and Goodwin, 1982] rather than that of the particle. What is passed on in sexual reproduction is on first count the gametes, and in the light of maternal constraints upon development [Cohen, 1979] [Tilney–Bassett and Abdel–Wahab, 1979] [Collins, 1977] that which is transmitted to the forthcoming offspring is a *morphogenetic field* [Goodwin, 1979]. For the moment we will not attempt to explicitly state what such fields (metaphysically) are. Nevertheless these fields interact with chemical agents as well as many other environmental agents, and in turn shape and reshape ontogenesis. What is transmitted "genetically" for human beings, for example, includes also cultural properties, even though such properties have a clearly Lamarckian mode of inheritance. But is is clear that within various social orders wealth and property is of a transmissible basis, and we find here an example of "downward causation" where the social environment, can through the mediation of a physical environment, have profound effects upon the biological realm. Poor diets and diseases are transmitted through families, as in many cases is a sociophysical environment. It is well known that many diffusable substances can be transmitted from the mother to the foetus [Zetterlund et al., 1977], but so are sociocultural properties. It is an obvious *petitio principii* to reject such effects because they are either not as "long lasting" as chromosome effects or because, as most evident in the case of socioculture, such effects are Lamarckian.

Once however non-Mendelian modes of inheritance are admitted the genotype–phenotype distinction becomes virtually impossible to draw, and theoretically useless. Where for example does one put C-type RNA viruses which may capture genes in somatic cells and if not infect the gametic cells, transfer such material to other organisms [Steele, 1979]. This is seen clearly in cases of the actions of slow viruses, such as those causing Creutzfeldt–Jacob disease, which has been incorrectly ascribed to the actions of Mendelian genes [Wahlsten, 1979, p. 450]. It is unclear precisely where one would place such mechanisms in the neo-Mendelian scheme of things. No matter, there is no rational compulsion existing for accepting the genotype–phenotype dichotomy and basic considerations of parsimony beg for its rejection. This scepticism can be made, I hope, quite compelling by viewing how in fact, once this distinction is rejected, the relationship between the genotype and phenotype is explicated.

We can distinguish two broad neo-Mendelian positions on the relationship between genotype and phenotype: *genetic programmism* and *norm-of-reactionism*.

Genetic programmism views the relationship between the genotype and the phenotype with reference to the metaphors of either a computer programme or an architectural blueprint, with a generous helping of information and coding terminology and automata theory. This metaphor is seen quite explicitly and non-trivially in Mayr's [1974] distinction between closed and open behavioural programmes, but stands more generally as an important part of the folklore of the genetic programmism position. Sometimes, as our introductory survey clearly illustrated, the genome is said to contain "information". As we also saw, Maynard Smith [1982, p. 50] has attempted to defend the central dogma of molecular biology by claiming that information, which is taken to be neither material nor energy, does not flow protein → RNA → DNA. Norm-of-reactionism by contrast takes the relationship between the phenotype and genotype, and between the phenotype and environment to be many: many relations such that no single phenotype corresponds to a unique genotype or vice versa. Ayala and Valentine [1979, p. 52] take the norm-of-reaction to be the different phenotypes which an organism with a given genotype may have in differing environments, and this is, they point out, a potentially infinite set. Consequently organisms "do not inherit a particular set of traits but rather a genotype with a very broad range of reaction". [ibid., p. 52].

Let us consider the norm-of-reactionism position first. The position despite initial intuitive plausibility faces an epistemological paradox, which lies in the empirical investigation of the function relationships. Let us note that Ayala and Valentine's claim that the cardinality of the set of various norms-of-reactions is potentially infinite (countably) is not merely a metaphor. If we consider the possibility of ontogenesis in organisms in space travel, in environments not yet experienced on the Earth, the claim is non-trivial. It is difficult to obtain evidence in the sense of getting precise measurements even for controlled breeding experiments under well known environments, and given the traditional characterization of norm-of-reactionism the norm-of-reaction as defined by Ayala and Valentine is empirically impossible to obtain.

The genetic programmism position faces conceptual and metaphysical difficulties. Here I will examine the concepts of a "programme", a "code", a "blue print" and "information". No application of any of these concepts as they have been used in material already cited is justified in my opinion.

What are "programmes"? Obviously the metaphor comes from computer science and so we turn to this science for enlightenment. Since the computer scientists are more concerned in general with problem-solving rather than the explication and defence of basic concepts we had best turn to the philosophers of computing science for explications. J. M. Brady [1977, p. 10] takes a computer programme to be "a written representation of an algorithm in a programming language". Burks [1979, p. 405] is in complete agreement with this definition. The concept of an algorithm may in turn be explicated by recourse to recursive function theory: an algorithm in an alphabet X (being a non-empty finite set of symbols) is an effectively computable function E the domain of which is the subset of the set of words of X and the range of which is also words in X. A function $f(x_1, x_2, \ldots, x_n)$ is effectively computable if and only if there is a mechanical decision procedure for determining $f(k_1, \ldots, k_n)$ given $k_1, \ldots, k_n$ [Mendelson, 1964, ch. 5]. Programmes thus appear to be abstract entities of a set theoretical nature [Boden, 1977, pp. 6–7] [Sloman, 1978, p. 103].

The programming metaphor breaks down logically and biologically. Logically and metaphysically programmes cannot be causal agents but at best can only describe mathematically biochemical processes. The description provided is however not a very informative one and is dependent upon a prior understanding of biochemical processes and contributes nothing as a measure of explanation and prediction. Biologically, organisms are self-organizing, self-repairing phenomena. No computers exist which are self-reproducng and self-programming, and if they did, they must in turn have a master-plastic programme initially supplied by the computer programmer. Analogies between computers and organisms are thus hardly strong and in many crucial senses must break down. For example, the computer programme does not itself build the hardware, and direct its own construction. Yet genes are taken to direct the construction of the "molecular machines" – organisms, which they are found in. Sheldrake [1980] is quite correct in taking the term "genetic programme" to be an "empty phrase", a materialist form of vitalism, open exactly to the same criticisms which that position faced.[5]

A *code* is a mapping of all possible symbols of the alphabet of X onto sequences of symbols of some other alphabet Y [Abramson, 1963, p. 46]. Lila Gatlin in *Information Theory and the Living System* [1972] has stated that the sequence of bases along the DNA chain may be regarded "as a sequence of symbols that store information in the same manner as a sequence of letters in any language" [ibid., p. 4]. Gatlin continues: "one could regard language as an ordered sequence of symbols with a definite meaning, as the basis of all life" [ibid., p. 4]. Worse: "The fact [is] that *DNA is a linear sequence of symbols*" [ibid., p. 4 emphasis added]. These remarks either confuse symbols with what they designate or else advocate a strange form of linguistic idealism. Worse, given Gatlin's assumptions, the use of code-talk is semantically invalid: by definition two alphabets of symbols are needed, and the sequence of bases along the DNA chain constitute only one such alphabet.

Closely associated with the metaphor of the computer programme is the metaphor of the DNA blueprint, based this time upon the architectual metaphor. Blueprints are models from which structured particulars can be built. D. S. Lehrman [1970, p. 34] has pointed out that the relationship between a blueprint and the entity to-be-built is one of topographical isomorphism and that for example, the represented ratios of lengths to widths are exactly the same. But this is simply not the case for the organism and the genome.[6] Yet even if it was, the position lands itself into metaphysical hot waters: blueprints don't build boats. Given neo-Mendelian assumptions of the genome as a central directing agency, what builds the organism?

The final concept requiring examination here is that of "information". We have found Maynard Smith [1982, p. 50] defending the central dogma on information-theoretical grounds, accepting that information is another non-material/energetic phenomena in the world. This metaphysics would seem to be consistent with statements made so explicitly and frequently that they must be interpreted literally, statements such as this: "The genetic information contained in DNA controls the development and metabolism of an organism through the processes of *transcription* and *translation*" [Dobzhansky et al., 1977, p. 24] (cf. also [Eigen, 1971, pp. 465–471]).

It would appear that received opinion has it that information theory and thermodynamics are intimately related [Mercer, 1981, ch. 2]. The familiar story told by Szilard and Brillouin [Brillouin, 1962] is that Maxwell's *Gedanken Experimente*, which was a *prima facie* challenge to the universal truth of the second law of thermodynamic fails: it can be answered by information theory, insofar that Maxwell's demon must expand energy to gather information about the velocity and position of the gas molecules. (Rejoinder: since the demon is a spiritual being it is not necessary that the demon obey physical laws).

The relationship between thermodynamics and information theory is through the concept of *entropy*. Information is negentropy, a negative term in the entropy of a system. The statistical form of the second law of thermodynamics is: $S = k \ln W$, where "S" is the entropy of the system, "k" is Boltzmann's constant and "W" is the number of microstates of the system. Shannon and Weaver [1949] it is also said, have reduced the concept of entropy to pure probability theory; an intuitive proof of this is given by Gatlin [1972, pp. 33–34]. If all microstates are equiprobable then the probability of each individual microstate is $p_i - 1/W$, hence $W = 1/p_i$ which by replacing in the $S = k \ln W$ equation yields the Shannon–Weaver formula of $S = -k \ln p_i$. Shannon–Weaver's formula is the expectant value of the Boltzmann term $-k \ln p_i$, thus $H = -k \sum_i p_i \ln p_i$. As an argument for the "reduction" of thermodynamics to information theory the argument is circular, since the derivation uses as a substitution instance the very formula which should be derived. But there are further problems.

The symbol "k" in the thermodynamics equation is Boltzmann's constant, $k = 1.38054 \times 10^{-23}$ joules per kelvin, a quite specific number, so "k" must be a constant symbol. The "k" however in the Gatlin formula is an arbitrary parameter, so "k" must be a variable. Further the logarithmic base is also an arbitrary choice; usually $k = 1$ and base 2 logarithms are used, giving $H = -\sum_i p_i \log_2 p_i$, although base e logarithms may also be used. Further the sole use of logarithmic functions is to ensure that information has additive properties (i.e. $I(a, b) = I(a) + I(b)$) and Brillouin [1962, p. 1] is explicit in his statement of this justification. However additivity can still be secured by any number of other logically possible mathematical functions, which in turn will give mutually conflicting values to our logarithmically defined function.

Gatlin's problems, as well as others', e.g. Reidl [1978], go metaphysically quite deep. In particular it is assumed that information measures the organization of a system. But the equation of order, information and negentropy on the one hand, and disorder lack of information and entropy on the other is, as Petersen [1975] has cogently argued, unsatisfactory. The equation of disorder with entropy means that exploding stars, parts of which have both high energy and momentum, would have to be classified as highly ordered, which is highly counter-intuitive. The equation of information and negentropy, made to solve Maxwell's problem, would mean that one could get a machine to function merely by having information about it [ibid., p. 84].

Perhaps even worse, a DNA sequence which is functional, and one which is redundant, both having the same number of Gatlin's "letters", logically must contain the same amount of information. But then this information which presumably has the property of self-identity is taken by neo-Mendelians to *both* direct and *not* direct cell activity. The contradiction follows immediately one accepts that the quantity of information has the property of self-identity. Therefore there are no properties by which one can explain differences of DNA function if we follow Dobzhansky et al. [1977, p. 24] and take the genetic information contained in DNA to control the development and metabolism of an organism. Nor for that matter do we explicate the obvious order present in the functional DNA sequence. Adding reference to the consciousness of an agent as Riedl [1978, p. 23] has suggested, will not improve matters as part of the current received wisdom of mechanistic molecular biology is that consciousness itself is ultimately physiologically reducible. One can thus only agree with Apter and Wolpert [1965] that measurements of the amount of information in organisms are highly arbitrary, and as the arguments above establish, theoretically useless for an understanding of biological processes. Consequently Maynard Smith's [1982, p. 50] defence of the central dogma of molecular biology collapses.

It is concluded that the remaining pillars of the metaphysical ideology of neo-Mendelianism collapse.

3.0 Beyond neo-Darwinism: The search for a new science of life

The philosophical critic's role is to loosen one's grip upon the ways and prejudices of received theoretical frameworks. Psychologically this is most effectively done not merely by negative criticism, from a purely logical stand independent of the issue of whether a satisfactory alternative is presented, but by outlining alternatives which may dramatically illustrate the fact that the ways and prejudices of old theoretical frameworks are far from being natural and inevitable.

The orientation most despised and dismissed in the text books in usually a sentence as being a mere historical curiosity, is *Vitalism.* Vitalism is taken by physicalistic orientated biologists as a form of magic, which was undermined by the synthesis of the first organic chemical compound, urea, and which has been steadily refuted by the titanic advances of molecular biology. The works of leading figures such as Hans Driesch (1867–1941) [1914, 1927, 1929] may be left unopened to gather dust on forgotten library shelves. However as Rupert Sheldrake [1981 (1)] has cogently argued, Driesch's position is far from the superficial trivia physicalist ideology would have it to be. Whilst Driesch followed the vitalists of all periods in taking matter to be organized by non-physicalist forces, he did not deny that organisms had material bodies, or even that physicalistic biology would enjoy considerable success. Its limits were however with properties such as the regulation of form and the physiology of ontogenesis. Driesch correctly pointed out that no purely physico-chemical machines existed which had the power of regeneration after arbitrary removal of parts (of course many higher organisms cannot be regenerated if certain vital parts are removed, but can usually regenerate non-fatal mutilations) nor the powers of self-reproduction.

These results suggested to Driesch a radically different alternative philosophy of the organism from the utilitarianism, atomism, and mechanism of Darwinism. First organisms were a real object open to study by biology and not merely the epiphenomenal product of biochemistry. In particular organisms possessed properties of *equipotentiality* (ontogenetic potential is not merely limited to those parts normally actualized) *equifinality* (the same ontogenetic end point may be reached by different starting points and by different pathways) and the *self regulation* already mentioned. In particular Driesch observed the reproduction of typical form after perturbations. The factor operating here he called "entelechy", a non-physical causal agent which organized physico-chemical processes during ontogenesis. The difficulties which have faced dualistic accounts of mentality arose also to face Driesch's position, perhaps psychologically, with *prima facie* more destructive force. There was the problem of explaining the interaction of entelechy and matter as well as the more urgent problem of explaining the nature of entelechy, none of which Driesch succeeded in doing. This meant that there were no new types of experiments to occupy the minds and times of experimental biologists, who anxious for the growth of knowledge or careers, would obviously choose a physicalist biology as a paradigm within which to work.

A second author D'Arcy W. Thompson [1945, 1961] in his classic work *On Growth and Form* whilst accepting the operation of natural selection as an eliminative force (the position accepted by the present author) rightly denied that every structure is an inherited adaptation. Thompson's book is a beautifully written catalogue of how direct physical forces may place constraints upon development. Form was taken to be ultimately explicable within the framework of the "Newtonian language of elementary physics":

> The form, then, of any portion of matter, whether it be living or dead, and the changes of form which are apparent in its movements and its growth, may in all cases alike be described as due to the action of force. In short, the form of an object is a "diagram of forces", in this sense, at least that from it we can judge of or deduce the forces that are acting or have acted upon it; in this strict and particular sense, it is a diagram – in the case of a solid, of the forces which have been impressed upon it when its conformation was produced, together with those which enable it to retain its conformation; in the case of a liquid (or of a gas) of the forces which are for the moment acting on it to restrain or balance its own inherent mobility. In an organism, great or small, it is not merely the nature of the *motions* of the living substance which we must interprete in terms of force (according to kinetics), but also the conformation of the organism itself, whose permanence or equilibrium is explained by the interaction or balance of forces, as described in statistics. [ibid., p. 11].

The limitations of this programme need not be extensively documented here, since my ultimate interests are with reconstruction now, rather than continuing critique. What is of interest to us here is the final chapter of *On Growth and Form* where Thompson put forward a general analytical approach to the study of biological form.

Thompson's approach consisted of taking a specified biological form as a standard of reference upon which one gave a description, or mapped this form on a system of Cartesian co-ordinates and then attempted to demonstrate that various forms of other organisms arose from a *continuous* deformation of the original co-ordinate system. A famous example frequently reproduced is that of the transformation of a *Diodon* or porcupine-fish into a sunfish *Orthagoriscus mola*. This is mathematically delightful, but faces the strong objection that Thompson has failed to give an account of a generative mechanism responsible for the deformation between organisms [Rosen, 1967, p. 90]. Most importantly however, it could be cogently argued that Thompson's work was little more than a study of *analogous structures*, which are not of high evolutionary interest. Despite the mathematical developments which have been made in extension of Thompson's position, the position will not be of much biological use without further considerations of taxonomy, phylogeny and ontogeny to establish homology of structure. Hence Thompson's theory of transformations can be of little use in establishing the exact *basic structure* of extinct forms or transitional forms, without vicious circularity.

Thus we have briefly outlined some interesting, but ultimately problematic positions of some dissents from neo-Darwinism (or Darwinism). Yet it could well be argued that the difficulties facing neo-Darwinism might well be avoided by giving neo-Darwinism a substantial metaphysical and theoretical facelift. In the following section I will explore this possibility.

3.1 Modifying neo-Darwinism: The addition of epigenesis

Epigenesis is a process which has five essential features [Kitchener, 1978, p. 144]. (1) Development proceeds in a series of causally related events, the earlier events producing the later events through interactions between genes, their products and cytoplasm and between cells and cell systems. (2) Development proceeds through a series of stages. (3) Development is a process of progressive differentiation from an indifferentiated homogeneous cell to a differentiated heterogenous biosystem. (4) Development is a process of increasing complexity and organization, in particular, organisms are integrated wholes, most of the parts of which are internally related to each other. (5) Developmental processes exhibit emergence insofar as there arise properties at later developmental stages which were not present at earlier stages.

In two well-known metaphysical images of epigenesis, the role of the environment as a causative agent is minimal. In Waddington's "epigenetic landscape" [Waddington, 1968] the environment may disturb *homeorhesis* for a short period of time before development is channelled back along the development canals (chreods). The genotype–phenotype relationship, still rigidly held in Waddington's position, is analogous he maintains, to the relationship between premises and conclusion of a deductively sound argument [Waddington, 1975, p. 217]. This is equivalent to the rigid preformationist position, that the phenotype is implicitly contained in the genotype. Søren Løvtrup's *Epigenesis* [1974] contains a more explicit statement: both preformationism and epigenesis can be reconciliated since both are necessary and jointly sufficient explanations of ontogenesis. This solution for Løvtrup amounts to denying any specific causal import of the external environment upon embryogenesis, hence "any difference prevailing within the embryo at any stage of development must be causally related to differences present in the fertilized egg" [ibid., p. 10]. The principle generative mechanism is a predetermined programme analogous to a computer programme "outlining a number of operations and a "preformed" set of data on which the operations are to be performed" [ibid., p. 14]. My criticisms of both positions should be readily anticipated by the reader.

By contrast to these positions *dialectical theories of development* [Riegel, 1975] [Lerner, 1978] give a "creative" role to the environment at the expense of accepting a very problematic set of metaphysical concepts. Lerner opts for what he calls "probabilistic epigenesis", which stresses a dialectical interaction between nature and nurture variables. This he contrasts with a "static interaction" conception, which views development "as arising from an interaction between sources which are unalterable by each other over the duration of the interaction period" [ibid., p. 5]. In the dialectical position "each respective source is dependent on the other for the quality of its influence" [ibid., p. 4]. Moreover, whilst physico-chemical phenomena are necessary for the development of any organism they "do not suffice in accounting for the phenomena characterizing the higher level" [ibid., p. 7]. Indeed by virtue of the third law of dialectics, that of the transformation of quantity to quality "at the level of the organism there is present – at one and the same time – the molecular and the epigenetic, the continuous and the discontinuous" [ibid., p. 8]. Levins and Lewontin [1980] have given a broad dialectical materialist interpretation of ecology which meshes well with the Riegel–Lerner position.

The developmental position however faces difficulties in squaring its dialectics with received Weismannian dogmas. If these are accepted, then the theory is internally inconsistent; if not, some form of Lamarckianism must be accepted [Kitchener, 1978, p. 156]. In general the main difficulty lies in the system of presupposed metaphysics – Marxian dialectics. Whilst Marxism makes relatively good social theory, its philosophy of nature is poor [Bunge, 1975].

The work of Riedl [1977, 1978] constitutes in summary a modification of the neo-Darwinist position through the addition of systems of feedback loops between the genotype and the phenotype. This he believes enables the neo-Darwinist position to account for a variety of phenomena such as homodynamy coadaptation, parallel evolution, orthogenesis, Cartesian transformation typostrophy, heteromorphosis, systemic mutation and spontaneous atavism, which Riedl believes that the mainstream neo-Darwinism position has failed to account for. I am not as confident as Riedl, that neo-Darwinism would be silenced by this list: there are extensive discussions of co-adaptation, orthogenesis and typostrophy. However phenomena such as *spontaneous atavism* are especially puzzling if one does not make the *ad hoc* claim that such phenomena must arise from genetic mutations. In spontaneous atavism reversion to a primitive type (in part) occurs. Examples are the occassional cases of three toed horses with a foot structure virtually identical to that of extinct ancestors such as *Merychippus* and *Mesohippus*, furred faces in humans and tailed baby humans. This would allegedly indicate the preservation of non-functional ancestorial patterns.

Riedl's theory of evolution takes as its generative mechanism "the systems conditions which link different levels of complexity to feedback loops of cause and effect" [ibid., p. 358]. This revision is necessary because of a philosophical problem facing neo-Darwinism: if non-adaptively neutral and non-maladaptive mutations are improbable, and since organisms are biosystems we must hope in most cases for systematic mutation. The *coincidence theory* which postulates that systematic mutation occurs through a mere coincidence is untenable. Not only is such an occurrence utterly improbable, but too many alterations per reproduction would have disastrous effects upon the stability of the biosystem. The *storage theory* postulates the loci changes are stored in the population gene pool, and accumulate altering the phenotype when these small alterations come together through hybridization. However these changes must be small so as to escape elimination, but if they are insignificant they will hardly be promoted by selection (and the barriers between "small" and "insignificant" are vague themselves). How can this paradox be resolved?

The answer for Riedl lies in the functional organization of the epigenetic system. The system is said to obey four principles: (1) the normative, (2) the interdependency (3) hierarchic and (4) traditional principles. The normative principle is "the multiplication of identical units of information" [Riedl, 1977, p. 395] and occurs at all levels of complexity in the biosystem. The "interdependency" and "hierarchic principles" would appear to be nothing other than alternative names for gene regulation and the operon model of gene regulation respectively. The traditional principle is "nothing more than an addition of the time axis to the three other principles". [ibid., p. 359]. The core of Riedl's position is this:

> Inasmuch as the genetic information determines functional structures, the functional dependency of the products determines the acceptable freedom of the genetic information. [ibid., p. 360].

Now a functional system of n "epigenetically independent units", would seem adaptatively impossible as the probability of successful adaptation would increase as a power function of the probability of each unit. Hence the units must not be epigenetically independent. He argues that by use of a model in which the mutation of any one genetic unit occurs the operator will reduce this probability, – to quote his own figures – from 10^{-12} to 10^{-6}, thus by 10^{6}. The phenotype here serves to restrict the "adaptive repertory of the genome" much like obtaining two sixes from a toss of two dice which accidently stick together, hence eliminating some other possible outcomes which might have occurred in the case of independent tosses, and also predisposing a certain set of possible outcomes over others (e.g., "end up" tosses are less likely than alternative outcomes).

Do these rather sketchy proposals resolve the difficulties which Riedl has presented for mainstream neo-Darwinism? I think not. The probability argument against neo-Darwinisn has only been moved to another level: now there is an allegedly "epigenetically independent unit", the operon, to which the difficulties which Riedl canvasses for the *coincidence theory* will apply. I imagine the strategy would then be to produce a *storage theory* for systems of operons. But can he? We can well imagine the storage theory working for structural genes, but how does the mechanism work for operons? Do they have a long term biochemical memory? If so what are the mechanisms involved? This is not a trivial question if one rejects the usual *ad hoc* mutational explanations of phenomena such as spontaneous atavism. It is important to note Riedl's own remarks about this phenomena:

> This phenomena [i.e. spontaneous atavism] is stronger than is usually thought, for it tells us that complete subsets of the 40-million-year-old

> patterns of the *Mesohippus* genome are still present. And since such an ancient structure can survive elimination only when it is absolutely necessary, we must assume that the leg of the modern horse cannot be developed without it. [ibid., p. 356].

It seems reasonable to me to propose that Riedl's position outlined above does nothing to explain preservation of patterns and the *Bauplän* of organisms. The mere existence of the proposed genotype–phenotype is necessary but hardly sufficient to account for Riedl's own list of explicandum phenomena. The solutions to the difficulties which have preoccupied us in both this chapter and its precursory chapter do not appear to be met by modifications of the received viewpoints. The search therefore for a more radical alternative is in order.

3.2 The neo-Lamarckian alternative

Ho and Saunders [1976] as we saw in the previous chapter criticized neo-Darwinism for its failure to account for the increase of complexity in evolution. They proposed that the addition of laws of organization would produce a satisfactory theory of evolution, although the presentation of these laws has yet to be done. In their 1979 and 1980 papers they came to reject most of the neo-Darwinist synthesis. Apart from well known empirical difficulties in actually documenting that certain minute variations are adaptive, both genetic changes and the extreme occurrence of natural selection are irrelevant for major evolutionary changes. With regard to the former claim, they follow the "pan-neutralists" in claiming that most mutations are adaptively neutral and that the resultant genetic differences are of no physiological significance. This claim is supported by evidence from the fossil record, where it seems that significant evolutionary changes occurred, when ecological niches were relatively empty and thus where competition was minimal (cf. also [Whyte, 1965]).

Neo-Darwinism is, they maintain, at its weakest in account for its central phenomena: *adaptation*. Here organisms are seemingly discarded, and the environment reduced merely to an agent of selection, a position which is dependent upon implicit Weismannian assumptions which they reject. Their alternative accepts and develops ideas of C. H. Waddington, minus his preformationism. Bio-systems exhibit the property of *homeorhesis,* the ability to return to a normal developmental pathway after perturbation, ensuring a stability of the epigenetic system as a whole. Perturbations which take the biosystem beyond its homoerhetic range usually result in the death of the organism, but if it does not, a relatively large change in the physical characteristics of the organism may occur. The environment is taken then to have an absolutely creative role to play in development: variation arises out of the epigenetic response to an environmental challenge. For a population in a given environment, the epigenetic system will be much the same, as the organisms are all members of the same species, as indeed will be the environmental challenges facing a population. This process is none other than the phenomena of "genetic assimilation" described by Waddington [1957]. A novel adaptive response arises in organisms exposed to a new environmental stimulus which during development becomes "canalised", such that the original response is deepened in intensity and regulated until an approximately uniform response results from the range of intensity of the specific environmental challenge. The response becomes assimilated after a number of generations and can even arise in the absence of the stimulus.

Waddington proposed that the mechanism of canalization of novel development responses could be accounted for through traditional neo-Darwinist mechanisms. Ho and Saunders being sceptical of the theoretical cogency of these mechanisms, advance their own: canalization and genetic assimilation may be due to *dauermodifications* and maternal effects, both of which have already been discussed. Rejecting the genotype–phenotype distinction, they accept genetic assimilation as a Lamarckian process [Ho and Saunders, 1980], not merely mimicking Lamarckian inheritance as they once thought [Ho and Saunders, 1979, p. 581].

The cogency of the position will depend upon how widespread dauermodifications are in nautral populations and whether such changes persist indefinitely, rather than being ultimately resisted by later generations. In this sense Ho and Saunders have put all their theoretical eggs in the one basket. If we are to accept that there are non-Mendelian heretitary mechanisms, there may well be a large number of evolutionary mechanisms. The somatic selection hypothesis of E. J. Steele [1979] is a recent example of a neo-Lamarckian mechanism which could supplement Ho and Saunder's proposal.

Before outlining the somatic selection hypothesis, we should note that both Ho and Saunders and Steele are critical of the usual textbook refutations of Lamarck which usually consist of experiments involving minor physical mutilations of organisms over several breeding generations, such as cutting the tails of mice. But such an argument is based upon a misunderstanding of Lamarck's position. Lamarck held that there is a fundamental tendency for organisms to increase their complexity. The mechanism for this is through the inheritance of acquired characteristics. These acquired characteristics are of two kinds: (1) instrumental modifications directly resulting from environmental action such as injury or multilation; (2) those arising from increased uses of bodily parts in response to environmental challenges. Lamarck did not propose that instrumental modifications are of central evolutionary importance at all. Hence the challenge made by an anonymous reviewer of Steele's book in *Nature*: "Why does a one armed woman not produce one armed children" [Anon., 1981, p. 631] is easily met: they are the wrong sort of acquired characteristic.

Lamarckianism has always been plagued by the problem of the lack of a satisfactory generative mechanism. Steele proposes that endogenous C-type

RNA tumor viruses may be one such mechanism. These viruses contain genes coding for the enzyme RNA-directed DNA polymerase, which enables the synthesis of DNA segments from an RNA template (reverse transcription) which is converted into double helic DNA by DNA-polymerase. It is proposed that C-type RNA viruses can capture mRNA segments, cross Weismann's barrier and either infect an ova or sperm, or travel as an independent particle in the sperm infecting the egg, and through reverse transcriptase a DNA copy of the viruses' transported RNA is made, and integrated into cellular DNA. These viruses possess properties such as the capacity to transduce host cells and to be able to enter the non-lytic cycles with host cells, as well as display both organ and species specificity, properties which make them favourable vectors for soma to germline gene transference. The theory predicts that evolutionary change will be much faster than predicted by neo-Darwinism as inherited effects are shown in the next generation. Steele appeals to the neo-Darwinist concept of natural selection as the mechanism which "fixes" such characteristics in the population. But there is a difference: widespread elimination of organisms is not necessary, since natural selection operates first amongst the population of somatic cells.

The main supporting evidence drawn by Steele is from immunology. This falls into two parts and critics have only attacked one part of Steele's argument. First there is the experimental research on mice which Steele has done with Gorczynski [Steele and Gorczynski, 1980] which indicates, they believe, that neonatally induced immunological tolerance via major histocompatibility antigens can be genetically transmitted to at least two generations of offspring. The Brent research group has failed to replicate the Steele–Gorczynski results [Howard, 1981] and have as far as I am aware failed in a second replication experiment. This does indeed cast grave empirical doubts upon Steeles position.

Nevertheless Steele's theoretical argument remains: experimental evidence from rabbit idiotypy studies indicates that a transmission of "individual-specific" antibody idiotype from immunized rabbits to their offspring occurs. Two terms are in need of definition. Antibody molecules consist of two identical heavy (H) polypeptide chains and two identical light chains (L): the idiotype is the specific LH antigen combining site. By "individual-specific" it is meant "(1) absence of a given idiotype in other antibacterial sera of rabbits from the same closed breeding colony . . . (2) absence of the designated idiotype in preimmune homologous sera. . . or in a pool or normal rabbit Ig" [Steele, 1979, p. 25]. Now, this phenomena, Steele alleges, is not adequately accounted for by either the *germline hypothesis* (all idiotypic specificities are encoded in the germline cells) or the *somatic selection hypothesis* (only some such specificities are encoded, the rest arise by the dual processes of mutation and selection). The former position assumes that every rabbit idiotypic pattern is a consequence of the species germline genes, but only a small portion of this plentitude are expressed in the parents, so there is no inheritance of acquired somatically variable idiotypes. Steele [1979, p. 26] rejects this explanation on the grounds that it conflicts with a large body of experimental knowledge which indicates that "in all *established* germline idiotype systems, viz. in mice, *all* individuals of the strain (or group of strains) express the designated germline idiotype" [ibid., p. 26]. He believes that the genetic differences between the high homozygosity of the mice and the lesser homozygosity of closed colony rabbits is insufficient to account for these results.

The somatic selection hypothesis is unfortunately not criticized by Steele in the same detail as the germline theory. It would appear that since the somatically variable idiotype appears in the F_1 generation, and since the germline theory is rejected, it is not possible for selection to occur here. Steele then offers an argument for the best explanation: a Lamarckian mechanism must be at work here. Critics seem to have forgotten in their haste to demolish Steele that the phenomena of an inheritable somatically variable idiotype requires explanation and even if Steele is wrong, current immunological theories have yet to explain this phenomena. Of course one may attempt to undermine the dilemma which Steele has produced for mainstream immunology. Mitchison [1980] has stated that the assignment of individuals to "individual specific" categories is doubtful. But he has given no argument in defence of the rationality of his doubts. What he would need to show is that either in a closed breeding colony of rabbits, one cannot detect the absence of a given idiotype in the antibacterial sera of a subset of rabbits, or else that the absence of a given idiotype in either normal rabbit Ig G antibodies or in preimmune homologous sera of such rabbits cannot be detected. But if Mitchison's argument was carried to its logical conclusion far bigger problems are thereby produced for immunological science than Steele's work has produced: why accept the standard methodology of individual specific assignments at all?

The critics however, who populate the pages of *Nature* are correct in pointing out the limitations of Steele's work for the revival of neo-Lamarckianism, and this is especially evident from a pan-neutralist position. Nevertheless Steele's work, even if virtually certain to be experimentally undermined, should open up research interest in the role of microorganisms, viruses, viroids and perhaps larger organisms as agents of inheritance. Certainly Ho and Saunders and Steele have given us no comprehensive and satisfactory answers to our major questions of concern. Regardless of this, they may be totally wrong and yet it still does not follow that mainstream biology is innocent.

3.3 Rationalist, Pythagorean and Structuralist approaches to biology

It will be a helpful aid in understanding the positions which I will discuss here to briefly review in very general terms, and terms prejudiced by the course of the argument of this work, the views of the pre-Darwinian rational morphologists. It is a true generalization to assert that for the classical rational morphologists organisms were internally related systematic wholes. In Cuvier for example, this systems view of the organism is illustrated by a number of "rational principles". First, *the principle of the conditions of existence* states that the parts of an organism must be related such that the existence of the organism as a whole, both in itself, and in its relationships with other organisms, is possible. This principle in turn was taken to support Cuvier's principle of correlation: given a sufficient knowledge of the structure of a part of an organism the whole may be reconstructed. Thus from the shape of one organ, one may well be able to determine the shape of other organs. Third, Cuvier held to a principle which might be called "the subordination of characters" which proposed that there are constraints placed upon changes in the form of organisms because some parts are so vital to survival. In Goethe, Geoffroy, the followers of Geoffroy such as Serres, Savigny, and Audouin, Oken and others, the same view of organisms may be traced [Russell, 1916].

Another principal concern of rational morphology was a search for rational principles of typical form. Goethe was one of the first to explicitly discuss the idea of the unity of plan, where he believed that the separate parts of structure where discoverable in all animals. The concern however with typical form is seen nowhere more clearly than in the law of parallelism of Oken, where every animal in its development passed through all stages of the animal kingdom, motivated by rational necessity. Von Baer, whilst empirically disputing the theory of parallelism or recapitulation in the strong form as developed by Oken, nevertheless distinguished between four basic types of spatial-structural plans of the organism; (1) the peripheral or radiate type, (2) the longitudinal type, (3) the molluscan type and (4) the vertebrate type. The typical form of organization determined the manner of development. From this von Baer formulated various laws of form, the following being offered as an example: *fundamentally the embryo of a higher animal form never resembles the adult of another animal form, but only its embryo* [Russell, 1916].

The tradition contrasts sharply with Darwin's position. Cuvier's conditions of existence, Darwin thought, were fully explicable by the principle of natural selection. For example, Darwin thought that the formation of a complicated organ such as the eye could readily occur if all the modifications were slight and gradual. This however is precisely the sort of claim which Cuvier would deny, and critics such as von Baer and Wigand [ibid., pp. 241–242] attacked Darwinism for its inability to explain the correlation of parts and typicality of form.

Webster and Goodwin [1981, 1982] follow these rational morphologists in criticizing their Darwinian foe's conception of the organism, a conception which is an unhappy and disharmonious mix of an empiricist atomist conception of the organism *per se* with a materialist revamp of idealistic holism which sees the organism as an "expressive totality" of "a central directing agency". Organisms are viewed as "structures" and exhibit the same three properties which organisms for Driesch exhibited: equipotentiality, equifinality and self-regulation. Further Goodwin [1976] takes organisms to be cognitive and co-operative systems:

> . . . cognitive insofar as they function and evolve on the basis of knowledge of themselves and their environment, co-operative insofar as the dynamic modes of operations of biological systems (growth, differentiation, rhythmic activity, etc.) are to be understood in terms of the co-operative (correlated) behaviour of particular processes. [ibid., p. 187].

Such a viewpoint may well represent poison to physicalists who have intuitions that "physicalism is somehow intrinsic to empirical science as we conceive it" [Khin Zaw, 1980, p. 52] but from the perspective of this work, we need not worry further about such physicalist throes.

The neo-Darwinist synthesis is taken by Webster and Goodwin to fail to present any satisfactory theory of the production and reproduction of biological form. They argue that: (1) the central directing agency metaphor of the genome is ill-founded; (2) theories of morphogenesis based upon this concept are problematic and (3) morphological invariance cannot be explained by the invariance of the directing agency (as is seen for example in selfish gene theories which dominate Sociobiology). The central argument supporting these three propositions is that there is no one–one correspondence between a given genetic or chemical type and a morphological type [Webster and Goodwin, 1982, p. 32]. They also wish to deny a weaker thesis of one–one token/token correspondence: "a *particular* genetic composition is not even necessary, let alone necessary and sufficient, for a particular morphology" [ibid., p. 33] assuming a fixed external environment. The arguments in turn for this conclusion can be readily summarized. First, there is a recessive "eyeless" gene in *Drosophila,* which after interbreeding of homozygous individuals for several generations, produces flies with normal eyes, although backcrossing indicates that the individuals are still genotypically "eyeless", thus in turn indicating that a re-organization of the gene complex has occurred to prevent the expression of this gene. Nor does invariance in genetic composition necessarily result in morphological invariance, since organisms with a wild type genotype can be of a mutant morphology, and organisms of a mutant genotype can be of a normal morphology, produced through environmental effects ("phenocopies"). Third in *Drosophila* studies, aristapaedia mutants are morphologically more variable than the wild type, with some individuals having one half of the body mediated by a bilateral symmetric axis, a normal phenotype, the other half a mutant phenotype. Whilst such empirical criticisms hardly constitute a knock down case against neo-Darwinism, placed within the framework of a general philosophical critique of neo-Darwinism, they are quite powerful.

What is Webster and Goodwin's proposed alternative? The paper calls for the outline of laws of form which are part of a generative theory of transformations. This proposal has also been made by Rashevsky [1961], R. Rosen and others of the so-called "relational biology" paradigm [Rosen, 1967, ch. 6]. The relational biological paradigm lacks however a unified metaphysical and mathematical approach to the understanding of biological phenomena. The recently proposed

"Constructional biology paradigm" [Danielli et al., 1982] proposes that a "holo-Pythagorean science based on Gaborean algebra" will be a powerful scientific tool for the understanding of biological processes. The approach has arisen from a linkup between theories in the neurophysiology of memory and morphogenesis and embryology.

K. H. Pribram's model of human memory [Pribram, 1971] is based upon the metaphysical image of the hologram. In a hologram each of the parts contains sufficient "information" to characterize the whole; translational invariances of images occur; there is a strong resistance to damage, and a tremendous storage capacity – all of which suggest a powerful image of the mechanisms of human memory. The organizing properties of experimental memory are non-local and the brain is a "decentered structure", precisely as Webster and Goodwin's organisms are. Moreover morphology and embryology can be mathematically described by harmonic functions (solutions of field equations) and their sequential transformations have holographic properties such that the whole can be reconstituted from the part.

These techniques are seen in B. C. Goodwin and L. E. H. Trainor's [1980] field description of the cleavage process in embryogenesis. The field description is based upon intuitions of similarity between the cleavage process and electron density distributions. The cleavage process is taken as a series of transformations as the result of an energy minimization principle, defined on a function representing a generalized surface energy given by a differential equation of the form:

$$(1) \quad E(\theta, \phi) = A\left[\left(\frac{\partial u}{\partial \theta}\right)^2 + \frac{1}{\sin^2\theta}\left(\frac{\partial u}{\partial \phi}\right)^2 + \beta u^2\right]$$

where A and β are constants representing various physiological properties of the cell surface. The minimization condition is expressed by the following integral equation:

$$(2) \quad \delta E = \delta \int_0^{2\pi} \int_0^{\pi} E(\theta, \phi) \sin\theta \, d\theta \, d\phi = 0$$

subject to the constraint:

$$(3) \quad \int_0^{2\pi} \int_0^{\pi} u^2(\theta, \phi) \sin\theta \, d\theta \, d\phi = 1$$

which must in turn satisfy the following Euler–Lagrange equation for the undetermined multiplier of (3), α:

$$(4) \quad \frac{1}{\sin\theta} \frac{\partial}{\partial \theta}\left(\sin\theta \frac{\partial u}{\partial \theta}\right) + \frac{1}{\sin^2\theta} \frac{\partial^2 u}{\partial \phi^2} - \alpha u = 0$$

the solutions for u being restricted to a real eigenfunction set of spherical harmonics:

$$(5)\quad u(\theta,\phi) \rightarrow Y_{\rho m}(\phi,\phi) = \sqrt{2}\, N_{\rho m}\, P_{\rho}^{m}(\cos\theta)\cos m\,\phi$$

The values of ρ are natural numbers and the m values for a given ρ range from $-\rho$ to $+\rho$, and u takes on the value of $\rho(\rho+1)$ and the surface energy of each characteristic cleavage state can be shown for a spherical harmonic solution $u = Y_{\rho m}(\theta,\phi)$ to be $A[\rho(\rho+1)+\beta]$ [ibid., pp. 769–770]. The characteristic states of (5) are limited by mitosis constraints on the number of cells, given by 2^p, where p is the number of cell divisions. Thus the following characteristic function may be defined:

$$(6)\quad 2^p = \begin{cases} 2m(\rho - m + 1) & \text{if } m \neq 0. \\ \rho + 1 & \text{if } m = 0. \end{cases}$$

Equation (6) is satisfied by many sets of (ρ, m) values so a choice is made given the principle of energy minimization of the lowest ρ, and the highest m value for a given ρ to remove a twofold degeneracy every second mitosis division. These constitute selection rules for the cleavage process. Morphogenetic fields are taken to undergo systematic transformations in conformance with the constraints imposed by the selection rules: "particular constraints expressed as selection rules generate specific form, different organisms arising from the operation of different rules within a domain of organized potential form which is the living condition" [ibid., p. 767]. The detailed task of interpreting the field descriptors by means of biological variables remains however to be done.

Constructionalist biology like any other field of inquiry rests upon metaphysical foundations. A holographic brain and a holographic morphology does not by logical necessity require a holographic universe. But if constructionists are to produce an explanatorily comprehensive programme integrated with other areas of human knowledge, the choice of a holographic universe is most natural. David Böhm [1971, 1973, 1980] has argued that relativity (special and general) and quantum mechanics "imply undivided wholeness, in which analysis into distinct and well-defined parts is no longer relevant" [Böhm, 1973, p. 144]. This metaphysical insight, Böhm believes, requires ideas of the implicate and explicate orders. Stated simply each spatio-temporal region contains implicitly, or "enfolded within it" a total structure of the whole, which may in the case of the hologram be *explicated* much as a television broadcast can be carried by radio waves and explicated into a visual image by the television receiver. What however in general carries the implicate order is an unbroken and undivided totality which Böhm calls the holomovement. This he says is "*undefinable and immeasurable*" [Böhm, 1980, p. 151]; it conforms to no particular order, and is not bounded by a particular measure.

The mathematical description of the implicate order is described in terms of *metamorphoses* rather than *transformations*, such that if E denotes a set of transformations under a given explicate order (D_i, R_i, R_o) (where "D_i" is a displacement operator, "R_i" a rotation operator and "R_o" a dilatation operator given specific interpretations in an n-dimensional mathematical space) for a metamorphosis M, E will metamorphose to E′ by: $E' = MEM^{-1}$. Whilst this appears to be nothing more than a similarity transformation, Böhm denies that implicate changes are merely geometrical transformations: "everything alters in a thorough going manner while some subtle and highly implicit features remain invariant" [Böhm, 1973, p. 157]. This would seem to indicate that Constructionalist biology, which in the main has made use of ideas of transformations, does not fit harmoniously with the holographic universe of Böhm.

Here is not the place to further explore Böhm's system beyond noting that given his account of the holomovement as an undefinable and immeasurable one-knows-not-what mathematization, involving autonomous subalgebras which are aspects of an undefinable "whole algebra" of the holomovement, hardly makes much sense for the reason that given the holographic metaphor the mathematical ineffability of the whole will be replicated in that of the parts. Hence it follows that any mathematical modelling of Böhm's position is impossible: the subalgebras must be ultimately undefinable. Böhm's system is, I believe, ultimately inconsistent, because the hologram simply is not a good metaphysical image of the undivided wholeness which we have seen in metaphysical systems from Zeno to Bradley. If there is undivided wholeness, then we have no "parts", period. Our world is like St. Augustine's God: an infinite sphere, the circumference of which is nowhere, the centre of which is everywhere.

If we attempt to give a holographic-implicate-explicate view of morphology arguing as Goodwin has done (private communication 19/11/82) that morphogenetic generation is a particular type of transformation from potential (implicate) to actual (explicate) contingencies determining the details of the transformation, metaphysical paradoxes arise. For example, what we would take as emergent properties of an organism and cell system as a systematic totality, if not literally "present", "enfolded" or "implicated" in parts, must be in some manner preformed in the parts to be explicated by a morphogenetic-field-holomovement. This however seems to restate with a vengeance a more complete form of the central directing agency myth, for each part now is one such director from which the whole can be "called forth". Indeed Webster and Goodwin accept that more than one form is possible for a given composition, such that the existing state of a system (S_1) determines a set of possible transformations (T_1) to the next state S_2. On the holographic metaphor, this many–many relationship is impossible because the whole must in turn be implicated in the series of successive parts $P_{n-1}, P_{n-2}, \ldots, P_1$ which make it up. Since one can unfold from a part p_1, the whole, it follows that either preformationism must be accepted where the wholes $w_1 = w_2 = w_3 = \ldots = w_n$, or else p_1 contains a set of mutually contradictory wholes, which in turn contain another implicate set of such whole, each whole in turn having its respective parts.

How can generative processes be handled mathematically? Especially, can the basic notion of transformation as the Constructionalists understand it be accepted without thereby accepting preformationism? I do not have a satisfactory answer to this question, but my intuitions are that the assumption that harmonic functions are primarily "holographically biased" is simply incorrect, that the formalism itself can be meshed with an alternative philosophy of nature. But the painful dilemma here is that this is what we are precisely searching for!

Further difficulties which face "Structuralist biology", and I think possibly the greatest and certainly the most interesting from a philosophical point of view, are of the rationality of the proposed methods of mathematical modelling. These difficulties are seen nowhere more clearly than in the mathematical biology of Thom [1972, 1975], and we shall conclude our study here of Structuralism with an examination of Thom's position.

Thom's text [Thom, 1975] is addressed to the problem of form in biology, thought, language and society. "Form" is admitted to be a conceptually slippery concept and Thom's attempts at explication are restricted only to topological considerations, all of which he admits are inadequate [ibid., p. 13 & pp. 153–154]. This however does not prevent Thom from identifying Science as the study of morphology. What is morphology? A morphology is understood to be a catastrophe set K, a set of qualitative discontinuities in the substrate space of observables M, and manifold "which is in general impossible to exhibit" [ibid., p. 7]. When a representative point M of the system meets K, a "discontinuity" in the system occurs which is said to be a *morphogenesis*. A real system is "an open connected set of regular points in substrate space whose closure is a conditional chreod" [Thom, 1972, p. 69]. This term is allegedly taken from Waddington's work. But for Thom, the structure of any system consists of certain "elementary chreods". Everyday life is said to have a morphology: the "catastrophe set of experience" is said to be "covered by a finite number of such elementary chreods" [ibid., p. 69]. Not even acts such as throwing a rock escape the chreod category: they are said to be *projectile chreods* [Thom, 1975, p. 306].

Structuralist biology, and indeed any structuralist theory at all, should attempt to do four things according to Thom [1972, p. 70]:

(1) to form a finite lexicon of elementary chreods g_i;
(2) to build experimentally the "corpus" of the empirical morphology;
(3) to define the "conditional chreods", the objects of the theory;
(4) to describe the internal structure of a conditional (or elementary) chreod by associating a mathematical object to it, whose internal structure is isomorphic with the structure of the chreod.

Perhaps a translation of this is, that Structuralist theories attempt to present formal mathematical models which exhibit the regularities of empirical morphology. If this is a satisfactory translation, then the concept of a chreod seems to be doing no theoretical work at all; indeed its seeming applicability to even throwing rocks strongly indicates that a concept such as this, capable of referring to everything in the universe, in actual fact refers to nothing.

We should also become suspicious of Thom's catastrophe foundation to Structuralist biology by a consideration of an important model such as the *predation loop*. Here the Riemann–Hugoniot cusp catastrophe is used as a model, and the predation loop is taken to be represented by a unit circle in the plane. According to figure 13.3 of Thom [1975, p. 298], the unit circle meets the bifurcation curve at points J and K, where K is the catastrophe point for capture, and J the point where an actor appears. However traversing the arc 123 of the unit circle leads Thom to conclude that predators in turn become their own prey! Surely this is the *reductio ad absurdum* of the model. No, says Thom, "the mind of the predator is dominated, alienated by the image of its prey", after which "it jumps from the surface corresponding to the prey to its own surface in an instantaneous *cogito*" [ibid., p. 299]. Even though this is nonsense, it does not even supply a plausible fairytale (despite Zeeman's attempt [1977, p. 654]) to account for how a predator becomes its own prey: if it is back on its own "surface", then we simply have restated the original problem.

Sussman and Zahler [Sussman, 1976] [Sussman and Zahler, 1978] have developed an extensive critique of the social and biological interpretations of catastrophe theory. Whilst their main target has been Zeeman, their criticisms are readily applicable to Thom's far more speculative orientation. The criticisms are that: (1) catastrophe theory does not provide an adequate mathematical modelling of social and biological processes; (2) catastrophe theory does little more than exhibit a mathematical structure which is only vaguely related to the allegedly modelled entities, perhaps only if the phenomena in question exhibit discontinuities. However a "discontinuous function cannot be said to explain discontinuous behaviour merely because it is discontinuous" [Sussman and Zahler, 1978, p. 138]. I would like to elaborate on criticism (1) here.

For a function f on n variables, a critical point is a point n-dimensional space where all the first order partial derivatives s vanish. Critical points are said to be *non-degenerate* if the matrix of the second partial derivatives have a non-zero determinant, and degenerate if this is not the case. Now degenerate critical points, or catastrophe points, do not in themselves represent points of discontinuity in a "functional sense" as there exists monotonic continuous functions defined everywhere in a closed interval [x, y] which are nowhere differentiable [Brouwer, 1967]. Thus one cannot take catastrophes such as the swallowtail catastrophe in a four dimensional phase space to paradigmatically represent "splits", "tears" or other forms of functional discontinuity [Saunders, 1980, p. 119]. It thus remains problematic how in fact one interprets physically, biologically and socially such singularities.

The second difficulty is that the spatial implications of the theory are such that frequently more than one type of catastrophe can be applied to the one morphogenetic process. Goodwin [1976, p. 212] who advanced this criticism, cites the example of the modelling of the process of gastrulation in amphibia by both the swallow-tail catastrophe and the cusp catastrophe. To which one might well say: "so what, examples of equivalent systems of modellings abound in science". But the point remains that if these mathematical constructs are to topologically represent concrete phenomena and process, we now ascribe mutually incompatible topological properties to the *one* concrete particular or process.

Hence catastrophe theory has limitations, and Thom's position hardly stands as an adequate foundation for Structuralist biology. The difficulties which we have seen which face Structuralism are both mathematical and metaphysical in nature. It remains to see how the position will develop within the next decade, and a premature assessment must be rejected. After all, Darwinists have had over one hundred years to get their own theoretical house in shape.

3.4 The hypothesis of formative causation

The survey of this chapter is concluded by a brief discussion of Sheldrake's [1981 (3)] hypothesis of formative causation. The position involves postulating that in addition to the traditional forces of physics, there exists one other force, the morphogenetic field. Whilst the field itself is affected by both electrical potentials and currents, as well as chemical substances, at least in developing organisms, Sheldrake is committed to proposing that morphogentic fields are primarily *sui generis* in nature. All material morphic units, including sub-atomic particles, atoms, molecules, crystals, organells, cells, tissues, organs and organ systems and organisms are ordered by the actions of various morphogenetic fields. The order here refers not merely to spatial organization, but also to the internal structure of the morphic units. Associated with each kind of morphic unit is a characteristic morphogenetic field: thus there is a morphogenetic field for protons, another for hydrogen atoms another for water molecules, another for proteins of a specific type, another for the gut cells of elephants. Consequently, there is a plenum of morphogenetic fields which populate the universe, standing in a one–one relationship with identifiable types. The mode of causation of such morphogenetic fields, Sheldrake calls "*formative causation*".

The action of morphogenetic fields is dependent upon the presence of already organized systems which serve as *morphogenetic germs*. A morphogenetic field becomes associated with a particular type of morphogenetic germ because of the characteristic form of the morphogenetic germ. The associated morphogenetic field contains the "virtual form" of the final physical system of which the morphogenetic germ is a part. How in fact such fields came to contain virtual forms is not explained by the theory, and is taken to be a brute fact about the nature of such fields, just as physicists must accept under pains of vicious infinite regression various brute facts in nature (i.e. whilst the values of some physical constants can be explained by "more deeper" theories, not all can). The morphogenetic field then orders events such that the virtual form is actualized. Upon actualization, the morphogenetic field will tend to stabilize the system and correct deviations from such stabilized states. The form of the specific morphic units is determined by the forms of past systems via the force of *morphic resonance*. Morphic resonance resembles energetic resonance in its selectivity of response, but is not a form of energetic resonance and nor does it involve a transfer of energy [ibid., p. 95]. Morphic resonance however, is analogous to energetic resonance in that it takes place between vibrating systems: "by morphic resonance the form of a system, including its characteristic internal structure and vibrational frequencies, becomes *present* to a subsequent system with a similar form; the spatio-temporal pattern of the former *superimposes* itself on the latter [ibid., p. 96].

As morphogenetic fields are non-energetic, there is no *a priori* reason to suppose that they must obey all presently known physical laws, and Sheldrake proposes on grounds of simplicity that morphic resonance is unattenuated by space and time, although morphic resonance is taken not to involve backward temporal causation; morphic resonance takes place only from the *past*. It follows then as a theoretical consequence of this position, that systems have the morphology that they have because of the similarity of organization of past systems: an organism has the characteristic form of its species because past members of the species had that form which is inherited by both the influence of genetic and morphogenetic field factors. Further the larger the number of similar past organisms, the "deeper" the chreodes will be reinforced, and the Waddingtonian developmental vallies, deepened.

From this position Sheldrake attempts to deduce a number of experimental consequences. One deserves citation here above all others; this is the changes in the rate of learning of new patterns of behaviour. If animals of an inbred strain are taught some behaviour, then there should be a tendency for the same *types* of animals to learn more quickly the same task, as these motor fields will be reinforced by morphic resonance. Thus, to clarify the use of the term "type" here, one should observe an increase in the rate of learning in both animals descended from trained ancestors as well as genetically related animals from untrained ancestors of *that environment*. Hence it is not necessary to claim that one experiment will have action-at-distance and across time effects on all animals in the universe of the same species. Thus an increased rate of learning in successive generations of both trained and untrained lines would support the hypothesis of formative causation; an increase in trained lines only, the Lamarckian position; and no increases, the neo-Mendelian orthodox position. Sheldrake argues that McDougall's experiments [McDougall, 1927, 1930, 1938] in fact support the hypothesis of formative causation.

The hypothesis of formative causation has only recently been proposed and awaits extensive experimental tests. The tendency which I have found from personal discussions with my local biological populace, is the ultimately question-begging response that the hypothesis of formative causation is ultimately a form of vitalistic dualism. This criticism can readily be met by a variety of responses, especially the response that new physical principles and physical fields have been introduced in physics itself to aid explanation, and the arguments given in the present work, establish that physics does not stand in a dictatorial place in the scheme of human knowledge.

Nevertheless the hypothesis of formative causation does suffer a major defect in its present form: it does not, as Sheldrake admits, explain how forms originate, and in general he is quite sceptical of attempts to support general evolutionary hypothesis about the origin of life [Sheldrake, 1981 (3), pp. 24–25]. This is taken in general to be a question of metaphysics, which is outside the scientific domain. This attitude is unacceptable to the present author. Morphology as a science of form *should* explain the origin of form. Sheldrake cannot escape this problem because morphogenetic germs must by definition have a morphology. If he is to give a non-circular account of morphology, the morphology of such germs must be explicable, without opening up a vicious infinite regression. This is not done.

4.0 Conclusion: The state of the argument

The surveyed orientations which may be taken to provide an alternative to the neo-Darwinist position are still in the early stages of their own ontogenesis. No approach is free of both metaphysical and theoretical problems, and the empirical and experimental development of all these orientations has only just begun. As most of these positions are of recent origin, some within only months of writing the draft of this chapter, no final assessment should be given. What is important is that there are reasons to believe that the orientation of the biological orthodoxy of neo-Darwinism and its promiscuous child, Sociobiology, is a fundamentally flawed one, and these reasons hold independently of whether a satisfactory alternative is presented or not. Further, the surveyed orientations if not straightforwardly ontologically anti-reductionist, are at least methodologically anti-reductionist and fit more satisfactorily the *Weltanschauung* of Systemic Emergent Realism, than the *Weltanschauung* of Physicalist Scientific Unificationism.

CHAPTER 6

Causal-explanatory reductionism III: Neuroendocrinological reductionism and the rationality of the foundations of feminist social theory

1.0 Introductory statement of the argument[1]

In this chapter I will examine the theoretical cogency of the position of *Neuroendocrinological Causal-Explanatory Reductionism* (NECER) and make observations about the cogency of sex-related cognitive difference research. This area is a forgotten one for male philosopher-scientists who are critical of many of Sociobiology's central theoretical proposals, but the field is small enough to be dominated by a circle of researchers who hold positions of power on the editorial boards of the major journals publishing material in this area. The result of this is that critiques of the field must be published in alternative journals and books [Rogers and Walsh, 1982, p. 269]. This power structure would be in itself an interesting topic for a sociology of knowledge examination.

My principal concerns here, however, are with both the cogency of (NECER) and the social implications of this position. (NECER) is viewed here as posing a direct challenge, not only theoretically, but politically, to the *Weltanschauung* of feminism. Here I will argue, that in terms of the supporting empirical evidence, the conceptual coherence of central theoretical concepts, the soundness of supporting arguments, the meta-scientific validity of assumed methodologies and to a more limited degree, the presupposed epistemologies and ontologies, that the positions are either incoherent, false or theoretically untenable. The alleged case against the socio-metaphysical foundations of feminist social theory remains to be cogently defended.

One of the fundamental assumptions of (NECER) is that there exist separate categories of behaviour, causally related, and generally peculiar to each sex, or to some sexual orientation such as homosexuality. It is this assumption which is central to the positions of Goldberg [1977], Rossi [1977], Money and Ehrhardt et al. [1968, 1972] and Dörner et al. [1975, 1977, 1978], and in general, sex-related cognitive-behaviour difference research. I will attempt to challenge this fundamental assumption (FA), by attempting to reduce to absurdity those positions dependent upon it.

(NECER) is, as I have said, a direct challenge both politically and theoretically to the *Weltanschauung* of Feminism. Politically, whilst "feminism" has no unifying political philosophy, at the very least any Feminist position should be concerned with criticising traditional "sex-roles". However, the positions of Goldberg especially, but Money as well, accept that traditional "sexual roles" are essential to the proper functioning of society. Money [Money and Tucker, 1975, p. 8] sees such roles as "the glue that holds a society together". Goldberg sees "patriarchy" as "inevitable" and an essential property of any human social order. Rossi, whilst disagreeing with Goldberg that "patriarchy" is "inevitable", nonetheless sees women as (non-trivially) biologically suited to the domestic sphere, and men as (non-trivially) biologically unsuited to this sphere. Dörner sees homosexuality as little more than a neuroendocrinologically caused disease, which stands in need of endocrinological correction, so that homosexuality cannot be regarded as an alternative and legitimate mode of human sexual expression. To classify some phenomena as a "disease" is not merely to make a factual statement about the phenomena, it is in addition to presuppose that the phenomena is undesirable, not a good thing to have, and in need of cure. Insofar as feminists have usually assumed that a socio-cultural model of homosexuality is valid,[2] and have rejected the biomedical model, Dörner's position brings this assumption into question.

Theoretically, Feminism seems also to have accepted three assumptions which together (NECER) maintains, constitute a *prima facie* difficult problem for the orientation.

(1) Feminist social theory has in general maintained that socio-cultural phenomena, relative to biological phenomena, exhibit *emergence* (an ontological thesis), and in addition, socio-cultural explanations are *explanatorily autonomous* (a methodological-conceptual thesis). These assumptions are also widely accepted as part of the sociological/culturological position, as I hope to have made evident in this work. Thus we find yet another challenge here to the Culturology position.

(2) The central research problems of Feminist social theory remain:
   (a) the explication of the nature of gender relationships with respect to social systems, and the theoretical and practical possibilities of change of such relationships;
   (b) the explication of the socio-cultural-political position of women in human societies, both past and present, with respect to the relevant social systems, and the consideration of the theoretical and practical possibilities of change of such existing positions and possibly the relevant social systems.

(3) Feminist social theory has also accepted that human socio-cultural formations have been "male dominated", that predominately positions of power, as defined by categories used *within* societies (what cultural materialists call "emic categories" [Harris, 1979]) have been held by males. There are

no, and have been no, matriarchal or sexually egalitarian social formations [Rosaldo, 1980, pp. 393–394]. Of course there have been hypotheses advanced, and conjectured examples, but the consensus seems to be this. To put it simply, but not falsely, women have in general tended to remain in the "private domain" of the family, having daily responsibilities of caring for children and running the "domestic economy", whilst men tend to have more extra-familial relations and responsibilities – or so the story goes [ibid, p. 394].

Given the assumption that all of these propositions are accepted, a *prima facie* problem arises: does the universality of "patriarchy" thesis conflict with the thesis of the explanatory autonomy of the social sciences? Is it the case that pure non-biological socio-cultural explanations of the central preoccupations of Feminist theory are inadequate, and that either pure sociobiological or sociobiological-cultural "mix" explanations are more reasonable? It is the meta-scientific interests of a sociobiological orientation to say "yes" to this question.

I will now outline a number of sociobiological arguments, based on neuro-endocrinological premises, which would seek to provide causal explanations of some of the central research problems of Feminist social theory and in one case a negative conclusion to the possibility of women's emancipation.

2.0 Initial outline of the (NECER) positions

At the risk of repeating material, let us point out from the outset, that the question of whether a theoretically justified scientific explanation of some socio-cultural phenomena can be given with reference strictly to biological or bio-physiochemical mechanisms (NECER), is logically distinct from the question of biological determinism. The thesis of biological determinism is that because of these mechanisms, for some behavioural modes of social organization, human agents "could not do otherwise", and that the mode of social organization cannot be other than it is. There are extensive conceptual difficulties in clarifying the definiens here [Lehrer, 1981]. I hope, however, to avoid that debate by making the generous concession to biological determinists that their thesis can be satisfactorily formulated.

The point, however, which needs to be made is this: *not all (NECER) are biological determinsts.* There are four positions which I shall consider here: (1) Goldberg; (2) Rossi; (3) Money and Ehrhardt; and (4) Dörner. Both Goldberg and Rossi embrace biological determinism in its stronger and weaker forms respectively. The strong thesis asserts that certain behaviours and modes of social organization are either "theoretically" or "practically" impossible to alter. A weaker form of the thesis asserts that it is practically very difficult to alter certain behaviours and modes of social organization. Money and Ehrhardt, and Dörner, do not embrace either thesis of biological determinism – indeed, Dörner's biomedical research on the neuroendocrinological causes of homosexuality is orientated towards what he believes is a "cure" of homosexuality. Nevertheless, these latter authors embrace (NECER) and this is my central target for criticism. If the thesis of (NECER) is false, then any biological determinist argument based upon (NECER) will have false biological premises and the advanced argument will be unsound.

2.1 *Goldberg and the inevitability of patriarchy*

Goldberg's *The Inevitability of Patriarchy* [1977], advances the thesis that patriarchy, male attainment and male dominance are (i) universal to human societies, past and present, and (ii) most likely "inevitable" given the neuroendocrinological nature of human beings. By "patriarchy", it is meant any social system in which a statistically significant higher proportion of males occupy upper positions in social hierarchies [ibid., p. 25]. "Male dominance" is the "mutually acknowledged" feeling that in dyadic or familial relationships, in terms defined by that society, males have the overriding authority [ibid., p. 37]. Finally, "male attainment" refers to the alleged fact that in past and present societies, males attain the highest status positions and roles, and are more motivated to strive to obtain them. These theses, Goldberg maintains, are statistical, and are not refuted by the existence of "exceptional women", but I will come to question this.

The universality of patriarchy, male dominance and male attainment are physiologically based: given hierarchical social structures and high status roles, males will obtain power, status and dominance and the legitimacy of this will be acknowledged [ibid., p. 72]. The causal relationship between the manifestation of these behaviours and the underlying neuroendocrinology of the individual is this: (1) "[d]ominance tendency results from the interaction between a foetally prepared central nervous system and the presence of endogenous testosterone" [ibid., p. 88]; (2) "dominance tendency" describes the "emotional-behavioural differentiation and is responsible for patriarchy, male attainment and male dominance" [ibid., pp. 65–66]. Goldberg claims that the way male dominance is conceptualized is irrelevant to his argument: it may consist of a lower threshold for the "release" of dominance behaviour, a greater propensity to learn dominance behaviour, a weaker male ego, amongst other possibilities [ibid., p. 66].

2.2 *Rossi's biosocial "facts" on parenting*

Alice Rossi [1977] has also claimed that sex hormones programme the brains of male and female foetuses to react differently to socio-environmental cues, when subjected to such stimuli. Gonadal hormones do this by affecting brain organization during foetal development. She distinguishes carefully her position from Goldberg's: what is genetically determined is "the ease with which the sexes can learn certain things" [ibid., p. 4], not what men can do relative to women. She asserts that: (1) women in all cultures have cared for the newborn and prepared food for cooking; (2) men can learn these skills, but are less capable of handling infants and preparing food than women and will need compensatory training to catch up with the innate sexual superiority of women in this respect. Hence, Rossi's position is a form of weak biological determinism by contrast to Goldberg's position. Further, it is of interest for its argument for a biological superiority of women in the specified respect over men, a kind of reverse sexism, but one indefensible nevertheless.

2.3 Money and Ehrhardt and the androgental amazons

The work of Money, Ehrhardt and co-workers since the 1960's has involved in part the study of the effects, if any, of prenatal androgen upon the developing brain and latter behaviour in social life, with special reference to the effects, if any, of abnormally high androgen levels upon the female brain [Money and Ehrhardt, 1967] [Money and Ehrhardt et al., 1968, 1972]. The claim of central importance to us here is the proposal that high levels of androgens in foetal life cause increased "tomboyishness", the choice of career over marriage [Money and Ehrhardt, 1972, p. 102], higher IQ's and with a negative effect upon maternalism [ibid., p. 245].

I am not particularly concerned with criticizing the methodology of Money and Ehrhardt, as this task has been already accomplished by other writers [Rogers and Walsh, 1982]. Further mention should be made however about these cited studies. In Money, Ehrhardt et al. [1968] 15 girls with Androgenital Syndrome (this and other terms will be defined in due course) served as test subjects and were contrasted by an allegedly random control group of 15 local school girls who were also alleged to be matched on variables such as race, socioeconomic background and age. In Money and Ehrhardt [1968] the group under study consisted of female children whose mothers had taken the drug progestin during pregnancy to prevent miscarriage, but which had the side effect of masculinizing the foetal genitalia. The group chosen for study did not have masculinized genitalia, so it was taken to be possible to control biases in treatment by parents, resulting from the masculinization of the infants' genitalia. The control used here was a group of females with Turner's Syndrome, a condition characterized by low to nil sex hormone activity and a resultant underdeveloped female morphology.

2.4 Dörner and the disease of homosexuality

Dörner et al. [1975, 1977, 1978] [3] argues that male homosexuality arises from a lack of "male-type" differentiation in the hypothalamic-pituitary axis of these individuals, resulting in a "feminization" of the brain. Dörner and co-workers are excited by this hypothesis, since presumably the phenomenon is observed in laboratory rats.

In a study of 21 homosexual men, with a matched control group of heterosexual men, all were injected with estrogen and their luteinizing hormone (L.H.) levels examined. L.H. is a gonadotrophic hormone secreted by the anterior lobe of the pituitary gland. In females it activates the estrogen producing tissues of the ovaries and produces ovulation and initiates the production of corpus lutenum. In males L.H. activates the androgen producing tissues of the testes.

It is alleged by Dörner et al. that in both homosexual and heterosexual men, the intravenous estrogen at first caused a decrease in serum L.H., but in the homosexual males a secondary increase in L.H. above the initial level occurred, which did not occur in the heterosexual control group. Since presumably such a phenomena would occur in women (statements of controls are unclear) Dörner hypotheses that a prenatal androgen deficiency in the homosexual group gave rise to a "female differentiation" of the brain.

These then are the four positions which I shall consider. All assume that sociological causal variables are either insignificant or non-operative, and this seems to be the case for Money and Ehrhardt's research as well, despite the prevailing view that they have developed a "balanced" bio-social theory [Rogers and Walsh, 1982]. To satisfactorily engage in critique of these positions a considerable amount of preliminary conceptual and biological "spade-work" is necessary: some clarification of central concepts and presentation of certain empirical and theoretical material. We turn to this task in section 3.

3.0 The conceptual and biological background

In this section consideration will be given to the following topics: (1) gender categories; (2) sexual categories: (3) the nature of hormone action; (4) the sexual differentiation of the brain.

3.1 Gender categories

I am in agreement with Gould and Kern-Daniel's claim [Gould and Kern-Daniels, 1977] that the terms "sex" and "gender" are frequently used inconsistently in the literature. Here I shall attempt to clarify these terms. Sex is a biological concept, gender a socio-cultural concept. Whilst the term "gender" has a biological definition (i.e. the set of individuals of the same species incapable of fertilizing each other), to eliminate further confusion the term is not used in this sense here.

One account of the concept of gender is this: gender is a characterizing function, which assigns human individuals to one or other of two mutually disjoint classes, a masculine class and a feminine class, on the basis of whether a set of either masculine characterizing predicates (emically defined) $\{M_1, M_2, M_3, \ldots M_n\}$, or feminine characterizing predicates (emically defined) $\{F_1, F_2, \ldots F_p\}$ may be truly ascribed of the individual's actions. The predicates truly ascribed, would constitute necessary and jointly sufficient conditions of class membership.

There are a number of problems facing such a position. The most serious problem is that it is possible that the emically defined characterizing classes are internally inconsistent and jointly inconsistent. For example if we care to reflect upon our own Western emic categories of "masculinity" and "femininity", we find problems such as this: if masculine individuals are premaritally heterosexually experienced, and feminine individuals are precisely those which are not, then an individual x cannot be masculine in this respect without depriving individuality of her femininity in this respect. The inconsistency could be removed by rejecting such predicates from the class, but such a move conflicts with the allegation that these categories are emically defined. Still, it might be said, this is only one example, and does not make a case.

Such a move will now however remove the problem of internal inconsistency. A strong case does seem to have been made by the Feminist tradition, that the Western emic categories of "feminity" and "masculinity" are incoherent. To be "feminine", is to possess properties which it is logically impossible for someone to have. The argument is that one cannot simultaneously be a "Madonna" and a "hot blooded whore", since it is a necessary condition of being a "Madonna", that one is precisely not a "hot blooded whore". Likewise for a host of other properties a woman must not be assertive, but must still "stick up for her rights"; she must not be intelligent, she will scare away the boys – but if she is unintelligent, she will scare away the boys anyway, and to scare away the boys, is unfeminine! The masculine individual must also be an impossible combination of James Bond, Einstein and Dennis Lillee, an alcoholic drug crunching machine and also a bronzed healthy athletic tower of muscle, with a brain on top of it all. The lists can be extended at will, and satirists among us will no doubt perform their own, more satisfactory *reductio ad absurdum*. If I am right in my claim that the characterizing predicates are inconsistent, it follows that under this account of gender, there can be no masculine or feminine individuals at all.

A second problem with this position is that it is theoretically possible that individuals will not be assigned to either class. Thus if individual i may be truly ascribed by $\{M_1, M_2, \ldots M_{n-1}\}$, then i is not masculine. If the question of logical consistency discussed above was overturned, then it is *logically possible*, individual j, an androgynous individual, could be truly described by $\{M_1, M_2, \ldots M_n, F_1, F_2, \ldots F_p\}$, which is inconsistent with the postulate of class disjointness. The problems existing for such an approach to gender ascription are therefore immense.

An alternative strategy is this: there are a set of "aspects" $\{A_1, A_2, \ldots A_n\}$

to which either masculine, feminine, or neutral ascriptions are made – an idea underlying Sandra Bem's sex-role inventory [Bem, 1976]. In Bem's inventory, masculinity and feminity are not two points of a single dimension, but are "two orthogonal dimensions" [ibid., p.51]. The inventory consists of twenty masculine personality characteristics, twenty feminine characteristics and twenty neutral characteristics. A person indicates in a score from 1 (always or almost always not true) to 7 (always and almost always true) how well the characteristic describes him or her. The student t-ratio for the difference between total feminine score points, and total masculine score points gives the subjects degree of sex-role stereotyping. If the masculinity score is significantly higher than the femininity score, the individual has a masculine self-image; if the feminine score is significantly higher than the masculine score, then the individual has a feminine self-image, and if the scores are approximately equal, then the individual has an androgynous self-image.

Masculine items include leadership qualities, "aggression", analytical thought, independence; feminine items include affection, compassion, gentleness, sensitivity to others needs; neutral items include conscientiousness, happiness, truthfulness. In this case it is meaningful to state that x is masculine with respect to some aspect A_1, but feminine with respect to yet another A_3, with respect to the prevailing culture which we examine. This seems a more reasonable conceptualization of gender, quite consistent with our daily usage, and at least coherent. It certainly avoids problems which infect other orientations, such as Parsons and Bales [1955], which associated "masculinity" with an "instrumental orientation" and feminity with an "expressive orientation", or alternatively the Levi-Strausian equation of male with culture and female with nature. It is an open question as to how the instrumental-expressive distinction can be drawn (perhaps the most "efficient way" of getting some "jobs done" may involve an "expressive orientation"); the equation of women with nature can only be made if one considers reproduction primarily as a biological act rather than a biosocial act, and women primarily as reproducer organisms, rather than intelligent, purposive and skilled social agents.

If an account of gender is accepted along the lines proposed by Bem, one of the fundamental assumptions of (NECER) is immediately rendered problematic. There does not exist separate "male" and "female" categories of behaviour. Indeed, it is *logically possible* and, in fact *true*, that some individuals, may be of a "feminine" gender category, even though they are males in the full sense to be specified in the next section; other individuals seem to be of a quite neutral gender, with their lives geared around abstract questions of truth, justice and goodness. (NECER) must show that the gender class of masculine individuals and feminine individuals, is co-extensive with respective classes of male individuals, and that the neutral class is empty. This is a task which we have yet to see accomplished.

3.2 Sexual categories

Organisms which are sexually classifiable, are those organisms which have a capacity to reproduce by the fusion of haploid nuclei, these being called gametes. In sexual reproduction, the production of an organism, numerically distinct from the parent occurs, through the processes of meiosis and syngamy. "Asexual reproduction" is defined (negatively) as any form of reproduction which is not sexual, and includes budding from a parent such as in hydras, cellular fission as in bacteria, protozoa, diatoms and green algae, and parthenogenetic reproduction in which ovum develops into the new individual without fertilization, as in aphids at some periods of the year and in rotifers.[4]

An organisms producing (either exclusively, or for a period of time) the smaller size gametes and in a generally more numerous quantity, or the parts of a single organism producing such gametes is called a male organism (for that time) or the male parts of an organism.[5] An organism producing (either exclusively or for a period of time) the larger size gametes, and in a generally less numerous quantity, or the parts of a single organism producing such gametes, is called (for that time) a female organism.[6]

Sexually reproducing organisms may be classified as either *monoecius* or *dioecius*; in the former category each individual produces both sperm and egg (which does not *necessarily* imply self-fertilization), in the latter an individual produces only one type of gamete. In the latter category, the concept of a primary *sex-difference* can be defined. It is merely the differences between male organisms and female organisms which are causal-developmentally related to the physical property of being a producer of small gametes or a producer of large gametes. *Secondary sexual characteristics* however, are characteristics that differ between the sexes, excluding the gonads and ducts and other organs for dispersing the gametes. An example in humans is the capacity of males (in general) to grow facial beards, and in females to develop breasts.

The embryonic gonads consist of two principle parts, an outer cortex and an inner medulla, the development of these regions being dependent upon the "genetic information" of the sex chromosomes. Inductors are produced by cells of the embryo, which cause the cortex of the gonads to develop ovarian tissue if the embryo is of genotype XX in humans and testes if the genotype is XY. However, this does not exhaust the possible number of human genotypes. According to Money and Ehrhardt [1972, p. 29] the karotype 45Y has not been found, as presumably the amount of missing x-chromosome "information" is lethal. However there are individuals with karotypes 47XXX, 47XXY and 37XYY, as well as the Turner Syndrome karyotype 45XO. There are also certain mosaic variations, in which some cells of these individuals have an extra chromosome, and other cells a missing chromosome. In general, in the absence of inhibiting environmental influences, if at least one Y chromosome is present the body type differentiates as male. Consequently we can define the concept of "genetic sex" as follows: an individual human x, has the genetic sex "ge-male" if and only if x has in each cell (or *predominantly* in most cells) at least one Y chromosome, otherwise x is of genetic sex "ge-female". If x has a mosaic variation, then x is of genetic sex "ge-mosaic".

It is also necessary to distinguish between gonadal sex type and external-genital sex type. Gonadal sex type refers to what sort of gonads the person has: "gon-females" have ovarian gonadal tissues, "gon-males" have testicular gonadal tissue. "Ext-gen-females" have vaginas, "ext-gen-males", penises and scrotums and testes. Once more there is no rigid dichotomy between these types: true hermaphrodites possess both ovarian and testicular tissue and have "ambiguous" external genitalia which are a continuum from more or less "female" to more or less "male". Consequently categories of "gon-ambiguous" and "ext-gen-ambiguous" are of theoretical use. Ge-males (46XY) suffering from androgen insensitivity syndrome due to an autosomal defect, have tissue which is insensitive to testosterone. They have an inner reproductive structure of gon-male, but are of an unambiguous ext-gen-female nature, and are identified at birth as females, and raised as such. Ge-females (46XX) may exhibit varying degrees of "male-iazation" of their external genitalia, due to exposure to excessive amounts of testosterone from external sources during pregnancy.

Finally, a distinction can be drawn between "hormonal males" and "hormonal females" ("ho-male"/"ho-female"). A "ho-male" is anyone having, over an extended period of time, a blood plasma concentration (nanogram/ml.) of androgens approximately equivalent to, or greater than the "normal level of an adult male". A "ho-female" is anyone having a blood plasma concentration (nanograms/ml.) of estrogen equivalent to, or greater than the "normal level of an adult female" over an extended period of time. The "normality" phrase is extremely difficult to explicate, and is best taken as a primitive and understood "preanalytically", to avoid the circularities in definition which seem to arise.

This completes our rather tedious outline of definitions. The journey has not been in vain, for we shall bring our finely honed distinctions to bear upon our opponent's theses in due course. Now we turn to our third set topic.

3.3 The nature of hormone action

Our aim is to understand what hormones are, and what they do. They are, after all, the central generative mechanism of (NECER): consequently such an understanding is necessary for the advancement of our general critique.

First, *what are hormones*? A characterization which views hormones as chemicals produced by various glands and transported via the blood to target cells, would lead to the false conclusion that there are no local hormones, and that auxin, which stimulates growth in plants is therefore not a hormone. This explicate is inadequate. Dörner [1978] on the other hand, defines hormones as:

> . . . intercellularly active chemical messengers that are produced in specialized cells and exert biological effects on other cells of the same

organism by acting locally (as local hormones) or on distant target cells (as systemic hormones) [ibid., p. 13].

However, this characterization fails to distinguish hormones from enzymes or co-enzymes – a great biochemical sin, since according to the biochemist L. Stryer [1975, p. 808] no known hormone is an enzyme or co-enzyme. Nor would adding a clause stating that these effects should have *direct effect* on the genome through, or as part of, a "regulatory complex" (as Raacke has proposed [1977, p. 144]), result in a satisfactory definition, since hormones also alter the permeability of cell membranes, which is not a direct genomic effect [White et al., 1978, p. 1184]. Nor does a chemical characterization seem satisfactory, since hormones are chemically diverse; oxytocin, insulin and the thyroid-stimulating hormone are either polypeptides or proteins, whilst estradiol, progesterone and cortisone are steroids.

Tentatively I propose that hormones are any chemical messengers produced by a cell or cell-system, which affect other cells in any of the following ways:

(H) (1) influencing the rate of synthesis of enzymes and proteins already being produced by the organism, or those proteins and enzymes, not previously synthesized by the organism in its development [Korner, 1970];
(2) influencing the rate of enzyme catalysis;
(3) influencing the permeability of cell membranes and
(4) influencing the transference of electrical impulses (neural-transmitter substance) which do not act as enzymes.

Enzymes we define as any protein which is a catalyst. Consequently, (NECER) must supply plausible hormonal causal mechanisms for disignated socio-cultural phenomena, involving activities (H_1)–(H_4). It has not, I will argue, satisfactorily fulfilled its biochemical obligations.

How in fact do hormones function? The received view of hormone action for protein and amino-acid derived hormones, is that such hormones affect specific receptors of the cell membrane of the target cell, which then stimulates adenyl cyclase, a membrane bound enzyme. The action of adenyl cyclase enables the manufacture of cyclic AMP (adenosine 3′s′-monophosphate) from ATP (adenosine triphosphate). Cyclic AMP has the power to alter certain biochemical processes within the cell. The hormone does not enter the cell, but serves to activate cyclic AMP which is a second messenger. Steroid hormones primarily affect gene expression. These hormones enter the cytoplasm of target cells, and once binding to specific receptors, then translocate to the cell nucleus, where it links to specific sites on DNA, where it alters the DNA template available for transcription, therefore influencing the synthesis of specific mRNAs. We shall have more to say about hormone activity shortly in our discussion of the sexual differentiation of the brain.

The hormones which we are interested in here are androgen and estrogen. These hormones are responsible for the development of secondary sexual characteristics.

"Androgens" are defined as any hormones which result in the maintenance and development of male secondary sexual characteristics; estrogens are any hormones resulting in the maintenance and development of female secondary sexual characteristics. Androgen in males are secreted primarily by the testes, estrogens are secreted in females primarily by the ovaries. The principle hormones of the testes are testosterone and dihydrotestosterone, and their impact upon the development of male secondary sexual characteristics, is strongly documented by studies of accidental pre-puberty castration, and intentional castration of young boys for choirs, which occurred in the past in a number of parts of the world. The voice of the subject fails to deepen, the penis fails to develop, fat distribution resembles that of the female, and hair fails to grow on the face, among other effects.

Estrogens consist primarily of three compounds: β-estradiol, estrone and estriol, the latter two being products of the former [White et al., 1978, p. 1230]. Estrogens stimulate growth and development of female secondary sexual characteristics. In males, estrogens may function to regulate the pituitary glands' secretion of FSH through hypothalamic inhibition [Briscoe, 1978, p. 39].

There is no known hormone qualitatively unique to either human sex. Rather hormone differences are quantitative: "males" being "ho-males" *necessarily* have a higher blood plasma concentration of testosterone (and other androgens) than "females", who are normally "ho-females". Indeed the formation of estrogens from testosterone has been so well documented [White et al., 1978, p. 1231] that it is a part of text book knowledge.

We should also note that one of the estrogenic hormones, estradiol, elicits the "male brain pattern" (we will investigate the meaning of this phrase shortly) in the newborn male rat [McEwen, 1976, p. 48] [Harris, 1970]. Testosterone secreted by the testes is converted into estradiol by enzymes in the target nerve cells.[7] Such a claim raises grave difficulties for simplistic theses such as Goldberg's (and we note immediately that most of Goldberg's claims about human patriarchial predispositions are based upon the endocrinology of the rat), for now, if we follow him and make the logical leap from rat neuroendocrinology to human neuroendocrinology, we must conclude that patriarchical predispositions are caused by "female hormones", and his thesis collapses. However, we shall in due course criticize Goldberg on independent conceptual grounds.

Nor is there any reason to believe that a simple monocausal relationship exists between hormone action and social behaviour. The pituitary gland is stimulated to secrete gonadotrophins by releasing hormones, these being secreted by the hypothalamus. Since the hypothalamus is part of the old cortex on the floor and sides of the brain of vertebrates, behind the attachment of the cerebral hemispheres, it follows that the central nervous system must be involved in regulating hypothalamic action. Indeed if the CNS and the hypothalamus did not act as a *system*, rage could be manifested, at inappropriate times, leading to the real possibility of extinction (e.g. rages during mating may prevent completion of mating). The effects of the social environment on testosterone levels have been documented [Eberhart et al., 1980] [Levin et al., 1967] [Rose and Bourne et al., 1969] [Rose and Gordon et al., 1972] and it is well known that in humans, stress and other mental variables can result in an increased secretion of adrenal steroids through an initial increased release of CRF and ACTH [McEwen, 1976, p. 55].

Consequently it does not follow in a situation in which a correlation between some social behaviour and hormone level is established that the inference: "therefore the hormones are the generative mechanisms responsible for that social behaviour" is sound. It is possible that the peculiar hormone level could be part of a causal chain initiated by the form of social behaviour. This elementary point, I will argue, seems to have been forgotten by authors with reductionist prejudices such as our cited opponents.

3.4 The sexual differentiation of the brain

Underlying (NECER) is a set of neuro-metaphysical theses. These may be summarized as follows:

(NMT) (1) Sex hormones have an "organizing effect" upon the brain of the developing organism such that the presence of androgen at critical periods of brain development effects the central nervous system in such a way that later in life, in the presence of androgens and the appropriate external stimuli, male-behavioural patterns are manifested [Dörner, 1978, p. 14].

(2) The absence of androgen at critical levels, and the presence of estrogens, will result in a more or less female organization of the brain, and the manifestation of female-behavioural patterns [ibid., p. 14].

(3) The action of sex hormones on the Central Nervous System is analogous to their action on genital tract tissue [Phoenix et al., 1968, p. 35].

(4) It is meaningful to speak of male and female brain and CNS organizations, being causally responsible for the manifestations of respective male and female behavioural patterns under the appropriate stimuli.

(NMT) would no doubt be regarded by experimental neuro-psychologists as an empirical question, which is philosophically unproblematic, and the statements of which is entirely clear in sense and reference. F.A. Beach [1971] on this matter has by contrast remarked that "the organizational theory was originally expressed and continues to be stated in such ambiguous and even contradictory terms that it is rapidly leading to a state of utter confusion and futile disputation" [ibid., p. 293]. If this is so, then (NECER) suffers an enormous theoretical set back from the outset. The arguments in favour of such a set back will now be given; the arguments in themselves will be both conceptual and empirical in nature.

What is a male-organized brain and a female-organized brain? How are they individuated and identified? What is an organized brain? How in fact does it differ from an unorganized brain? Can there be functional but unorganized brains? Or perhaps, all brains are functional that are living and operating by "information processing" and other such control activities? If this is so, is not the concept vacuous? Answers to these questions are difficult to consider, since authors such as Dörner whose major theoretical theses depend upon the cogency of these considerations, never explicate these terms. Let us suppose however that we have some non-vacuous account of "organization", or better yet for (NECER), let us ignore the plethora of metaphysical problems lurking within this domain of inquiry. What now is a male organized brain, and a female organized brain?

Given an intuitive characterization of organization and (NMT)(3), we must of necessity maintain that if organism x has a male brain organization, then it is logically impossible for x to simultaneously have a female brain organization. This is so because male and female structures are derived from the same undifferentiated tissue, and to possess a *complete* penis, scrotum and testicles, as well as a *complete* clitoris, vagina and major and minor labia, would require two or more spatially distinct objects to be spatially coincident, which is logically impossible. At best, an organism which is a true hermaphrodite, can possess *incomplete*, and hence ambiguous genitalia. By (NMT)(3), we must then conclude that male and female brain organizations, are mutually exclusive as well, insofar as complete male sexual behavioural patterns (i.e. mounting, intromission and ejaculation responses) and female sexual behavioural patterns (i.e. lordosis) occur.

It seems, that the only coherent way in which male and female brain organizations can be defined, is behaviouristically; that is, x has a male brain organization if and only if x has the capacity to engage in a male sexual behavioural pattern, the causal antecedents of such a pattern being the characteristic neuronic organization of x's brain. The definition of "female brain organization" is analogous. Thus, we cannot by this account identify male or female brain organizations without examining first in our inquiry what behavioural pattern the organism exhibits. From these initial inquiries, we may then ascertain what precisely the underlying neural differences between male and female brains in fact are, and should be able to identify in an examination situation, the male organized brains from the female organized brains.

But at this point a difficulty arises in explaining how bi-sexuality is possible. The phenomena of bi-sexuality is firmly established as existing in human beings, and there is good evidence to believe that it is present even in the rat. Beach [1971, p. 264] describes a male rat which had a high competence as a masculine copulator, yet was observed to have an average lordosis quotient well within the range of estrous females (cf. also Stone [1924]). Obviously, if bi-sexuality exists, then it must follow on (NECER) assumptions that such individuals must have both a male and female brain organization. But if so, this means that (NECER) is inconsistent as (NMT)(1)–(4) form an inconsistent triad. According to (NMT)(3) we would expect that there could not simultaneously be a male and female brain organization in the one individual. But by our definition of male and female brain organization, and the strong evidence of bi-sexuality our supposition is contradicted. But the very methodological point of introducing this hypothesis in the first place was to explain what were taken to be discrete behavioural patterns, uniquely correlated with the sex of the individual. Since ge and ho males, and ge and ho females, can exhibit the behavioural patterns of the opposite sex, (NMT) remains inconsistent.

This is not to deny that there is no perfectly plausible, empirically supported neurally based sex differences. These are precisely those aspects of the brain controlling the male or female "sexual aspects" of the organism. But such an account, when conceptually unpacked, is not very informative at all: male and female brains are precisely those brains which males and females have. If the organizational hypothesis was vaguely defined and incoherent, this claim is logically true and vacuous.

Do we have reason to believe that even if some sort of brain organization did occur, that androgens are responsible for this? Many think so. According to Gerall [1966, p. 367]:

> . . . an androgenic substance . . . 'organizes' or induces the undifferentiated sexual neural areas to develop into a center regulating male behaviour. In the absence of such a substance, the intrinsic tendency in mammals is for the relevant neural tissue to be organized into female behaviour controlling regions.

Beach in response to this position produces two photographs of sections of the optical tracts of an amphibian, which clearly show the disorganization effects of androgen [Beach, 1971, p. 283]. Further evidence against the position is the previously cited study of the masculinizing effects of estradiol [Harris, 1970].

If androgens did have the alleged organizing effects which (NECER) claims, a number of theoretical predictions could be deduced. If male sexual behavioural patterns were dependent upon endogenous testicular androgen during critical periods to the extent that (NECER) believes, then if androgen is lacking during these critical periods, male sexual behavioural patterns would not be manifested. Further, we would expect that the administration of female hormones would have no effect upon such behavioural patterns. As Beach [1971, pp. 258–267] has observed from his first hand research, and from a summary of the literature, neither of these deductions are justified for the rat. Whilst we cannot soundly infer that these conjectures are thereby false for human beings, it nevertheless, places a strong burden of evidential support upon the shoulders of (NECER). We must constantly bear in mind that what is not being questioned is that any neuroendocrinological sex differences exist at all. Rather our interest here is whether those differences exist which our opponents allege exist. In the following sections, I will argue that their case is not made.

4.0 A critique of the positions

In section 3, we explicated some central concepts of the debate, and presented a biological case against (NECER). In this section, I will develop specific arguments of a philosophical and methodological nature against the cited positions.

4.1 Against Goldberg

Goldberg's text is entitled: *The Inevitability of Patriarchy* but the arguments of the text do not rationally support this thesis. If we grant that patriarchy is universal to human societies, at present and in the past, it does not follow that the cause of hierarchical social systems is based upon the neuroendocrinological nature of human beings. Indeed, given Goldberg's argument, hierarchical social systems cannot be causally explained by reference to this mechanism. We recall that his thesis was that patriarchy, male dominance and male attainment are universal and allegedly inevitable, as given hierarchical social structures and high status roles, males will obtain power, status and dominance, and the legitimacy of this will be acknowledged. Males have a "power" to dominate under hierarchical social structures. But note what Goldberg has smuggled in here: we are to accept that hierarchical social structures are already in existence for males to take control over. Therefore Goldberg does not account for these initially assumed hierarchical social structures by his neuroendocrinological hypothesis: he presupposes them. The general argument is then viciously circular. Therefore the thesis of the inevitability of patriarchy fails: even if males were neuroendocrinologically driven under hierarchical social systems to dominate, this does *not* support the claim that they will act in the same manner under a different social system. Goldberg has no non-question begging argument to this conclusion. Consequently his thesis disintegrates: patriarchy is not established by his arguments as being "inevitable".

Goldberg could escape this conclusion by arguing that a non-hierarchical society is either theoretically or practically impossible. But he has given no argument for this, nor do I know of any good arguments for this position [Nielsen, 1981]. It is concluded that Goldberg's claims of the inevitability of patriarchy, male attainment and male dominance have not been supported.

Let us consider the question as to whether, *given* a hierarchical social structure, Goldberg has supplied a plausible generative mechanism to account for the existence of patriarchy, male attainment and male dominance. Goldberg's causal explanation is that endogenous testosterone causes the development of a foetally prepared central nervous system, which in turn causes male dominance tendency, and such a power in turn causes patriarchy, male dominance and male attainment. However when we examine Goldberg's definition of dominance tendency, it is circular: it is just those emotional and behavioural differences resulting from physiological differences, which *cause patriarchy, male attainment and male dominance* [Goldberg, 1977, pp. 65–66]. Removing the problematic causal clause we obtain this definition: dominance tendency is the physiologically caused (by endogenous testosterone), emotional and behavioural differences between males and females – a definition which certainly casts its net wide! Indeed, so wide as to ensnare itself in a vicious circle.

It seems that the action of testosterone in a sufficient quantity must prewire the foetal CNS of the male, predisposing him to patriarchal activities. Now Goldberg must state that it is the *level* of testosterone which does such a wiring, in particular that very level which ho-males possess. But now a grave problem arises. Testosterone in such a level results in the "male-brain" and male secondary sexual characteristics. Having a male brain results in male dominance tendency. But for some individual to be classified by Goldberg as a male at all, the individual must be all of these: G = ⟨ge-male, gon-male, ext-gen-male, ho-male⟩. An individual not a ho-male, such as ge-males suffering from androgen insensitivity syndrome, is not therefore classifiable as a "male". Therefore it cannot be the case that testosterone in a "sufficient level" is the *cause* of any behavioural differences in any non-platitudinous sense, as it is analytic to the concept of maleness that such a testosteronic action on the CNS has occurred. And if it did not, then the "male" would be a "female", as in the case of androgen insensitivity syndrome individuals, A = ⟨ge-male, gon-male, ext-gen-female, ho-female⟩. We are looking for a generative mechanism to account for why it is that *males* so extensively hold positions of power in so many societies.

It is true but uninformative to say that they do this *because* they are males, just as it is true, but uninformative to say that hydrogen and helium have different chemical properties because they are different elements, and different elements have different properties. Goldberg's problem is that he has chosen a mechanism which is responsible for an individual being in G above at all. Consequently he merely restates the original question, and does not tell us why males once individuated by their characterizing properties should so extensively hold positions of power in so many societies. All Goldberg can ultimately say is that they do this because they are males. Recall that he cared little about what way male dominance was conceptualized, and he listed a plethora of mutually contradictory mechanisms [ibid., p. 66]. This scientific irrationalism fits well with my claim that Goldberg's position is circular and devoid of explanatory power.[8]

Ideas of gender are irrelevant to Goldberg's argument he maintains [ibid.,

p. 81], pointing out that a muscular person can lift weights, regardless of whether he/she sees him/herself as belonging to a specific gender. The strength analogy is an interesting one to pursue. First if we grant that a women had this power (as the 1981 European Body Building Championships at Wembley Conference Centre London demonstrated) in a specific social environment, she may never exercise this power at all by simply not engaging in weight lifting practices. Indeed there has been some suggestions that the old secure belief that men have a greater arm strength than women, may be a result of society encouraging greater activity on the part of the average man through labour [Douglas and Miller, 1977]. Likewise males exercise their alleged dominance tendency through either intentional or unintentional acts within a social environment. The elements of intentionality cannot be eliminated from the analysis, since quite a lot of men oppose dominance, patriarchy and hierarchical social structures. Therefore dominance tendency cannot be like a petrol soaked cloth which necessarily ignites in the presence of a set of causally sufficient *external* conditions. Rather the set of causally sufficient conditions *must include the condition that males want to continue to play dominance roles.* And *why* they want this, is not explicable merely by reference to one's neuroendocrinological nature: it requires also a sociological analysis. Necessarily (NECER) does not supply this, because if it did, it would follow that a set of neuroendocrinological conditions is not causally sufficient for the manifestation of the behaviours which we are concerned with here.

The next counter-argument is also applicable to Money and Ehrhardt's claim of a possible causal relationship between androgenital syndrome and "tomboyish" behaviour and high IQ in their sample of girls. Now some females have normal endocrinological natures, but exhibit dominance tendencies: perhaps some feminists, women politicians and business women. Now not all of these women could have had a specially prepared foetal CNS as the testosterone at a ho-male level, if it could act on the CNS would also "male-ize' the external genitalia (for by *hypothesis* the testosterone is circulating in the blood stream, and by definition, so called neural organizations are defined with reference to the manifested behavioural, morphological and physiological effects) and this would certainly have been detected. Now consider this argument:

(1) Testosterone is found in males in a causally sufficient level for the development of ho-maleness and is the *general* generative mechanism responsible for male dominance tendencies, patriarchy, high IQs etc.
(2) Testosterone is not found in any normal female (ho-female) in causally sufficient levels.
(3) At least some normal women (i.e. a growing class of "exceptional women") exhibit dominance tendency, high IQs etc.
(4) If x exhibits dominance tendency, high IQs etc., then the general generative mechanism responsible for this is testosterone, in particular the actions of testosterone on a foetal CNS.

But from (4) it follows that:

(5) If x is a female and x exhibits dominance tendency, high IQs etc., then the general generative mechanism responsible for this is testosterone, in particular the actions of testosterone on a foetal CNS.

But from (2) and (3) it follows that:

(6) In at least some females, testosterone is *not* a cause of dominance tendency, high IQs etc.

But from (5) and (6) it follows that:

(7) In at least some females, testosterone is a cause of dominance tendency, high IQs etc., and is not a cause of dominance tendency, high IQs etc.,

which is an outright contradiction. Since the argument is formally valid, one or more premises must be rejected by my opponents. (1) is their assumption and they accept (2). (4) must also be accepted by them, as it is the only generative mechanism postulated. Since (5), (6) and (7) are logical implications of earlier premises, this leaves only (3) to be rejected – my assumption. Now I submit that (3) is most likely true, constituting an aspect of contemporary common sense knowledge, and in any case substantiated by a previous given argument.

Perhaps one would be inclined to argue that testosterone causes dominance tendency, high IQs etc. in men, but in the growing numbers of feminists, business women and so on, this is a result of intentional, social processes which are emergent relative to their neuroendocrinological nature. This utterly *ad hoc* move, would maintain that testosterone is like a stimulant, which gives males the edge over females, just as taking a sleep prevention tablet may give one an edge against someone who is trying to stay awake simply by unaided will power.

This exit cannot be made, for the stimulant analogy breaks down. It is not the current blood plasma level of testosterone which is taken as the causal agent, but the effects of this on the foetal CNS. The hypothesis is about causal activities in the past, not the present, so the highly likely consequence of testosterone doses to many subordinate housewives failing to produce business women (and yielding only breast degeneration and facial hair) is irrelevant.

A second argument which can be advanced against the stimulant analogy is this:

(1) If testosterone is a stimulant for dominance tendency in males, but not in females, then in an hierarchical social structure, its effects will *tend* to be manifested in all males.

(2) In an hierarchical social structure, at least some males (ho-males) do not *tend* to be stimulated for dominance tendency at all, but on the contrary are stimulated not only to fight against the dominance tendency of other males, but to fight against the hierarchical social structure itself.

Therefore,

(3) The dominance tendency stimulation effects of testosterone do not *tend* to be manifested in all males under a hierarchical social structure (from (2)).

Therefore

(4) Testosterone is not a dominance tendency stimulant (from (1) and (3)).

Of these premises (3) and (4) are logical implications of earlier premises; this leaves (1) and (2) contestable. (1) would seem to be accepted by Goldberg, but it could be rejected in favour of a statistical thesis which maintains that testosterone is a dominance causing mechanism only for some men, and not for others. But to do this is little more than to reject testosterone as a *general* generative mechanism. Further it is to raise extreme conceptual difficulties, as was illustrated by an analogous argument earlier, in explaining how some homales are in fact motivated to fight dominance tendencies and patriarchy. Presumably we do not wish to argue that testosterone predisposes one to mutually contradictory behaviours. Further to argue that in some males, namely the "radicals", the testosteronic mechanism is inoperative is an utterly *ad hoc* hypothesis, and one necessarily false, for "radical" *males* are by definition, homales, so that any testosteronic mechanism must have been operating. Thus to weaken premise (1) is not to escape either falsity or contradiction. This leaves (2) to contest. (2) I submit is true, as illustrated by the male critics of patriarchy, whose actions are consistent with their arguments. We would need very strong counter-arguments, none of which have been given, to reject (2). Since the argument can be seen to be formally valid, the case against Goldberg and Money and Ehrhardt stands.

4.2 Against Rossi

With regards to Rossi's position I will concentrate solely upon her specific theoretical claims, having dealt already in depth with the bio-philosophical assumptions underlying her position.

As we recall from our summary of her position, Rossi claims that men are less capable of caring for infants and preparing food. But this claim meets some difficulties. First, from a methodological point of view it will require studying the respective caring capacities of males, and their respective cooking capacities. But such extensive comparative studies have not been done. What we do know is that many single parent families with males as the sole parent have reared children, and many fathers in the classical nuclear family have bottle fed children, changed nappies, washed, and nursed babies to sleep. Single men have had to cook, and many fathers cook some meals if the wife is ill or working. Further many men are first rate chefs as a glance at the local phone book will tell one. These are commonplace facts. They indicate however that it is nonsense to say that men will need "compensatory training" to clean dirty nappies and cook a meal: they will simply need *training* if they have not done these things before. It is not at all clear what could possibly count as *compensatory training*. In usual psychological discourse, to compensate is to offset some deficiency by developing some other characteristic (cf. Oxford English Dictionary). This however is precisely not what is proposed here. We see no new characteristics being developed by males to boil potatoes.

Rossi may well argue that women have special biological advantages which men do not have: a set of unlearned responses to infants. This may be the case insofar as an infant crying at feeding time, may stimulate the secretion of oxytocin, hence causing nipple erection. Perhaps, and more controversially, mothers may unconsciously cradle the infant with the left hand, so the heart beat may sooth it (by fooling the infant into thinking that it is still in the womb! – *and* if it can even hear the heart beat at all), but this does not support the claim that women have a "special advantage" and are therefore better carers because of that capacity. First, it is true that men do not have functional breasts (excluding hormonal treatments or surgically constructed breasts), but nothing follows from this about the respective care of the parties, as women can just as easily bash babies as men. Second, if we grant the cradling claim, this would seem an utterly minimal advantage: if heart beats are so soothing, men could perhaps consciously place babies on their left side (and the average distance from the babies ear to the beating heart of the male, may well be less than that average distance from the babies ear on a breast to a female heart), or set up a stereo system with recorded heart beats, and so on. In short we find that Rossi has offered no satisfactory arguments to support her chief allegations.

4.3 Against Money and Ehrhardt

Once more I will concentrate upon philosophical and methodological issues, regarding Money and Ehrhardt's chief theoretical claims, which were previously summarized. Further, by the term "methodological", I am not primarily interested in studying and debating the merits of Money and Ehrhardt's modes of data collection [Rogers and Walsh, 1982]. The points to be made here can be made independently of the consideration of Money and Ehrhardt's failings in that respect.

What is lacking from (NECER) is some plausible neural model which canvasses how in fact choices, beliefs, reasoning and modes of social action can be caused by high levels of androgens in foetal life and what the appropriate neural circuitry corresponding to, or identical with, such *prima facie* mental episodes actually looks like. But to solve this problem, not only requires taking up some position in the mind-body problem debate, but requires as well, a very precise theory of beliefs and choices with readily, empirically operationalizable criteria of identification and individuation. Needless to say, these issues are difficult, vague, controversial, and at present unresolved.

Regardless of this matter, Money and Ehrhardt's claims face internal difficulties of their own. It is proposed that high levels of androgens in foetal life caused increased "tomboyishness", increased IQ and general male orientated values, beliefs and behaviour. Now with regard to the IQ question, even if it was true that the androgen affected girls had a mean IQ greater than the mean IQ of the controls, nothing of interest follows from this claim unless we had good reason to believe that the mean IQ of males as a class, was higher than the mean IQ of females as a class. That the IQ tests have been standardized to eliminate such sex differences counts as a matter of logic against Money and Ehrhardt's claims: it would follow that their IQ tests were not standardized, hence methodologically invalid.

When we conceptually unpack the category of "tomboyism", we find grave difficulties. One criterion of "tomboyishness" is the choice of male-type clothing, especially jeans rather than decorative dresses [Money and Daléry, 1976, p. 366]. However it is clear from surveying the clothes actually worn by the young that jeans whilst a utilitarian garment, are frequently very decorative, and worn by both sexes. This is also verified by surveying popular teenage orientated radio stations' advertisements, as well as newspaper and magazine advertisements. Thus it is unclear how in fact a meaningful distinction can be drawn between the utilitarian style of clothes and decorative clothes. Cross culturally, matters are much worse, for the uniforms of soldiers of many nations has both a utilitarian and decorative aspect: in many so called primitive societies decorative dress is an important aspect of battle strategy, used to strike fear into the hearts of the enemy and to protect themselves from the perils of battle. Further, Japanese Samurai did not wear "pants" as we conceive of them.

This failure to clearly explicate central theoretical concepts permeates Money and Ehrhardt, and their co-workers' work. From the previously cited early studies, right up until Ehrhardt and Baker [1974], "a high expenditure of physical energy" has been taken as a male-related behaviour. But this criterion is inadequate, since it must assume that minding a house is not an activity which involves "a high expenditure of physical energy", which is a false assumption [Rogers and Walsh, 1982, p. 273]. It also follows from this assumption that male academics are not involved in male behavioural activities. The absurdity of the position is evident.

To sum up my claims: the categories employed by Money and Ehrhardt and co-workers are too fuzzy and ill-defined to allow any general conclusions to be drawn except for a strong theoretical scepticism about the cogency and scientific meaningfulness of their work.

4.4 Against Dörner

I turn now to Dörner's claims. Dörner's position is perhaps the most carefully documented, with years of research on his part, but nevertheless it suffers from severe conceptual and epistemological problems. Consider Dörner's [1978] definition of a "female organization of the brain", which homosexuals are alleged to have:

> . . . a neuroendocrine predisposition for male hypo-, bi- or even homosexuality and a strong evocability of a positive oestrogen feedback action [ibid., p. 14].

This definition is, unfortunately for Dörner, viciously circular: homosexuals have a "female organization" of the brain because of their homosexuality – a statement which is utterly useless as a scientific explanation.

I remarked earlier that the positive estrogen feedback relation does not support the claim that the hormone levels caused the homosexuality: Dörner makes without justification a biased choice about the direction of causation. Indeed from the mere fact of a correlation, we cannot justifiably conclude that the nature of homosexual life caused that particular hormonal level either, without further evidence. I will attempt now to show that the hormonal explanation is untenable for further reasons.

It is fallacious to conclude that male homosexuals are effeminate [Meyer-Bahlburg, 1977] as any fierce Azande warrior would tell you [Harris, 1980, p. 494]. It is therefore reasonable to believe in the existence of both macho homosexuals and macho bi-sexuals. But if this is so, Dörner's central theoretical conjecture is false. Further, the difficulties which bi-sexuality raised for the brain organization theory also face Dörner's position. We need to postulate two mutually inconsistent modes of brain organization of the one brain, which is incoherent on any intuitively plausible account of "organization", and conflicts in any case with the previously outlined assumptions of (NECER). If the "female brain organization" of homosexual men arose through foetal developmental deficiencies of androgen, *and was presently of causal relevance*, why for example did not a general bodily "female-ization" occur with, say, lack of facial hair growth capacities, fat distribution resembling the female form, and perhaps breast development? Of course Dörner did not observe this because the homosexuals were male: H = ⟨ge-male, gon-male, ext-gen-male, ho-male⟩. We would expect from Dörner (1) a neurophysiological account of the nature of "female brains" contrasted with "male brains"; (2) an empirical demonstration that homosexuals have "female brains" – none of which he supplies. We have then nothing more to conclude that it is logically impossible for *homosexual males* to have a female brain organization, because they are males. Dörner's position is then incoherent.

It might be said in criticism that we *can* imagine a woman's brain in a man's body. Perhaps this is what Dörner means. That is, we must now assume that not merely is hormonal specificity determined at the gonadal level, but the specificity is determined at the brain level. So even though the same hormones (such as L.H. and FSH) trigger gonadal activity in both "males" and "females", under this thesis, the message to the gonads is not merely "get functioning", but "ovulate" or "make sperm". But if so he moves now from the frying pan of contradiction into the blazing fires of theoretical absurdity. For now we have a body which is male guided by CNS and hormonal information appropriate to a female body. It is highly unlikely that a homosexual could function sexually at all, given the brain control of such functions (from hypothesis), for "confused messages" would constantly be received: "testes start ovulating"! After all, L.H. surges cause ovulation to occur [McEwen, 1976, p. 52], so we would expect, to follow Dörner's line of reasoning, that the testicles of our subjects would ovulate. But many homosexual men do function sexually and do not ovulate; we have a presumption of proof in favour of this claim. Therefore the conjecture is unreasonable. But we must ask: how could one possibly get a "female brain organization" with a male body as the result of a natural developmental process?[9] This does not seem theoretically possible, for somehow the rest of the individual's body must be testosteronically affected whilst the CNS and brain is not, and yet, if this is the case, what then controls the production and regulation of testosterone given the hypothesis of its brain control and regulation, and what is the "blocking mechanism"? The hypothesis is too fantastical to believe, and a swift use of Occam's razor is justified.

4.5 General criticism of (NECER)

Underlying (NECER) is the assumption that various social behaviours can be adequately explained by the postulation of hormonal action upon the foetus' central nervous system. But how do we know this? What gives us reason to believe in Dörner's sample of homosexuals, that the cause of their homosexuality was a hormonal event in ontogenic foetal history? I have argued that appeals to present hormonal levels and the present brain organization fail to offer needed support. Unless the causal effects of foetal hormones are observable in neural structures, known to be so alterable, one is left merely with a group of homosexuals, housewives or patriarchs, who have since birth acted and reacted with physiological and social mechanisms. Gone are the foetal hormones, just as the original "pre-biotic soup" of the earth has now gone. If we were unethical Laplacian demons, perhaps we could trace the entire ontogenic-social history of various individuals to test the conjecture (which may then be false). However, we are not magical beings and cannot do this, assuming we can even initially know what we are looking for in viewing this individual's brain in complete detail (which is unlikely given the lack of a detailed neurological theory). Nor would an appeal to twin studies solve this epistemological problem, for the sample problem would arise for each twin.

It might be thought by the empirical psychologist and neurologist, that the above conclusion could not be correct. Can we not diagnose prenatal genetic sex by fluorescence microscopy of amniotic fluid cells, as well as detect any testosteronic concentration differences between genetic males and females? Indeed we can [Dörner, 1976], but this will not support Dörner's position. What is needed, is a knowledge of the ontogenic history of an actual identified homosexual. Given the length of human life (and the morality of the situation), no individual has been examined from foetal life to adult homosexual practices.

(NECER) is plagued by epistemological, conceptual, methodological and theoretical problems. Consequently this programme fails to give adequate reasons for the abandonment of *sui generis* sociocultural explanations of the relevant social phenomena.

5. 0 Scepticism about sex-related cognitive differences

To this point in the chapter, the presented critiques of the positions introduced in section 2 have been given. It will be useful to close with a brief discussion of some philosophically relevant aspects of the topic of sex-related cognitive differences, and especially those aspects of the topic which may be related to my opponents' positions.

The significant empirical findings appear to be these: (1) Females have a verbal performance superior to males [Guilford, 1967] [Maccoby (ed.), 1966]; this appears to be in most areas of verbal functioning. (2) Males are allegedly superior to females in visual-spatial perception [Maccoby and Jacklin, 1974]. (3) Despite similarities in early acquisition of quantitative concepts and performance similarities up to high school, males excel at mathematical reasoning at high school and later at university and in adulthood [Maccoby (ed.), 1966] [Maccoby and Jacklin, 1974].

With each of these issues, debate could be made as to the methodological and statistical adequacy of design. First, there are statistical problems which sex-related cognitive difference research shares with the rest of statistical-empiricist orientated social science with regards to statistical methodology [Morris and Henkel (ed.), 1970]. However, there are some peculiar problems which this methodology brings to this field of study. Sherman [1978, pp. 17–19] points out that there is a bias to assume that failure to demonstrate a sex-related difference means an inadequate procedure, rather than that the null hypothesis is probably true. Contrary to good Popperian metascience, conclusions with negative results are less likely to be published, which makes biases for the report of sex differences in cognition exceedingly stronger [ibid., pp. 18–19]. It also seems that by contrast to many areas of psychological research, studies in this field show an extensive range of methodological flaws [Fairweather, 1976]. Sherman has given another breakdown of these flaws [Sherman, 1978, pp. 19–36]:

(1) numbers of studies ignore other studies where no sex-related differences were reported;
(2) numbers of studies commit elementary statistical blunders, such as correlations significant for one sex, but not the other, are interpreted as providing significant sex-difference data;
(3) there are a number of instances of researchers' own data showing no significant differences between the sexes, where theories accounting for sex differences are advanced in the same paper!

The field of research thus exhibits strong biases, just as its cousin, racial IQ research does (which hardly seems possible to deny now, after the exposure of "howlers" and possible deliberate lying by Burt [Evans and Waites, 1981]). Both fields have been shown to have quite unsightly histories [Shields, 1975], full of characters who have used science as a tool for dragging the Negro and Women into further degradation: these sciences were born in sin. Waving the is–ought distinction, researchers usually attempt to drive a wedge between acts of truth-seeking and acts of policy-recommendation and ethical assessment. However *even* if we accept the "factual/evaluative knife", this does not imply that the theories in question are not *ideological* in the common sociological sense of legitimizing sectional interests. The question of truth is irrelevant to this question. Sex difference research is ideological, insofar as it seems biased towards providing justifications (which need not be at all "philosophically rational") for discriminations against *either* men or women with regards to various positions of status and power, although the principle discrimination seems primarily, Feminists have argued, to be against women [Hubbard and Lowe (eds.), 1979]. Why should we be interested in such questions as to whether men have a greater mathematical ability than women, rather than say whether men can solve the Rubik's cube faster/slower than women? Because the latter is trivial, and it is trivial because *it precisely has no social consequence*. Consequently, given the ideological ramifications of such work, its *a priori* biases, a sceptical attitude should be maintained towards this research programme as an argumentative presumption.

Space does not permit one to challenge every research paper supporting the existence of sex-based cognitive differences. I will however comment upon some of the theories advanced to account for such alleged differences. The inadequacies of these theories cast further doubts upon whether such differences exist outside the researcher's imagination and own a priori biases.

The theories advanced to account for the sex differences are: (1) the greater male variability model; (2) the X-linked inheritance model; (3) the gout model; (4) hormonal models and (5) brain lateralization models. Here I shall not discuss point (4) further.

The greater male variability model cannot be stated rigorously, since the model *itself* is not rigorously stated! Males allegedly have greater genetic variability than females, and for this reason there are more male geniuses and mental defectives. Presumably Darwinian sexual selection was to occur, with females selecting those males with the most successful characteristics so this variability would be gradually diminished (assuming no "dominant idiot X-linked genes"), but alas, the conclusion of this is unclear in its relevance to the relevant sex-related cognitive differences. The model of greater male variability is consistent with both male inferiority or superiority to females in any respect, or with the null hypothesis, since the variability considered is between males and does not justify an inter-group inference. Further, the mere fact of variability does not in itself present any generative mechanism which could be taken to account for the relevant sex-differences. The model is therefore devoid of explanatory power, even if we accept its assumptions, which have been questioned since the time of Karl Pearson's paper *Variation in man and woman* [1897].

The X-linked inheritance model is simply that the various high potential abilities such as spatial ability and mathematical problem solving ability are X-linked, that is carried on a recessive gene on the X-chromosome [Stafford, 1961] [Wittig, 1976]. It follows that as sons never receive their father's X-chromosome, father-son correlates should be zero and as females receive one chromosome from each parent, father-daughter and mother-daughter correlates will be greater than zero, but father-daughter correlates will be higher than mother-daughter correlations, as a daughter may receive either of her mother's X-chromosomes, but her father's only X-chromosome. However, accepting the questionable genetic assumption that complex human abilities are the sorts of things which can be carried on recessive genes, the model faces disconfirming evidence. In a study by DeFries et al. [1979] of 672 father-son pairs for spatial ability scores, the correlation was 0.33 in a Caucasian group and 0.26 in a sample of Oriental fathers and sons, with an average correlation for other family relationships of 0.33 for the Caucasion sample and 0.25 for the Oriental sample. Williams [1975] in a study of 55 father-son pairs on a Wechsler Arithmetic Subtest, obtained a correlation of 0.31 which is inconsistent with the predictions of the X-linked model. Other studies [Guttman, 1974] [Bouchard and McGee, 1977] [Loehlin et al., 1978], do not confirm the X-linked model.

Theoretically, the X-link model fails to account for the relevant sex-related cognitive-differences. If "high potential" is carried on the X-chromosome as a recessive, then it would allegedly imply that more males than females would exhibit such potential. It does not explain *why* females allegedly have a greater verbal ability than males. Second, according to the basic reasoning behind sex linkage, recessive genes on the X-chromosome express themselves in males when they have one allele on their single X-chromosome, on females when they have the allele on both X-chromosomes. On this basis, it is taken to follow that more males will exhibit this characteristic than females. One argument has it that the probability of inheriting two X's is the square (Pr(X). Pr(X)) of inheriting one, but the reasoning here is fallacious if we infer from this, that more males than females will thereby possess this characteristic, for we ignore the probability of drawing a Y, Pr(Y), as we are considering XX and XY individuals. If the recessive gene is on *all* X-chromosomes (and recessiveness is a relationship between genes, not a property of a gene), then without further information about the sex ratio biases, all we can say is that it is equiprobable whether one obtains either a male (XY) or a female (XX)(the sample space here for births) and if these abilities are X-linked, insofar as one obtains either sex, one draws the respective characteristics. If we draw an X for a male, which necessarily we must as they are of karyotype XY, the characteristics is *ex hypothesi* manifested. But the characteristic must, by virtue of its recessiveness, have the alleged characteristics manifested when they have the allele on both X-chromosomes – and since they are of karyotype XX, necessarily the characteristic is manifested *on the assumption* that the respective gene is present on all X-chromosomes.

If my assumption is rejected, then the above argument collapses. However if it is not present on all X-chromosomes, it does not *necessarily* follow that more males than females in a population will manifest these qualities, but is a matter for empirical inquiry as to whether this is the case. Note further that if this theoretical argument is cogent (and if the reader should recognize that this is controversial), then the assumptions upon which our previously cited empirical criticisms were made, will be brought into question. Since I am arguing as a sceptic here, I will interpret by arguments such that the empirical argument proceeds on (NECER)'s own genetic and probabilistic assumptions. It is therefore not incoherent to canvass both arguments, but is merely a good example of a sceptic hedging his bets, both bets which (NECER) must call.

The gout hypothesis offers an excellent example of fallacious neo-Humean reasoning about causal relationships. Hyde and Rosenberg [1976] argued that gout, a metabolic disease which results in high levels of serum uric acid, and its eventual painful salt deposition in joints or the kidneys, has frequently been correlated with "genius" and gout is not as extensively found in women. It is taken that a causal relationship exists here, from the uric acid blood levels (acting as an "intelligence stimulant") to the "genius" or high intelligence level. However, not all "geniuses" have suffered with gout, and nor is there any existing neurological model which could explain *how* gout could act as a relevant causal agent. How in fact do "intelligent stimulants" work? The conjecture is difficult to fit into any plausible metaphysical view of intelligence, and alternative hypotheses such as that no direct causal relationship exists between gout and intelligence, but that the correlation arises through some third party mechanism, such as diet, stress etc. should be sought. This would enable us to explain two obvious facts:

(1) many non-intelligent males suffer from gout
(2) many intelligent females do not.

Even if the gout hypothesis is coherent, it fails to explain the specific sex-related cognitive-differences of our interest. Consequently this model as well is devoid of explanatory power.

The final model which I will consider is the brain lateralization position. Cerebral commissures have been performed to prevent inter-hemispheric spread of epileptic seizures. The operation consists of severing the corpus callosum, a set of nerve fibres connecting the left and right hemispheres, consisting of about 200×10^6 nerve fibres. However connections of both hemispheres to the lower brain regions remain, so therefore given no damage to the sense organs and their connections to the hemispheres (e.g. the optic chiasma, the bundle of nerve fibres connecting each retina of the eye to the visual cortex) an experiment such as the following can be performed. An object is flashed onto the left (or alternatively the right) visual half-field, is seen by the right hemisphere (or alternatively the left). The right hemisphere, a separately functioning brain system, can detect the object, but whilst being capable of recognizing words, commands and relating words to pictorial and diagrammatic representations, it is incapable of stating verbally what has been seen in speech or writing. The left hemisphere does not have this handicap: its linguistic ability is not significantly impaired by commissurotomy, but by contrast to the right hemisphere, it is deficient in spatial ability. This seems to be the received opinion on the matter [Eccles, 1980].

Both hemispheres are also self-conscious, that is it possesses a developed awareness of self, insofar as both hemispheres recognize photographs of themselves (as well as photographs of family, friends and political figures) [Sperry et al., 1979]. At this point some difficult questions in the philosophy of mind can be raised: if we are to get anywhere initially with our central thesis, let us decide for the moment not to ask the question, "how many minds and how many persons?", although this question will be of relevance later.

"Brain lateralization" is a term which is seldom given a precise explication. Authors seem to mean that either the left or the right hemisphere *specializes* in some activity. Puccetti [1973, p. 341] has however offered this definition: "each hemisphere is able to undertake different, even conflicting discrimination tasks at the same time". Now this characterization is consistent with a *non-specialized* hemispheric arrangement such as this: at t_1 R.H. makes the paradigm case of a spatial discrimination and at t_2, the L.H. the paradigm case of a linguistic discrimination and at t_3 the hemispheres reverse discriminations. Hence the Puccetti definition does not capture our preanalytic intuitions of what lateralization involves. Perhaps this is because our pre-analytic intuitions in this respect are in drastic need of modification. In either case, it is far from clear, what in fact the sense and reference of the term "brain lateralization" in fact is, and any logically satisfactory theory if consistent with our pre-analytic intuitions, will require a detailed and precise account of the concept of "specialization".

With this preliminary material we can turn to an examination of the principle two brain lateralization models advanced to explain alleged sex-related cognitive-differences: the Levy–Sperry model and the Buffery–Gray model [Levy-Agresti and Sperry, 1968] [Levy, 1972, 1974] [Buffery and Gray, 1972]. The Levy–Sperry model states that women, like left handed males, are more likely to be non-hemispherically lateralized for verbal and spatial functioning. Maximal verbal functioning is alleged to occur when the L.H. is specialized for verbal functioning, the R.H. for spatial functioning. Women's inferiority on spatial tests is due to a lesser degree of hemispheric lateralization. The Buffery–Gray model alleges that women's superiority on verbal tests is an indication of greater hemispheric lateralization. The two models are logical contraries, since both cannot be true, but both could be false.

Now the Levy–Sperry model runs into grave difficulties, insofar as the assumption that maximal verbal and spatial function is obtained through greater hemispheric lateralization, is inconsistent with the alleged data it is advanced to explain: namely that women also have superior verbal ability. I see no way of countering the criticism without modifying the model (we assume the data is valid)[Star, 1979(1), p. 121, 1979(2), p. 66]. The analogical reasoning, from left handed males' alleged bilateral language representation and poor spatial ability, to claims about women, is theoretically dubious because women are not left-handed males, and the basis for the analogy is non-existent. The claim that left handed males show a spatial deficit in functioning is empirically questionable [Sherman, 1978, pp. 126–128], as is the claim that left handers have bilaterally less specialized hemispheric language representation [Marshall, 1973]. Consequently the Levy–Sperry model is empirically and logically questionable.

The Buffery–Gray model alleges that left cerebral dominance for verbal function is established earlier in girls than in boys, hence females will have a less bilateral spatial function than males, which is taken to be the *cause* of females' alleged lesser developed spatial skill. In males linguistic and spatial abilities are specialized in the left hemisphere, spatial abilities in the right hemisphere. Spatial tasks require a gestalt perception, where as verbal tasks do not.

All parts of the Buffery–Gray model have been strongly criticized. Fairweather [1976] is sceptical of currently existing evidence which is taken to support the claim that L.H. language dominance is established earlier in girls than boys, but this is not the major empirical problem. As Sherman [1978, p. 115] points out, there is no independent evidence presented which *shows* that bilateral representation of spatial function is associated with superior spatial abilities, and there is evidence to the contrary [Waber, 1977] [Coates, 1974]. Sherman does not however go far enough in her criticism in my opinion: what is theoretically wrong with both the Levy–Sperry and Buffery–Gray models are that they are not *explanatory* models. It is not *explained why* hemispheric lateralization or bilateralization should result in either superior or inferior linguistic ability, because there is no neural model presented at all which predicts these results. On the Buffery–Gray model, females' less bilateral spatial function causes their spatial capacities to be less poorly developed than it is for males. But this reasoning is circular *unless* it is shown on an independent basis other than merely that of the data of the explanadum, that such a causal relation does hold. This has not been done, and the claim in any case seems highly likely to be false.

I mentioned earlier that comment is needed on the received view of the commissurotomy patients with regards to the philosophical question of how many minds and how many persons there are, insofar as this relates to our thesis. Now it might be thought that what answer one gives to this question is irrelevant to the question of cognitive-differences. However consider Puccetti's [1973] claim that not only do the commissurotomy patients have two minds each of which is a distinct person in the one body, but since it is absurd (for reasons Puccetti gives that need not be discussed here) that a brand new mind has been created, this dual personality existed *before* the commissurotomy, and exists presumably in all of us as well, both *males and females*. If Puccetti was right then the brain lateralization models of cognitive-differences between the sexes is incoherent. We cannot speak anymore of L.H. and R.H. lateralization of functions, because the L.H. and R.H. are really two brains, of two separate people [Puccetti, 1977]. I am not prepared to argue this question here in such limited space: my point remains that if Puccetti is right, then the brain lateralization models of sex-related cognitive differences have been philosophically refuted.

It is time now to state my conclusions from this study of sex-related cognitive differences. First, my confidence in the basic explanandum of the field is low. The field contains *a priori* biases towards the "production" of such confirming data. The field is highly ideologically laden: it is unclear as to why these questions, are asked, except insofar as such questions are of interest in a sexist society. Second, all the considered biological theories are either theoretically inadequate, semantically incoherent, or empirically false. I have not considered social explanations of alleged sex-related cognitive differences, since I am sceptical of the coherence of these questions and the legitimacy of the field of inquiry, and second, the present chapter has been concerned with the task of criticizing biological theories. It is logically possible that all social accounts of the alleged sex related cognitive differences fail. Then, either such phenomena are inexplicable and veridical, or else non-veridical (assuming we have exhausted all practically possible explanations). I would then maintain that such phenomena are mostly inexplicable because they are non-veridical.

6.0 General conclusion: The state of the argument

It is maintained as a general conclusion that (NECER) faces philosophical, theoretical and empirical problems which rationally justify its rejection.

One issue remains outstanding, this being the widespread acceptance amongst feminists of the very assumption which fire Goldberg's book: namely the verity of the universality of patriarchy. Now I have argued that even the strongest "biological determinist" position, that of Goldberg [1977] fails to show that the attainment of a non-sexist society is not possible.[10] Yet this is surely the main ideological motive of such positions, and since Goldberg's position fails to achieve this, our interest in the universality of patriarchy should therefore lessen. The falsifying counter-example, we hope, may be soon supplied from the ideological ruins of the present social order.

The question may then be asked: "what importance is the origins question (i.e. of patriarchy) to feminist social theory anyway?" Two answers are usually given to this question: (1) it is important politically because the limits of social engineering need to be known to gauge the probability of success; and (2) it is of abstract theoretical value to obtain a general theory.

The first reason does not provide adequate justification. I would maintain that biological parameters place no limits upon the possibility of building non-hierarchical non-sexist societies. There are simply no good arguments for the position. Second, given that we have such knowledge, perhaps the best way of testing the respective positions is to actually attempt the experiment of building such a society. It is certainly a good bet: freedom for all if it succeeds, and if it doesn't then I am wrong, and then I suppose we'll have to put up with our fate like we put up with death, and other poor, nasty and brutish things.

The second reason assumes that there can be a universal theory of patriarchy. However even if the phenomena is universal, it does not follow that the phenomena is mono-causal in nature. It is indeed necessarily parsimonious to assume so, but parsimonious explanations are not necessarily true.[11] All monocausal accounts of patriarchy have failed. Perhaps this is because there can be no *general* theory at all, because patriarchy is little more than a "historical accident", which has been socially perpetuated and reproduced because human beings have not yet made a satisfactory attempt to eliminate it. This question can no longer be seriously excluded. I propose that given the plausibility of this thesis and the failure of the inevitability of patriarchy argument, the origins question is not worth addressing at all as a socially relevant question. Energies are best devoted towards drawing the plan, and building an alternative social order.

CHAPTER 7

Causal-explanatory reductionism IV: Ecological Sociobiology and cultural materialism

1.0 Introductory statement of the argument

In general, critics of Sociobiology have been preoccupied with various genetic models and have devoted proportionately less energy towards a critical examination of the ecological premises of the Sociobiological argument. One reason for this is that Sociobiologists themselves have been preoccupied with the genetic components of their research programme. Stephen Emlen [1980], however, has addressed what he takes to be the second major research frontier in Sociobiology: to explain why one form of society exists under one set of ecological conditions, and another form of society under yet another set. Emlen states:

(EPT) There is a second major research frontier in Sociobiology – one whose primary concern is understanding the *forms of social organization* that are found in nature. Why does one form of society exist under one set of ecological conditions, and another form exist in another?

Building on over a century's accumulation of natural history studies, workers in the last decade have begun developing models that allow a better understanding of the type and structure of animal societies. By integrating behavioural and ecological theory, it has become possible to begin to predict various aspects of social organization on the basis of a small number of ecological variables. The types of questions being asked include: When do we expect an animal to live socially as opposed to solitarily? Under what conditions will an animal be territorial and defend an area or a resource rather than be nomadic or share the resource? If social, what determines the optimal group size, and does this vary for different activities (foraging versus breeding, etc)? What are the methods of resource acquisition and how are resources divided among members of the population? What is the

> form of the mating system, and what are the options for the accumulation of multiple mates? How and when do animals disperse, and what is the effect of this upon the kin structure of the population?
>
> In essence, this side of Sociobiology attempts to predict the basic elements of the form of animal societies. [ibid., pp. 127–129]

I shall argue, that with respect to human ecology, Emlen's position is not at all new to social theorists. Marvin Harris [1974, 1979] has developed a much more rigorous ecological research programme, complete with an epistemology and social ontology. Whilst Harris remains a sharp critic of a genetic orientated Sociobiology [Harris, 1979 chp. 5], the position of *Cultural Materialism* would seem, it will be argued, to provide a precise explication of Emlen's intuitions of the form a human ecological Sociobiology should take.

The thesis of this chapter is that both Emlen's human Ecological Sociobiology (and in certain respects, the programme for non-human animal behaviour) and Harris' cultural materialism are philosophically (i.e. conceptually and epistemologically) and theoretically flawed. Consequently it remains to be demonstrated by defenders of these positions that they are rationally tenable research programmes.

2.0 Central theoretical presuppositions of Emlen's Ecological Sociobiology

According to Emlen Ecological Sociobiology is concerned with understanding, predicting and explaining aspects of social organization on the basis of a limited number of ecological variables [Emlen, 1980, p. 127]. As he says:[1]

> Its basic premise is that given certain ecological constraints ("problems"), there exist a limited subject of adaptive social organizations ("solutions"). This is because alterations in group size, foraging patterns, territorial defense, etc. that are maladaptive (e.g., in terms of effective predator defense or efficient resource use) initiate negative feedback in the form of decreased energy intake, deterioration of physiological condition, and, ultimately, reduced fecundity and survival. Such ecological feedback is a stabilizing influence, shaping certain features of social structure to maximize the ability of the participants to cope efficiently with the constraints imposed by the environment. [ibid., p. 143]

This sentiment is also expressed by Jack W. Bradbury [1980] who states:

> One of the more widely accepted concepts in Sociobiology is the notion that social systems are determined by, or at least limited by, the distribution of crucial resources. [ibid., p. 189]

Thus we may state as the first tenet of Ecological Sociobiology:

(ES_1) Forms of social organization are "solutions" to ecological problems, i.e. social systems are both determined and constrained by ecological variables.

Further whilst recognizing that human beings have the greatest amount of behavioural plasticity of any species, human beings are also proper subjects for investigation by Ecological Sociobiology. Emlen states:

(ES_2) But it is my belief that the resulting form of social organization will converge upon that which is most energetically efficient and predation – secure (adaptive) to its individual members, regardless of the degree of genetic determinism or cultural plasticity of the behaviours. [ibid., p.137]

So much then for the general tenets of Ecological Sociobiology; now to put some theoretical and empirical flesh on these abstract bones. We shall now outline some of the problems which Ecological Sociobiology addresses and the conceptual machinery employed.

In general Emlen's programme is little more than an extension of well known biological optimization models, especially foraging Strategies [Maynard Smith, 1978 (1)]. A common assumption in virtually all of these models is, as Maynard Smith notes:

> . . . that the way in which an animal allocates its time among various activities (e.g. consuming one prey item rather than another, searching in one kind of patch rather than another, moving between patches rather than continuing to search in the same one) can vary, but the efficiency with which it performs each act cannot . . . . Thus the models of foraging so far developed treat the phenotype set as the set of possible behavioural stategies, and treat structure and locomotory or perceptual skills as constants contributing to the state equations (which determine how rapidly an animal adopting some stategy acquires food). [ibid., p. 33]

Biological "solutions" to ecologically posed problems are usually taken to be a "maximum solution"; as John R. Krebs [1978] states:

> . . . animals will, as a result of evolutionary selection pressures, tend to harvest their food efficiently, so if we can work out in theory the decision rules which would maximize the animal's efficiency, these rules ought to predict how the predator makes its choices. [ibid., p. 23]

The reasoning behind this is simply the fundamental neo-Darwinist insight that in the long run, "natural selection favours individuals with the highest fitness" [ibid., p. 23], and thus natural selection will favour "efficient predators". "Efficiency" here could mean a maximization of food intake or a minimization of the fluctuations of food intake or the maximization of food intake over some specified time period. Thus testing usually involves evaluating which of a plausible set of solutions are being adopted, but this is not necessarily easy to do from an empirical point of view, especially if the population is lagging behind a changing environment. Yet before passing onto some empirical examples, we should note well Maynard Smith's words [Maynard Smith, 1978 (1), p. 52]:

> The role of optimization theories in biology is not to demonstrate that organisms optimize. Rather, they are an attempt to understand the diversity of life.

This is to say, contrary to the impressions which one gains from some ecologists, such optimality models are *causal models*, telling us how in fact certain *prima facie* biological phenomena arise.[2] Maynard Smith is quite right to take this interpretation of foraging stategies. If he did argue that the role of optimization theories in biology was to show that organisms optimize, the argument would be obviously circular: the various decision rules spoken of by Krebs are derived from a set of premises consisting of empirical assertions about the specified biological phenomena, as well as the Panglossian optimality assumption, so that the advanced argument contains the problematic conclusion already as a premise.

There are three important predictions of foraging theory with respect to the matter of optimal diets. Given a number of different food items with a specific food value and handling time from which an animal can choose, then according to Krebs [1978] predators should:

> (i) prefer more profitable prey, (ii) be more selective when profitable prey are common, (iii) ignore unprofitable prey which are outside the optimal set regardless of how common they are. [ibid., p. 29]

The model is modified to account for problems raised by recognition of prey timelags, and incorrect identifications [Pyke et al., 1977]. If a high proportion of foraging time is spent in rejection, low profit ranking prey move into the optimal set [Krebs, 1978, p. 29]. In such a situation, assumption (iii) is inapplicable.

In Werner and Hall's [1974] study of a sample of ten blue gill sunfish (*Lepomis macrochirus*) feeding on different size classes of *Daphnia* the diets observed are taken by Maynard Smith to be in good agreement with the predictions of the model [Maynard Smith, 1978 (1) p. 44]. Different size classes of *Daphnia* were mixed and the sunfish allowed to feed. The number of prey of each class obtained by the sunfish was ascertained by an analysis of stomach contents after foraging sessions. Relative "profitability" was taken to depend upon the size of the prey classes. It was found that at a low density class mix, the sunfish ate the prey of the three classes according to how often they were encountered; at an increase in the density of the class mix from 20 of each class to 350, it was found that the sunfish ate at a statistically significant higher level, the most "profitable" size class of prey. In Kreb's et al. [1977] study of the feeding of the Great Tit *Parus major*, as with the sunfish, unprofitable prey (in this case small, by contrast to large pieces of worm meat) were not completely ignored. However, contrary to the predictions of optimal foraging, the proportion of unprofitable prey taken was not completely independent of the encounter rate: this was taken to indicate that the Great Tits required time to recognize small prey.

Yet another optimal foraging model which will be of interest to the following discussion of Emlen's paper is a model of the exploitation of patchily distributed food. Optimal foragers should prefer to forage in the most profitable areas, and only forage in less profitable areas when the availability of the more profitable areas is low. Consider a function f(T) representing the average cumulative food gain (f) as a function of time in a patch/area (T). Kreb's diagram [Krebs, 1978, pp. 42–43] takes the optimal predator to stay at a patch long enough to maximise the slope of the line AB (f′(T)) a tangent to the curve f(T). But from Krebs diagram (a) it is evident that the *slope* of the curve AB is not maximized at the value x = Topt. Rather for values of f(T), such that x < Topt, f′(T) is much greater, until the tangent becomes undefined at x = T = 0.

Regardless of this, we may still claim that this model predicts that the predator should not stay in a patch when it would be more profitable to travel to another one.

Emlen [1980] has outlined some further uses of ecological optimization models. For example, whether an animal "will be solitary or gregarious" [ibid., p. 129] is taken to be explicable by such models. There are a number of inescapable costs associated with group living including: (1) restriction of suitable habitat: (2) types of predation pressures; (3) dispersion pattern of resource base; (4) impact of disease and parasites and (5) the importance of "social information" to learning for adaptation to the environment. Emlen states:

> Group living is expected only under circumstances where advantages accrued through sociality more than offset these potential liabilities. [ibid., p. 129]

The same optimizational questions may be asked of phenomena such as territoriality and mating systems, and the answer would be given by a cost–benefit analysis: territoriality for example would be expected when (1) a resource in the area is worth defending and (2) when the associated costs of defence are low [ibid., p. 131]. A form of "behavioural scaling" would thus be predicted. It is believed by Emlen to be confirmed by studies of Hawaiian and African nectar feeding birds, who do not exhibit territorial behaviour when flowers are scarce and what nectar there is, is low, but at more abundant levels, benefits from territoriality exceed territorial costs. Territorial behaviour is energetically adaptive at such levels, but at even higher levels of abundance of flowers and nectar, territoriality may once again disappear since costs of territoriality once again exceed benefits [ibid., p. 33] [Gill and Wolf, 1975]. Energetic dependability is also taken to be of importance in understanding the evolution of mating systems. Ecological factors causally determine dispersion of mates and critical resources in space and time and phylogenetic factors causally determine the degree of emancipation from parental care. These factors interact to determine the degree of monopolization of mates.

Before surveying Emlen's comments on the applicability of such ecological considerations to human society, some consideration of the idea of this "environment" to which organisms must adapt must be given and our previous considerations in chapter 4 should be recalled. The ecologists H.G. Andrewartha and L.C. Birch [1954], Browning [1963], Maelzer [1965] and Andrewartha [1971], have been concerned to explicate the idea of the environment of a "typical animal". In Andrewartha's [1971] classification the environment has five components: (1) resources; (2) mates; (3) predators, pathogens, aggressors; (4) weather and (5) malentities. This position has been developed by B.S. Niven [1980, 1982] to encompass the interaction of the typical animal with these five components and to populations of animals. Without repeating her formalization here, we may say that the environment of an animal A at time t is defined as the union of the centrum of A at t and the web of A at t. The centrum corresponds to the "environment" as conceived by Andrewartha [1971]. The web is the set of Maelzer modifiers, these being objects in the environment linked to animal A only via some other object or objects, such as for example a measured amount of light-energy in a volume surrounding the photo-sensitive cockroach *Periplaneta americana.* It is assumed throughout this discussion of Ecological Sociobiology that some such account of the "environment" along these lines is accepted. The extension of Niven's formalization to the environment of a human being is a quite difficult matter, for we would need at the onset some intuitively clear account of the concept of a "typical human being".[3] Classification of the "environment" also cannot be made without the solution to a number of age-old philosophical problems. For example, some position must be taken on the ontological status of mathematical entities for example, for surely these are a part of the "environment" of mathematicians just as much as material objects. There are many conceptual difficulties which will arise in extending this position to complex human socio-cultural environments, and defining this latter concept is an open task not attempted by Ecological Sociobiology.

With this caution, let us review Emlen's remarks on the plausibility of Ecological Sociobiology with regards to both human behaviour and human society. These remarks are summarized in the following passage [Emlen, 1980, p. 137]:

> In species whose environments are highly variable or unpredictable, and in species that show tremendous cultural plasticity (including, but not exclusive to, humans) we expect this phylogenetic inertia [i.e. the more stable the long term environment, the more plausible it is to suppose the genetic determinism of behaviour] to be less pronounced and the potential for rapid behavioural change to be greater. But this does not mean that human societies are fundamentally different from all other species and are free from ecological constraints. Most human societies that retain a close connection to their environment are still faced with ecological "problems", and we still should expect a limited range of resulting "solutions". Knowing whether the social organization is arrived at through long-term genetic change by natural selection, through individual trial-and-error learning, or through the cultural transmission of optimal stategies of resource utilization is not the crux to understanding the potential importance of ecological Sociobiology. An organism that has the cultural flexibility to adapt its social organization to changing ecological pressures merely has the capacity

of arriving at a more optimal social organization more rapidly than one that is locked into the slower process of evolutionary change in gene frequencies dictating changes in its behaviour.

These theoretical propositions are illustrated by a consideration of the social structure and ecology of several groups of Basin-Plateau Indians. The Western Shoshone live in an environment which has as its main food resources grass seeds in the summer and piñon-pines in the winter, both of which are unreliable resources, neither worth defending, nor energetically dependable. Their social structure is taken by Emlen to be, as predicted by Ecological Sociobiology, nomadic (primarily consisting of family groups) and non-territorial. Contrast this with the social organization of the Paiute Indians who have more reliable and abundant grass seed resources, and who are settled primarily in permanent villages along rivers and who defend this land as territory.

Can this analysis be applied to modern industrial societies? Emlen is committed to saying that this is the case since Ecological Sociobiology is to be taken as an explanatorily comprehensive paradigm. It is at this point that we turn towards a critique of Ecological Sociobiology.

3.0 A critique of Ecological Sociobiology

Applying the reviewed models to modern industrial societies is recognized by Emlen as a problem, as such societies consist of elaborate "artificial environments". The difficulty humanity faces here is "our frequent inability to apply the principles of ecological sociobiology in attempting to understand the complexities of modern society" [ibid., p. 141]. Given this self-confessed limitation of Ecological Sociobiology it is more than a little difficult to understand Emlen's claims that from an increasingly precise ability to predict the fine structure of non-human/encultural animal social organization, that one can shape "social organization" by ecological manipulation [ibid., p. 141]. This "social organization" is, we must assume, human social organization, for if it is not then claim that "the possibility of such manipulations among human societies raises fundamental ethical questions" [ibid., p. 141]. But if this is so, Emlen contradicts himself: such fine structured predictions if not a matter of magic or good luck, require a theoretical basis which his former statements necessarily exclude.

The difficulties in even defining the socio-cultural environment of encultural beings are enormous as was noticed earlier. The development of Ecological Sociobiological models from optimization theory to account for the basic form of contemporary monopoly capitalism is a task yet to be done. For example, whilst there is at least a strong intuitive case to be made that the Basin-Plateau Indians are engaging in foraging strategies, it is difficult to apply the analysis to techno-agricultural production of wheat crops. For one thing, with computerized technology, the assumption common to all foraging models, that the efficiency with which various activities of obtaining food cannot vary [Maynard Smith, 1978 (1) p. 33] is false. Indeed, this assumption is necessarily false because the very point of technology is to obtain this very efficiency. Therefore all ecological optimization models accepting this assumption (and from Maynard Smith's survey, this includes virtually all currently developed models) are inapplicable. Such models assume the universal validity of the ecological determinism of neo-Darwinism. Organisms are no more than passive adaptors to a constraining environment. This image is not substantially altered by admitting that organisms may adopt a number of "solutions" to environmental "problems" for the image of the organism in its environment is still one of passive adaptation whatever "solution" is adopted, to a constraining environment. Only if the recognition was made that organisms may substantially alter their environment would the charge of ecological determinism be unfounded. But in the area where such interaction is most evident, the formal conditions for the applicability of general ecological optimization models is not satisfied.

This is not to say that human beings and other encultural beings are not influenced by their ecological environment. To say that is to speak an obvious falsehood. The Ecological Sociobiologist cannot be permitted to support their position by such arguments. To reject the claims made that a consideration of ecological variables is sufficient for the prediction and explanation of the "basic elements" of human societies, is not to proclaim the impossibility of human ecology as a science any more than a rejection of neo-Darwinism involves a rejection of biology.

Thus we may accept that human societies are constrained by ecological variables to wide and varying degrees. The case of the Western Shoshone Indians is hardly a typical view of the relationship between human beings and their ecological environment. Nevertheless, even if we can predict that such a society will be nomadic by a fairly uncontroversial style of argument (e.g. if they were not nomadic, given their climate and the difficulty in growing crops they would die) we fall far short of obtaining the understanding and prediction of the basic forms of organization and structure of Western Shoshone Society. We may be able to predict that the nomadic groups will be small, but it does not necessarily follow that these should be family groups, purely from ecological data alone. Nor for that matter do we find ourselves able to predict much of interest about the culture of the Western Shoshone: they obviously don't put space stations into orbit, but how in fact are their groups governed, what is their view of the cosmos and so on? These problems are compounded for technological, "rationalistic", science-orientated societies. But I need not labour the obvious in pointing out that Emlen fails to predict anything of interest about these societies: as I have argued above, he seems to have confessed this himself. But the extensive quotes made from the paper indicate that despite culture, *and indeed because of culture*, human Ecological Sociobiology remains a powerless explanatory paradigm. Therefore Emlen's paper is internally inconsistent and his major theoretical claims unsupported.

We recall that Emlen has claimed ((ES_2)) that forms of social organization, including human organization will converge upon that which is most energy efficient and predatation secure. To simplify this question let us restrict the issues to the question of energy efficiency. Now if this claim is to be meaningful, energy efficiency must be measured independently of relative considerations. That is to say, if we argued that capitalist societies are actually, under their specific "environment", the most energy efficient form of social organization, we are involved in painfully circular reasoning: capitalist social environments are necessarily as efficient as capitalist social environments. An independent characterization might be given by thermodynamic considerations, considering how much entropy is created through misuse, of how much available food, chemical and mechanical energy is wasted through mismanagement. From this perspective it is evident that capitalist social orders converge upon maximal wastage rather than to efficient use of energy and environmental management [Wheelwright and Stilwell, 1976, p. 177–228]. Electrical energy is spent on bill boards for advertising goods such as cigarettes, which may in fact lower one's inclusive fitness by riddling one's body with cancer. Our cars are quite inefficient, and it is widely recognized that much unburnt fuel is emitted in the exhaust fumes of the petrol driven internal combustion engine. True, this may be a technological limitation given mass production techniques. It is sufficient however to point out that capitalist societies involved energy wastages in many facets of life to refute Emlen's thesis.

Yet is Emlen's thesis of energy efficiency even true for the Basin-Plateau Indians? Now the Western Shoshone Indians, given that they live in the previously described environment, eating the grass seeds, and do not choose to migrate elsewhere have no choice at all but to live in nomadic groups given various laws of physics and biochemistry. Suppose that they did not live in nomadic groups. Then given shortages of food either they immediately starve or a single party or many groups must travel further and further each day until (1) either travel distances are so great that the village starves or (2) the village does not starve, but so many groups are now travelling that we might as well classify the society as nomadic anyway. It is not informative to state that this is the most energy efficient strategy for the Western Shoshone to adopt: it seems to be the only one which could be adopted granted all of Emlen's assumptions. The Paiute have no reasons to adopt the Western Shoshone form of social organization, but given their ecological base, since the Western Shoshone survive, the Paiute could *ceteris paribus* live a nomadic life of abundance. In neither case is it evident that these social orders are approximations to Emlen's energy efficient Nirvana in any thermodynamic sense. This would require empirical research which has not been done. Further from an *a priori* perspective it is evident that neither society is a maximally efficient energy user, for it is easy to construct a *Gedanken Experimente* where at least some energy inefficient activity is eliminated or rationalized.

So much then for our considerations of human Ecological Sociobiology. Let us ask: does Ecological Sociobiology provide us with a set of evolutionary principles which enables us to explain the diversity of life? This was the requirement which Maynard Smith [1978, (1) p. 52] set down for this position, rather than the attempt to demonstrate that organisms optimize. The latter assumption does not stand open to non-circular justification within the framework of optimization theory, but rather is an *a priori* metaphysical hypothesis. As Maynard Smith recognizes in response to such criticisms from Lewontin [ibid., p. 39] the optimality assumption does not come under test, but rather a specific set of auxiliary hypotheses of a particular model. Maynard Smith's position is that every existing ecological optimization model has been falsified as a *complete* explanation of all the available data; my position is that the explanations which we have reviewed do indeed fail to provide satisfactory explanations of evolution and the diversity of life. If one claims that this is unimportant, then given the above argument against the claim that we are attempting to demonstrate that organisms optimize, it becomes totally unclear what precisely one is attempting to do with optimization models.

Whether or not blue gill sunfish *Lepomis macrochirus* in actual fact feed on *Daphnia* in accordance with the predictions of the optimal feeding model of Werner and Hall [1974] is an empirical matter. How in fact such optimal feeding strategies come to be adopted is another matter. As Krebs [1978, p. 23] puts it, talk about "stategies" does not involve reference to mentality: it only means that "an animal is designed to follow certain rules" [ibid., p. 23]. Clearly this designing is done by natural selection. Krebs takes natural selection to favour individuals with the highest fitness. He identified optimizers with such individuals, and that natural selection will favour optimizers follows as a matter of deductive logic. In this we do not however have a mechanism for the generation of such optimizers. True, if they exist natural selection will favour them. We are however attempting to explain, as Maynard Smith stated, how in fact such phenomena evolved. The optimization account begins with the acting organism in its environment, and does not tell us how the former arose.

Consider Emlen's example of the Ecological Sociobiological account of the causal conditions for an animal to be either social or solitary. As this is described it seems much like a straightforward empirical question of balancing up the costs of being social, and the costs of being solitary. An animal is, at a period of time either social or solitary, just as an animal is either of a sexually reproducing species or an asexually reproducing species. Emlen's question about sociality vs. solitary behaviour is a precise analogue of Williams' question [1980]:

> My analysis is based on the assumption that the important evolutionary question is: What would be gained by an individual, in an otherwise sexual population, who cheated by eliminating meiosis and fertilization from its production of an offspring, but remained otherwise the same?" [ibid., p. 372]

Williams is asking a question which is nonsense from the perspective of developmental biology, and logic. What does it mean to say that such an individual remained "otherwise the same?" Surely such cheating must involve physiological changes of a quite radical nature – especially if the organism was a male. Further unless we believe in magic, some biological generative mechanism must be postulated for this transmutation, regardless of how profitable cheating actually is. Emlen argues that if the benefits of sociality are greater than the costs of sociality, an animal will be social rather than solitary. But an animal is necessarily *either* social or solitary, and for us to even talk about the relevant costs and benefits we need to know what the animal in fact is. But if we know from the outset what the animal in fact is, either social or solitary, then Emlen's cost–benefit analysis is useless as an explanatory and predictive device as the method is intrinsically circular.

To compare cost and benefits of some organism x being either social or solitary, we need to know the respective costs and benefits of that organism x being either social or solitary. But an organism is either social or solitary, it doesn't stand in an economic limbo outside of evolutionary time making up its mind which is the best bet to take. Hence it becomes logically impossible to ascertain the opposing costs and benefits, for to strip an organism of its sociality is to significantly alter the physiological and behavioural properties of the organism, so that the organism loses its self-identity. Questions such as "Why are orang-utangs the only solitary primates?" are much like the question: "Why is it that hydrogen has the atomic composition that it has?" The answer is that elements with just that atomic composition are classified as hydrogen according to current theoretical practices, and if "hydrogen" in some logically possible world had the atomic composition of "helium", then we would simply switch labels in our communication. There is in short nothing here to explain, only to justify [Causey, 1977]. Consequently Ecological Sociobiology, despite *prima facie* empirical credibility, has set itself some impossible questions to answer.

Am I claiming that social behaviour is not open to evolutionary examination? Most definitely not. Rather by the above *a priori* arguments, it is maintained that no mere cost–benefit argument as outlined by Emlen can succeed in answering his posed questions. This leaves Ecological Sociobiology with only the traditional neo-Darwinist mechanisms to account for the evolution and diversity of life. These, it has been already argued do not succeed. We have no more to say; we turn to our second set task.

4.0 Harris' cultural materialism: An exposition

Harris [1979, ch. 5] remains a strong opponent of a form of Sociobiology which seeks to explain human social life by means of the theoretical principles of neo-Darwinist evolutionary biology. Harris maintains that all his own programme of *Cultural Materialism* shares with Sociobiology is a commitment to "the general epistemological principles of science" [ibid., p. 119] but Sociobiology as defined is "capable of explaining only an insignificant proportion of human Sociocultural differences and similarities" [ibid., p. 119]. Harris' argument is simply that genotypes never account for all the variations in behavioural and cultural phenotypes [ibid., p. 121]. It is an argument almost universally accepted by social theorists critical of Sociobiology. Yet Harris' *Cultural Materialism* is, I will argue, quite consistent with the form of Sociobiology which Emlen wishes to construct: an Ecological Sociobiology, and in the human sphere at least, may be taken to be a philosophically sophisticated reconstruction of Emlen's basic insights.

What is Cultural Materialism? For an initial orientation let Harris speak in his own words:

> In its commitment to the rules of scientific method, cultural materialism opposes strategies that deny the legitimacy or the feasibility of scientific accounts of human behaviour – for example, humanist claims that there is no determinism in human affairs; and it opposes the currently popular attribution of the malaise of industrial society to too much rather than too little science. Cultural materialism, with its emphasis upon the encounter between womb and belly and earth and water, also opposes numerous strategies that set forth from words, ideas, high moral values, and aesthetic and religious beliefs to understanding the everyday events of ordinary human life. Aligned in this regard with the teachings of Karl Marx, cultural materialism nonetheless stands apart from the Marx-Engels-Lenin strategy of dialectical materialism. Condemned by dialectical materialists as "vulgar materialists" or "mechanical materialists", cultural materialists seek to improve Marx's original strategy by dropping the Hegelian notion that all systems evolve through a dialectic of contradictory negations and by adding reproductive pressure and ecological variables to the conjunction of material conditions studied by Marxist-Leninists". [ibid., p. IX]

The result of doing this is to achieve a "science of culture" which has not been achieved by alternative research strategies such as "cognitive idealism" and "dialectical materialism". Harris I believe, has indeed made some telling criticisms of such strategies, but it is my aim here to undermine Harris' own "strategy".

A theoretical thread running throughout Harris' main work *Cultural Materialism: The Struggle for a Science of Culture,* is that approaches other than Cultural Materialism are not merely problematic, but are methodologically and metascientifically *unscientific*. Harris is not content to opt for a broad characterization of "Science", equating it with simply "Knowlege" (by contrast to false or unjustified belief) but seeks to construct a position whose "epistemological assumptions are rooted in the philosophical traditions of David Hume and the British Empiricists" [ibid., p. XI], the chief task of which "is to create a pan-human science of society whose findings can be accepted on logical and evidential grounds by the pan-human community" [ibid., p. XII]. Thus the definition of "Science" and the definition of "Research Strategies" are important parts in the epistemology of Cultural Materialism [ibid., p. 5] and the basic philosophical point of Harris' programme would be lost if difficulties of an insurmountable nature existed here.

Recognizing that there is a distinctive thing called "Science", which is distinct from other epistemic enterprises [ibid., p. 27] commits Harris to supply criteria of demarcation of Science from non-Science, and criteria of distinguishing one "Research Strategy" from another. These as all contemporary philosophers of science know, are difficult questions. Harris' response after summarizing material, better summarized elsewhere [Suppe (ed.), 1977], gives us a two and a half page discussion of these issues. Still, this might be a very profound discussion, so let us examine this material.

By a "Research Stategy" Harris means "an explicit set of guidelines pertaining to the epistemological status of the variables to be studied, the kinds of lawful relationships or principles that such variables probably exhibit, and the growing corpus of interrelated theories to which the stategy has thus far given rise" [ibid., p. 26]. Cultural Materialism shares with the rest of "Science" an "epistemology" which "seeks to restrict fields of inquiry to events, entities, and relationships that are knowable by means of explicit, logico-empirical, inductive-deductive, quantifiable public procedures or "operations" subject to replication by independent observers" [ibid., p. 27]. Depending upon whether Harris wishes to follow the empiricism to which he pays lip service, *Metaphysics* and *Epistemology* could probably be squeezed into this scheme: they had better be, because one refreshing insight which permeates Harris' work is that metaphysical and epistemological issues are of crucial importance to empirical research, influencing the choice of problems, the method of inquiry and so on. If this was not true, then much of Harris' book would consist of nothing but irrelevant discussions. That Harris contrasts "Science" with Occultist Religion and the musings of "aggressive fanatics and messiahs", confirms the view that "Science" now is being equated with the domain of "Knowledge".

If this is the case a problem for Harris now arises. On the one hand he wishes to produce an account of "Science" which sees it as "a unique and precious contribution of Western Civilization," [ibid., p. 27] without denying that other civilizations have contributed to both scientific knowledge and knowledge in general. But Harris' account of a "Science" and a "Research Strategy" is far too broad to precisely supply the elixir for the demarcation criterion, once lusted after by empiricist philosophy of science. Natural-Rationalistic Theology involving the use of philosophical arguments and empirical considerations qualifies, although a Religious *Weltanschauung* appealing to revelation, will not. But perhaps Harris could grasp the counter-argument strongly by its horns, and digging his feet into the ground accept this implication. If this is done, does

Harris thereby produce a satisfactory metascientific foundation for Cultural Materialism? I think not.

The difficulty is that given Harris' cited account of scientificity he fails to supply a satisfactory demarcation criterion for the individuation of Research Strategies. The set of "guides" to the epistemological status of the variables of particular theories and the interrelationship between them, is insufficient to distinguish between any "Scientific" theories, or more generally "paradigms" (taken in the intuitive yet murky sense in which that term is constantly used). Both Cultural Materialism and dialectical materialism may produce theories which are restricted in scope to events, entities and relationships, which are knowable by means of "explicit, logico-empirical, inductive-deductive, quantifiable public procedures" [ibid., p. 27]. Yet, as cited earlier, Harris wishes to demarcate his own position from alternative strategies, for he accepts Laudan's claim that all evaluations of research traditions/strategies must be made within a *comparative context* [Laudan, 1977, p. 120]. Therefore most of the "basic ingredients" in the epistemology of Cultural Materialism are missing.

So much then for general epistemological considerations. In the remaining part of this section I will attempt to outline Harris' position in specific detail suspending criticism until section 5. The Cultural Materialist position may be summarized by a consideration of the following material:

(1) The Mental/Behaviour Distinction
(2) The Etic/Emic Distinction
(3) The Principle of Emic Mystification
(4) The Principle of Infra-structural Determinism and the Methodological Priority of Infra-structural Variables.

(1) *The mental/behaviour distinction*

We must clearly distinguish between "what people say and think as subjects and what they say and think and do as objects of scienfific inquiry" [ibid., p. 29]. Thus any adequate science of social life must distinguish between mental and behavioural fields. The mental field Harris takes to be "all the thoughts and feelings that we human beings experience within our minds" [ibid., p. 31]. The behavioural field consists of "all the body motions and environmental effects produced by such motions, large and small of all the human beings who have ever lived" [ibid., p. 31]. To describe such "motions" adequately, it is not necessary to refer to "what is going on inside people's heads" [ibid., p. 31], that is, to mental episodes. Cultural Materialism is committed to *explanatory behaviourism,* that there are non-trivial explanations of socio-cultural phenomena, making no reference to mental variables at all.

I am not going to discuss in detail the reasonableness of Harris' mental/behaviour distinction, beyond pointing out that Harris' acceptance of the distinction stands in conflict with a general consensus existing within the philosophy of action: that behaviourism is a moribund doctrine, at least in the simple minded form in which it is stated by Harris. Harris maintains that behaviour, that is bodily motions and effects, can be described independently of the mental states of people. Now this is true insofar as *human actions* are equated with *bodily events*. However no mere description of my various bodily events serves to provide action descriptions. There is a qualitative difference, Wittgenstein [1953] pointed out, between my hand "going up" and when I "lift my hand". Not all cases of the behaviour of hands "going up" are actions. Actions would seem to be intimately related conceptually to mental events, although it remains an outstanding problem to specify precisely what this relationship is. Consequently Harris' distinction between mind and behaviour cannot be as sharp as he would wish. It is indeed possible that Harris could rebut this argument by denying that action categories exist at all. If so, the above argument would beg the question against this radical move. But such a counter-intuitive denial would only be justified if Harris' research programme was philosophically tenable, and it is the negation of this proposition which I believe to be demonstrably true.

(2) *The etic/emic distinction*

The etic/emic distinction is an epistemological distinction concerned with the criterion of the truth of descriptions made by anthropological observers. Harris has not given a formally precise explication of the distinction, but one has recently been outlined by Oakes [1981, p. 2].[4] Let "Fa" be a sentence describing the property of some particular in society S at time t, such that "a" designates the particular p and "F" designates the property k. Then "Fa" is an emic description if and only if there is a criterion of truth C used by members of S at t, such that "Fa" is true if and only if the ascription of k to p satisfies C. "Fa" is an etic description if and only if there is a criterion of truth C* of the social sciences, such that "Fa" is true if and only if the ascription of k to p satisfies C* but need not necessarily satisfy C. An alternative explication is this: let "L" and "N" be propositional variables, and let the symbol "True_D" mean "true for the assigned elements of the domain of the respective propositional variable". Let C and C* be interpreted as above. Then the propositions in the domain of "L" are emic propositions if and only if:

$$(1) \quad (\forall L)(C(L) \leftrightarrow \text{True}_D(L))$$

and the propositions in the domain of "N" are etic if and only if:

$$(2) \quad (\forall N)(C^*(N) \leftrightarrow \text{True}_D(N)).$$

Informally, the hallmark of emic operations is their ability to explicate the natives' viewpoint, and the test of the adequacy of such statements is the ability of such operations to generate statements which the natives themselves accept as true. The hallmark of etic operations is their ability to produce statements which are scientifically rational and testable, frequently using categories unknown and unintelligible to the natives under examination. Thus an emic explanation of some sociocultural phenomena is an explanation, the explanans of which contains statements involving essentially emic variables, that is descriptions of the natives' beliefs, reasons, meanings, values or theoretization, whereas the explanans of an etic explanation does not involve such reference. However it does not necessarily follow from this that all etic explanations are "nonintentional" as Oakes [1981, p. 5] takes Harris to mean. Rather etic explanations need not make reference to the intentionality of their objects to be satisfactory from a scientific point of view. Behavioural-etic explanations do however eschew all reference to mentalistic variables in the explanans.

It is very important to note that etic $\neq$ objectivity and emic $\neq$ subjectivity. To restrict a sociocultural science primarily to emic considerations is not to embrace relativism or scientific irrationality. It is worth quoting Harris to substantiate this point, since Harris has a definite tendency to forget his own warning. Consider first Harris' statement [Harris, 1979, p. 45]:

> Partisans of idealist strategies seek to subvert the materialist effort by claiming that "all knowledge is ultimately 'emic'" . . . . The allegation is that in the name of demystifying the nature of social life, the observers merely substitute one brand of illusion for another. After all, who are the "observers"? Why should their categories and beliefs be more credible than those of the actors? The answer to these questions is entirely dependent on whether one accepts the scientific way of knowing as having some special advantages over other ways of knowing. To deny the validity of etic descriptions is in effect to deny the possibility of a social science capable of explaining sociocultural similarities and differences. To urge that the etics of scientific observer is merely one among an infinity of other emics – the emics of Americans and Chinese, of women and men, of blacks and Puerto Ricans, of Jews and Hindus, of rich and poor, and of young and old – is to urge the surrender of our intellects to the supreme mystification of total relativism.

This passage equates emic views with subjectivism and relativism: if Harris' Cultural Materialism is ultimately one emic view among others, then "total relativism" is, he believes, inescapable. Yet we should contrast this position with his earlier remarks:

> To be objective is not to adopt an etic view; nor is it subjective to adopt an emic view. To be objective is to adopt the epistemological criteria discussed in the previous chapter by which science is demarcated from other ways of knowing. It is clearly possible to be objective – i.e. scientific – about either emic or etic phenomena. Similarly, it is equally possible to be subjective about either emic or etic phenomena. [ibid., p. 35]

In this passage the above equation is denied. Since no argument is given by Harris that emics = relativism, it is more reasonable to believe the above position.

(3) *The principle of emic mystification*[5]

Despite the above interpretation, Harris maintains that emic explanations are theoretically flawed, as the native's conception of his/her world represents "a profoundly mystified consciousness" [Harris, 1974, p. 5]. The native's consciousness" is generally ignorant of alternative social orders, blinded by fears of the unpleasant facets of the human condition, and furthermore the social inequalities and the corruption of the native's society generally escapes notice due to the effects of social conditioning. Consequently the explanatory power of emic categories is low, if not zero [ibid., p. 6]. The theoretical significance of this principle to Harris' programme will be discussed in section 5.

(4) *The principle of infra-structural determinism and the methodological priority of infra-structural variables*

The next and most important piece of material which needs to be summarized is Harris' cultural ontology. Human societies for Harris consist of an *infra-structure*, *structure* and *super-structure*. The infra-structure consists of first the *mode of production*, which is made up of the "technology of subsistence", "techno-environmental relationships", "ecosystems" and "work patterns". In addition it consists of the *mode of reproduction*, which in turn consists of the "technology and practices employed for expanding, limiting and maintaining population size" [Harris, 1979, p. 52]. This would include "demography", "mating patterns", "fertility", "natality", "morality", "nurturance of infants", "medical control of demographic patterns", "contraception", "abortion" and "infanticide".

The *structure* of a society comprises the *domestic economy* – consisting of "family structure", "domestic division of labour", "domestic socialization", "enculturation", "education", "age and sex roles", "domestic discipline", "hierarchies" and "sanctions"; the *political economy* consists of "political organization", "division of labour", "political socialization", "class", "caste", "police/military control" and "war" [ibid., pp. 52–53]. The domestic economy is concerned with the organization of reproduction and basic production within domestic settings such as camps and families. The political economy is concerned with the organization of reproduction and production within larger social groups such as tribes, cities and empires.

The *super-structure* consists of ideational phenomena, such as art, rituals, literature, science and Cultural Materialist theories of society. Corresponding to the etic infra-structure, etic structure and etic super-structure are:

(1) an *emic infra-structure* consisting of "ethnobotany", "ethno-zoology", "subsistence lore", "magic", "religion" and "taboos";
(2) an *emic structure* consisting of "kinship", "political ideology", "ethnic and national ideologies", "magic", "religion" and "taboos" and
(3) an *emic super-structure* consisting of "symbols", "myths", "aesthetic

> standards and philosophies", "epistemologies", "ideologies", "magic", "religion" and "taboos" (yet again!)[ibid., pp. 54–55]

Harris makes a simplifying assumption of lumping together the mental and emic components [ibid., p. 54].

The central tenet of Cultural Materialism is the principle of *infra-structural determinism*. Harris states this principle as follows:

> The etic behavioural modes of production and reproduction probabilistically determine the etic behavioural domestic and political economy, which in turn probabilistically determine the behavioural and mental etic super-structures. [ibid., p. 56]

Harris is silent on the relationships between first the etic structure and etic super-structure and the behavioural and mental emic super-structure. Harris accepts a principle of the methodological priority of infra-structural variables: only after infra-structural variables have been excluded may one consider other variables. He also takes this principle to imply that there is a methodological priority of etic and behavioural variables over emic variables. It is relatively easy now to derive a contradiction from this: would the etic super-structure have methodological priority over the emic infra-structure? From the principle of the methodological priority of etic variables, we would say "yes", but from the principle of the methodological priority of infra-structural variables, we would say "no". We thus have derived a contradiction.

Why should a methodological priority be given to etic and infra-structural variables? The former question is already answered by Harris' principle of emic mystification. The latter question is answered by Harris' view of human nature. The methodological priority of infra-structural variables rests upon the existence of two "iron laws":

(1) *The Eco-Thermodynamic Law*: "human beings must expend energy to obtain energy" [ibid., p. 56]

(2) *The Malthusian Principle*: "like all bioforms, our ability to produce children is greater than our ability to obtain energy for them" [ibid., p. 56]

Harris has an argument to support these claims. Once more, for demolition purposes we let him speak in his own words:

> Infra-structure, in other words, is the principle interface between culture and nature, the boundary across which the ecological, chemical, and physical restraints to which human action is subject interact with the principle sociocultural practices aimed at overcoming or modifying those restraints. The order of cultural materialist priorities from infra-structure to the remaining behavioural components and finally to the mental super-structure reflects the increasing remoteness of these components from the cultural nature interface. Since the aim of cultural materialism, in keeping with the orientation of science in general, is the discovery of the maximum amount of order in its field of inquiry, priority for theory building

logically settles upon those sectors under the greatest direct restraints from the givens of nature. [ibid., p. 57]

This completes our exposition of Cultural Materialism. I am not concerned with discussing case studies of the applications of Cultural Materialist principles, since Harris' programme is demonstrably incoherent. We turn now to that demonstration.

5.0 The case against Cultural Materialism

We attempt to systematically undermine the position of Cultural Materialism, beginning with the etic/emic distinction.

(1) *Against the etic/emic distinction*

Oakes [1981, pp. 9–11] has argued that Harris violates his own methodological tenet in supplying emic solutions to etic problems. This is illustrated by many examples, but one of the most impressive is Harris' explanation of Aztec Cannibalism [Harris, 1977]. My interest here is not, however, in how consistent Harris is in following his own metatheoretical principles, but whether in fact these principles are valid. My primary interest here is whether Harris' etic/emic distinction is coherent.

Critics have attacked the etic/emic distinction on the grounds that all knowledge is ultimately emic, as Harris' etic category is the emic category of Western science, and have concluded that the distinction is both misleading and unnecessary [Oakes, 1981, pp. 15–18] [Fisher and Werner, 1978, p. 198]. Critics have argued that Harris' use of this distinction is primarily to justify the methodological principle that etic statements can be controverted by emic statements, without etic statements being falsified, a principle which also gains support from the principle of emic mystification. We will consider this latter principle shortly. Our immediate concern is to consider the implications of the claim that all knowledge is ultimately emic.

From a philosophical point of view, the epistemological use to which Harris puts this distinction is not justified. Yet from a semantic point of view the distinction may still be drawn: etic descriptions consist entirely of terms which are the observers', and etic explanations involve essential reference to at least one etic category; emic descriptions consist entirely of terms which are the natives', and emic explanations involve essential reference to at least one emic category. On this account etic and emic explanations are not opposed, and indeed an explanation with one etic category and one emic category, is both an etic and an emic explanation. Most explanations in anthropology are both etic and emic, as abandoning one's own culture *completely* and going "completely native", is not only difficult, but would produce an incomprehensible journal article! On this account, the etic/emic distinction has no epistemological significance, and whilst perhaps of use in some theoretical matters, cannot have the epistemological import which Harris' programme requires.

I have spoken already at length about the relationship between emics and relativism. Oakes [1981] whilst rejecting Harris' position seems committed to such relativism. Following Weber he argues that (1) metatheoretical values of social inquirers influence both the fields of inquiry and definition of problems. I accept this thesis. He then asserts: (2) "suppose that the domain of axiology exhibits an ultimate and irreducible polytheism" [ibid., p. 16] such that this sphere is "irrational: a sphere of permanent contradiction" [ibid., p. 16] then: (3) "The diversity and variability of the metatheoretical frameworks of the sociocultural sciences, *therefore*, are consequences of the diversity and variability of the values to which they are committed" [ibid., p. 17, emphasis added]. Consequently: (4) "there are no culturally universal and historically invariable cognitive norms which occupy the status of *a priori* or transcendental presuppositions of all sociocultural research [ibid., p. 17].

One would have thought that the conclusion of this argument would have been a thorough-going irrationalism, given that the sphere of axiology is riddled with "permanent contradiction". But Oakes argument is open to extensive critique in other respects: (3) is not deducible from {(1), (2)} as there could be a diversity of sociocultural research programmes, *even if* the domain of axiology was rational, namely by a proliferation of inadequate orientations or because such a proliferation was financially profitable. Therefore it does not follow that the proliferation of research programmes is necessarily due to metatheoretical value differences. The real weakness in the argument is (2) which is merely an assertion: "suppose that . . ." and offers no reason for acceptance. But there is some reason for rejecting it: Oakes and Weber themselves accept certain metatheoretical values which are so central to the process of arguing and explaining that talk of their rejection is nonsense. For example, one such value is that demonstrably inadequate positions such as Cultural Materialism should be rejected – otherwise why bother criticizing Harris at all?

It is concluded, that in rejecting the theoretical significance of the etic/emic distinction as proposed by Harris, one is not necessarily committed to accepting relativism. Both Harris and Oakes are wrong in this regard.

(2) *Against the principle of emic mystification*

We note immediately that if Harris' etic/emic distinction is rejected, then the point of the principle of emic mystification is also lost, at least as Harris wishes to use this principle unless he wishes to accept scepticism. This is just as well, as it saves Harris from the expense of further internal inconsistency: between the principles of emic mystification and the methodological priority of infrastructural variables. The former condition entails that emic explanations are never reasonable, the latter, that there is at least one condition where they are legitimate [Oakes, 1981, p. 6].[6]

Harris is right in warning us not to accept as the sole truth the native's account, even if we are reasonably sure he/she is not lying. The native may well be wrong about social reality. Yet on the other hand the native may well be right, and the social scientist wrong. Who is right and who is wrong is not a question which can be, as far as I understand it, decided by conceptual analysis and *a priori* argument alone and is utterly contingent upon the questions asked. Therefore the principle of emic mystification must be wrong.

(3) *Against the principle of infra-structural determinism and the methodological priority of infra-structural variables*

Oakes [1981, p. 14] has argued that Harris' principle of infra-structural determinism is epistemologically or self-referentially inconsistent. The argument is an old and fallacious one, namely that in accepting this principle as reasonable we are therefore deprived of reasons for accepting it, since one accepts the principle because of etic causal processes rather than because of emic reasons, therefore the principle cannot be rationally accepted. The well known rejoinder to this argument, that one accepts the principle of infra-structural determinism because of arguments, but that this whole process is in turn causally determined by infra-structural variables is not considered by Oakes. I have argued elsewhere that despite the importance of self-referential arguments in philosophy, those advanced against determinism fail [Smith, 1983 (5)].

The problems with the infra-structure/structure/super-structure distinction are conceptual in name: how do we in fact reasonably draw these distinctions? We may illustrate the problems which arise here by briefly considering a set of well known problems facing "economic determinist" versions of Marxism. Such versions propose that a society (all societies?) have an economic base which bears a determinate causal relationship to a political and cultural super-structure. A sharp distinction is usually drawn between *forces of production* and *relations of production*. At various points in his writings Marx lumps into the category of forces of production, science, trade and population, taking the relations of production to be the legal expression of property relations. But this is hopeless, since relations of production are alleged to also determine legal relations on this model, so circularity arises. Productive relations are intimately related to legal relations which are concerned to enforce rights of property. Therefore legal relations are, contrary to Marx, part of the relations of production. A problem exists in excluding many relations which seem to fall into the category of the relations of production, collapsing the base/super-structure distinction. Poulantzas [1973] and Hindess and Hirst [1975] have lumped all these items into an all-embracing category called the "mode of production", but this strategy as Parkin [1979, pp. 7–8] has cogently argued, robs the concept of its intended explanatory power: the mode of production simply becomes a metaphor for the social structure of society itself – if not society itself.

There are admittedly many ways in which economistic Marxists may escape this conceptual maze. One favourite strategy is illustrated by a paper of J. McMurty [1973], which defines the forces of production as anything which can make a use value, the relations of production as power relations of expropriation between human agents and productive forces, whilst property relations are based upon rights of legal title. But it seems on this account that difficulties arise in keeping the ideological super-structure from collapsing into the base. Abstract entities such as scientific theories, computer programmes and mathematical entities, now fall into the forces of production category. Perhaps one could accept this, as after all, Marx himself recognized in volume one of *Das Kapital* that human production is distinguishable from the production of the bee by its theoretical aspect. Yet to do so, is to seriously weaken the thesis of materialism and to depart very significantly from the past mechanistic models of base-superstructure.

Now such a problem as discussed above will plague Harris' materialism, since many phenomena of the infra-structure such as technology, *conceptually presuppose* phenomena of the super-structure such as science. In this case, there cannot be a technology without scientific knowledge since technology is "applied science" and "the science of the study of the practical or industrial arts". Further, there is an irreducible intentional aspect to technological innovations: the goal of technology is to change the world in accordance to the intentions, aims and plans of the innovator. Ironically the term "technology" appears within the definiens of the terms "mode of production" and "mode of reproduction". Therefore Harris' scheme faces circularity as well.

One may be unimpressed by these *a priori* and *conceptual* considerations and maintain that Cultural Materialism does not view human agents as mindless automata [Harris, 1979, p. 59]. Rather the primacy of the infra-structure means that thoughts "cannot assume a material social existence unless appropriate material conditions for their social acceptance and use are also present" [ibid., p. 59]. The recurrence of inventions, indicates, Harris believes, "that when the infrastructural conditions are ripe, the appropriate thoughts will occur, not once but again and again" [ibid., p. 59].

Now even if this was true, it is not a solution to the problem of definition raised above; only a redefinition could dissolve this problem if the argument was sound. Yet is Harris' point correct? The conclusion is far from "inescapable" as Harris believes it is, for even "cognitive idealism" can explain the recurrence of inventions easily: human beings are not "mindless automata" and thinking along the same lines in the same domain of discovery is bound to produce multiple discovery of inventions. This point can be developed further and directed against Harris' justification of the principle of the methodological priority of infrastructural variables. Even if we accept both the eco-thermodynamic law and the

Malthusian principle, acceptance of the methodological priority of infra-structural variables is not placed upon one as a logical commitment. Human beings also obey, we suppose, the laws of quantum mechanics, but we have yet to have any interesting quantum mechanical explanations of human sociocultural phenomena such as the Russian revolution or the Cuban missile crisis. Nor have we had any purely etic explanations (as defined by Harris) of any human sociocultural phenomena. The Malthusian principle, whilst true, fails to explain even the "population explosion". The mere *ability* to produce more children than our *capacity* to obtain energy for them, does not explain why these abilities and capacities are operating. Indeed, even if there were zero population growth in the world with totally abundant food, the Malthusian principle would still be true. It is a basic existential parameter facing potentially geometrically increasing rates of organism production with an ultimately finite universe and is far too general to act as a theoretical foundation for the sociocultural sciences.

Harris' concern with the practical elements of life is a worthwhile concern. But he wishes to maintain, as we saw in the above paragraph, that theoretical and ideational phenomena are ultimately determined by the infra-structure. Now rather than supply counter-examples to this claim, it is best to point out where the fallacy of reasoning lies. The mere fact that thoughts require a material social existence, only indicates that such conditions are *necessary conditions* of one's theoretical activity. Necessary conditions are not causes, and it is such causes which Harris requires to construct a sound argument. (It is a necessary condition for me to write this, that I exist, but that I exist, is not the cause of my writing.) Thus Harris is right to note that the eco-thermodynamic law and the Malthusian principle are true. This does not supply the slightest support for the claim that the actual phenomena which he himself outlines as belonging to the infra-structure such as "work-patterns", "medical control of demographic patterns" and "contraception" are not in themselves dependent upon super-structural phenomena such as theories of contraception, work-arrangements and so on, even though such abstract entities must be "socially embodied". It has not been explained why this claim conflicts with either the eco-thermodynamic law or the Malthusian principle.[7] Harris therefore has no good argument in support of the principle of the methodological priority of infra-structural variables, a principle already argued to be incoherent.

If Harris' Cultural Materialism is a cogent scientific enterprise, we need some idea of the general generative mechanism by which infra-structure determines structure, which in turn determines super-structure. But what is this relation of "determination"? Harris has told us that it is "probabilistic", but he says nothing else, and this phrase is as vague as the economic determinists' "determination in the last instance". To start with, I take it Harris is *not* proposing that the phrase "probabilistically determines" just means "causes" consistent with say Suppes' [1970] probabilistic theory of causality. Not only is Suppes' position open to counter-examples [Otte, 1981], but we are left with the same problem only this time formulated in causal-probabilistic terms. Here we must say something like this:

(PD) Infra-structure causes structure = df. the conditional probability of that structure existing given the infra-structure is greater than the absolute probability of that structure coming into being (on its own).

On Suppes' analysis "a causes b" = df $Pr(b/a) > p(b)$ and this is nonsense for the Cultural Materialist as structures do not exist independently of infra-structures.

Alternatively Harris might mean that it is probable that the infra-structure determines the structure. Now if this probability is *a priori* or subjectivist, it must in turn be justified by the theoretical power, simplicity and cogency of Cultural Materialism as a research programme. But since this probability is part of the defining features of Cultural Materialism its value cannot be obtained without opening up a vicious infinite regression: we need to know already what these probability estimates are before the cogency of Cultural Materialism can be assessed. If on the other hand the probability is *a posteriori* then it is up to experience to substantiate some given probability estimate. But if the previous philosophical and theoretical criticisms hold, such hopes are abortive – such is the nature of philosophical criticism.

This completes my philosophical critique of the principal tenets of Harris' Cultural Materialism. In the concluding sub-section, I would like to advance some empirically orientated arguments against Harris' principle of the methodological priority of infra-structural variables.

5.1 The principle of the methodological priority of infra-structural variables: Some empirical considerations

One should note that the principle of the methodological priority of infra-structural variables is a *methodological principle* not an *ontological principle*. It places a rationality constraint upon permissible types of scientific explanations. It is sufficient to refute this principle to take phenomena from the infra-structure and give structural or super-structural explanations of them. Such explanations may not be known to be true, but they must be plausible and legitimate explanations from an intuitive point of view. Thus one must show that Harris' principle fails as a methodological principle by ruling out from our consideration as *initially* credible explanations such structural or super-structural explanations. I will take two infra-structural phenomena – human population growth and famine – and attempt very briefly to sketch structural/super-structural explanations of these phenomena. Such explanations are not merely plausible but will be supported by empirical evidence. They may therefore be true. Harris' methodology precludes these explanations from initial consideration – so much the worse for Harris' methodology.

One popular explanation of the current "ecological crisis" is the spectre of too many people. Population growth is taken on the one hand to be the cause of famine since, as we have noted, Malthus was certainly right that organisms have the capacity to produce more offspring than can be satisfied by available energy and spatial requirements given a fixed technology and resources. Garrett Hardin [1974] sees the present world population situation aptly summarized by the metaphor of a lifeboat:

> Metaphorically each rich nation amounts to a lifeboat full of comparatively rich people. The poor of the world are in other, much more crowded lifeboats. Continuously, so to speak, the poor fall out of their lifeboats and swim for a while in the water outside, hoping to be admitted to a rich lifeboat, or in some way to benefit from the "goodies" on board. What should the passengers on the rich lifeboat do? This is the central problem of "the ethics of a lifeboat". [ibid., p. 561]

Hardin's answer is to let them die. This resource scarcity situation is caused by the human tendency to over-populate, and giving food and aid to these poor will only exacerbate the problem. It could perhaps even threaten the stability of the lifeboat of the rich. It is best, Hardin concludes, for some fortunate minorities to survive and act as the trustees of civilization rather than for all of humanity to perish [Hardin, 1971, p. 1297].[8]

It is not Hardin's thinly disguised reactionary politics which concerns me here [Vandermeer, 1977] but the theoretical assumptions existing behind this view.[9] An innate tendency to over-populate most certainly qualifies as an etic explanation of the "population problem" and "famines". There is, however, an alternative viewpoint.

An alternative non-etic explanation of the population problem would deny that there is any innate drive within human beings to produce as many offspring as possible within a given environment. This is taken to be refuted by the "rich lifeboat" countries which whilst having great richness have declining birth rates approaching ZPG even though it is biologically and socially possible to produce more children. The argument is hardly a conclusive one against Sociobiology, but placed within the context of an extensive foundational critique of Sociobiology, the argument has considerable strength. As such the existence of the lifeboats and their relative richness and poorness is in need of explanation and cannot be taken as a given. An alternative explanation might take population growth and famine as significantly conceptualizable and explicable as an aspect of social inequality, imperialist plunder and class relations. Note that it is not necessary to rebut the principle of the methodological priority of infra-structural variables to claim that *every* famine including those before the existence of capitalism can be explained in this way. It is sufficient to show that since the existence of capitalism there have been a significantly large number of famines caused by such social factors. Even if the immediate causal agent is a pest such as the brown rat destroying food crops in some underdeveloped land, the question still needs to be asked why, if this phenomena is not as extensive in the developed countries, pest control measures are not made. If this is through a general state of poverty, this state of poverty must in turn be explained. By such a chain of causes one may still trace the initial problematic effect back to a capitalistic cause.[10]

The documentation of the imperialist impact upon what is now called the underdeveloped world is quite extensive ([Amin, 1977] [Baran, 1957] [Barratt Brown, 1974] [Frank, 1971] [Magdoff, 1970] [Turner, 1973], to name only a few sources. Here is, in outline a structural account of at least some famine: with the emergence of capitalism came developments in either biological or chemical technology, increasing plant and animal yields, or increasing the yield by vastly increasing the area of land per acre. The latter strategy was not made without a significant cost to the original owners of the appropriated land in Africa, Asia and South America. What clearly occurred was a stagnation of agriculture under colonial rule due to a failure of both capital and labour investments as well as technological inputs into domestic agriculture. The natives were too busy working under gun point, cash cropping and the European bourgeois, too busy extracting a surplus to care about such matters.

Thus, it is hardly surprising that we find throughout the colonial periods an intensification of famine with a famine occurring in India only a short period of time after her conquest and domination. Significantly, Europe suffered only one major famine comparable to the ravages of India (thus it is evident that we are not committed here to claiming that all famines are "caused" by Capitalism). Significantly, food production in the Third World countries gained an improvement only in the post-colonial periods under the land reform and investment of the National Governments. Yet there are obvious problems of famine remaining even though there has been an upward production trend of food in the underdeveloped world. This is still slightly faster than their population growth rates, given the relative lack of technological inputs into agriculture which has occurred by contrast to the developed world. Wealth remains, however, concentrated in the hands of a small proportion of the population, serving to increase the severity of the problem especially when such large wealthy land owners receive a disproportionately large share of technological assistance [Mampilly, 1976].

It is further no surprise to realize, as the 1976 Report to the Club of Rome, *Reshaping the International Order*, and other sources [George, 1976], indicate, that the West's intake of "biologically expensive" animal protein is excessive. In terms of per person per year grain, for example, whilst directly consumed grain in India is 348 lbs. and in Canada 202 lbs., total consumption of grain per person per year via meat producers fed upon grain, increases the latter figure to nearly 2,000 lbs. To eliminate extensive dependence upon high level food chain organisms such as live stock given egalitarian distributions go a significant way towards meeting world hunger.

As it stands, however, the poor countries are negatively advantaged in food commodity markets at the international level. Hassan [1979] has argued that a Weberian concept of class, is applicable here: the poor stand in a "class situation" determined by the international "market situation" and thus are vulnerable to chronic food deficits. Having basically labour-intensive economies and high mortality rates due to food and medical deficiencies necessitates the production of many children, especially males, to serve as family labour, and to provide for the parents in old age. Is it not reasonable to do this in a social world where it is unlikely that anyone else will look after one? It follows, according to this point of view, that the solution to the double headed Medusa of over-population and famine is an alteration of the social system producing these difficulties.

Enough has been said to give the reader an understanding of the style of non-infra-structural argument which may be employed to account for two infra-structural phenomena on Harris' scheme. The argument may well be sound: it is at least plausible and not one which should be given an initial "second best" rating as the principle of the methodological priority of infra-structural variables implies. Hence we have further reason to reject Harris's principle.

6.0 Conclusion: The state of the argument

Two forms of Ecological Causal–Explanatory Reductionism has been considered here: these are Emlen's Ecological Sociobiology and Harris' Cultural Materialism. Both positions, it has been argued, are subject to extensive philosophical and theoretical problems. Neither orientation, it is concluded, supplies a satisfactory foundation to the sociocultural sciences.

CHAPTER 8

Reductionism and cultural being: Beings, agents, mentalities, persons and societies in the universe

1.0 Introductory statement of the argument

One all to frequently finds in intellectual works a far from subtle form of dishonesty: the writer after leading the reader through the graveyard of the systems of others, attempts an extensive intellectual synthesis by which he/she not only resolves the problems posed in the work itself, but allegedly stands as a shining and proud specimen of intellectual originality and lasting testament of the (implicitly stated) genius of its author. What follows are but some footnotes to the critique of Sociobiology and the PSU orientation and some necessary clarification of the Culturologist position.

In sections 2 and 3 our concern is with the conceptually slippery concept of *human nature*. Our position is advanced by consideration of Mary Midgley's recent defence of the concept of fundamental human nature in her book *Beast and Man*: *The Roots of Human Nature* [1978]. It is argued in sections 2 and 3 that Midgley's position is open to severe conceptual objections which vitiate her account. Scepticism is also levelled at other existing attempts to explicate the idea of an innate human nature and in particular Chomsky's innateness hypothesis [Chomsky].

It is argued that the concept of innateness is one which is devoid of explanatory power, and thus that concepts of human nature which presuppose it, are also methodologically flawed. Nevertheless, there is room in a *Weltanschauung* for the idea of *human nature*, however imperfect our present conceptions of it may be. In accordance with the realist philosophy of science of Harré [1970] and Bhaskar [1978], the essence of humanity can be specified, but only by appeal to a "Total Science" encompassing both the natural sciences and the social sciences, all of which is placed within a metaphysical *Weltanschauung*. Clearly the present chapter cannot present any such detailed account of human nature: the critic presupposes that a satisfactory theoretical biology and social science exist. That they do not, has been the argument of this work.

If this work is to be more than an extended argument for theoretical scepticism, some clarification of the Culturology position is necessary. In section 4, a very tentative and provisional sketch is given of a Culturology position. This point cannot be strongly enough stressed: it is my position that only given substantial progress in total human science, and most importantly in metaphysics, can the "crisis of the social sciences" be resolved. All that is possible here, is a preliminary clarification of terminology and strategies of argument. This uncertainty will obviously make this chapter hermeneutically difficult: in the face of our outstanding problems this is unavoidable.

2.0 Culturology and models of human nature

Petryszak [1978, 1980 (1), 1980 (2)] has convincingly argued that all social theories make reference to some thesis of the nature of human nature and that this reference is not explicitly justified. The idea of human nature to which Petryszak refers is that of a human "innate biological nature". I maintain that not only are assumptions made about the nature of human nature in this sense, but assumptions are also made (and seldom explicated and justified) about what will be called "the metaphysical nature" of human beings and societies. The "epistemological crisis" problem cannot be resolved without detailed, rigorous and rationally justified responses to questions such as: (1) what is the ontological nature of human beings, cultures and societies; (2) what are socio-scientific theories (3) what is the relationship between interpretation and explanation in such theories and (4) what is the role of critique in the social sciences? Fay and Moon [1977] are correct in my opinion in maintaining that neither the "humanistic" nor the "naturalistic" orientations satisfactorily resolve these questions. What alternative direction does the present author suggest, or is this work yet another stroll through the graveyard of the systems of others, leaving us ultimately cognitively shipwrecked, dazed, cold and lost?

In the following sections I will comment upon the question of the rationality of assumptions about "human innate biological nature", in an attempt to clarify and defend the Culturologist position. Then, by way of further clarification of the Culturologist position, I shall address the question of the ontological status of the social entities to which Culturologists refer in scientific explanations. The work is concluded by stating very briefly the role which *critique* in the social sciences should take, which is a position readily derivable from my previous claims that the incorporation of all aspects of intellectual inquiry into a plausible metaphysical *Weltanschauung* is of extreme importance. The answer is simply – trivially simple – that the social scientist engages in ethical debate insofar as this is possible, according to the system of ethics which he/she incorporates into his/her metaphysical *Weltanschauung*; the detailed development is, however, an enormously difficult task which should be the object of an entire research tradition, rather than one researcher. But we do not address the multitude of questions under the broad heading of the "nature of socioscientific" theory. I do not believe that significant metascientific progress can be made in this area without the resolution of major issues in epistemology and metaphysics, and unfortunately we resolve none of these issues here. But let us climb as much as we can of this icy peak here.

Thus we turn immediately to our first assigned question about what, if anything, the expression "human innate biological nature" refers to. The terms in need of analysis, at least in this context, are "innate" and "nature". Let us tackle the concept of "nature" first, and select a recent author who accepts that there is a basic biological human nature for study: Mary Midgley's *Beast and Man* [1978].

2.1 Midgley on human nature

For Midgley the view that there is a basic biological nature is summarized simply by saying that human beings have "instincts". She compares and contrasts this position with the "blank paper view" which sees human beings as having no "instincts" at all and thus being "infintely plastic" so that "all comparisons" with non-human animals with respect to their mental and social aspects must be irrelevant [ibid., p. 3]. Watsonian behaviourism and the Existentialism of Sartre [1958] are variations upon this theme, she maintains. (This latter point is asserted with very little textual analysis, but we need not pursue the specific point further.)

The term "instinct" is taken by Midgley to be synonymous with the terms "drive" or "programme"; instincts are "dispositions", a set of "causal properties" [Midgley, 1978, p. 51]. Instincts may be either "closed" or "open". *Closed instincts* are genetically fixed behavioural patterns, "programmed" in every respect where there is no place for learning at all. Examples of this are, she believes, the web spinning activities of spiders and the bee dance. *Open instincts* on the other hand, are behavioural patterns which are not fixed in every detail, there being a gap which is "filled in by experience" [ibid., p. 53]. Cats are said to have an open instinct to hunt. The instinct is not an encoded stereotype behavioural programme but a general tendency to behave in certain ways: to stalk prey, climb trees and spring, among other activities. Cats are able to improve their hunting techniques with experience and they may learn from other cats.

Open instincts, she argues, are said to be characteristic of higher animals, particularly man. The paradigmatic case of this in human beings is the "instinct of aggression". This, she defines as "a general tendency to attack members of one's species" [ibid., p. 54]. Thus to say that it is a part of human nature to be aggressive is to say that human beings have the power to attack other human beings without being taught how to do this, although such attacking abilities can be improved through cultural experience, through learning how to use new weapons of destruction.[1] The nature of human beings is then defined as the class of all open and closed instincts. Midgley admits that human beings have few closed instincts beyond reflex actions. She maintains, we should also note, that the "savagery" seen in human beings, from wife beating to the nuclear arms race, cannot be satisfactorily explained by Culturology, since Culturology cannot explain *why* the phenomena of aggression arose for the human race as a whole, and why it is so hard to change these states of affairs when other culturally caused phenomena are not so difficult to change.

Without for the moment explicitly criticizing Midgley's theoretical claims we should note that no explicit and detailed discussion of "natures" occurs in her book. True, I have said that Midgley appears to accept that the nature of human beings is the class of all open and closed instincts of such human beings. However such a class hardly serves to distinguish human beings as a kind from other logically possible kinds. Giving free rein to the philosopher's imagination we can well imagine on planet x organisms utterly unlike human beings except for being truly described by just those open and closed instincts which human beings possess. Hence at best Midgley could claim only to have presented an account of the innate biological nature of the human beings rather than a more general-metaphysical account. Even here, the concept of "nature" is put to little use: we can correctly paraphrase her position by stating that human beings have open and closed instincts $OI_1, OI_2, \ldots OI_n, CI_1, CI_2, \ldots CI_n$.

If I am right in this claim, I do not in this respect score any points against Midgley's position. Rather, it is claimed that we could approach the whole question of "human nature" raised in her book without consideration of questions of essentialism. Yet, I believe that consideration of such questions will, in fact, aid us in our analyses in this and following sections. Thus, let us depart from our consideration of Midgley's position to pursue this issue, returning from our tangent at the first opportunity.

Essentialist doctrines are usually dichotomized into two broad positions. On the one hand is the position of *individual essences*, on the other, the position of *general essences* [Parsons, 1979, p. 73]. According to the first thesis "some or all objects have characteristics (or properties) which are so intimately associated with the object that nothing else *could* . . . have precisely those characteristics without being that object" [ibid., p. 73]. According to the second thesis, "certain characteristics" are singled out as "being necessarily true of certain objects" [ibid., p. 74].

Both positions have been regarded as problematic by many philosophers, such as the claim made by Quine [1961, pp. 139–157], that quantified modal logic leads to essentialism, which has been taken as a matter of major philosophical concern for modal logicians [Linsky (ed.), 1979]. However one set of objections to essentialism seems answerable once one clearly distinguishes between *de re* and *de dicto* claims [Brody, 1973].[2]

The problem of trans-world identity is not as easily resolvable: what does identity across possible worlds mean? One claim, made by Chisholm [1967], is that there is for each object a set of properties which it has in all possible worlds in which it exists which no other object in any possible world has, which is distinct from that object. The central line of criticism of this proposal is that if this set contains properties that can be had by more than one object in a given possible world, then it is logically possible for two distinct objects (spatially distinct) to have all their properties in common, whilst if only one object in a given world can have the specified properties, it becomes difficult to specify any such properties [Brody, 1973, p. 358].

Brody's strategy is to reject the idea that if an object has a property essentially then it has that property in all possible worlds. His alternative account is this: "a has P essentially just in case a has P and, for all possible future worlds in which there is some O_1 identical with a, O_1 has P" [ibid., p. 359]. But the difficulty with this, is that if one takes a Leibnizian account of identity ($x = y =$ df $(\forall f)$ $(Fx \equiv Fy)$[Brody, 1980, p. 8], then for all *permissable* extensional or intensional properties of the domain of the predicate variable F, these properties are essential. This is so because for any property of a, a has P, and by the Leibnizian account of identity, O_i has P for any properties. But surely then the thesis of essentialism is logically vacuous if all properties then become essential properties.[3]

It seems that any account of essentialism which accepts Leibnizian accounts of identity will be subjected to triviality and alternative positions face problems of trans-world identity. My position is that Leibnizian accounts of identity should be abandoned. The skeleton of the argument is this: (1) this account of identity makes it especially difficult to solve other metaphysical problems such as spatio-temporal continuity; (2) there are independent logico-conceptual arguments which can be advanced against the position [Routley, 1980 (1), pp. 96–119] [Griffin, 1977]. What the alternative position is remains an outstanding problem,[4] but it is not a problem *peculiar* to my position.

Taking this non-full-indiscernibility account of identity as primitive (and indeed unexplicated), we would say that the *individual essence* E_1 of the object O_1 are those properties such that if O_1 did not have E_1, then O_1 would be counter-factually, an object O_2 such that $O_1 \neq O_2$ and if O_3 had E_1 then $O_1 = O_3$. The general essence of certain objects, of a kind, are those properties G which objects of that kind exemplify, and which would not be regarded as of that kind if they did not exemplify those properties.

Before concluding, some words should be said about trans-world identity. What identity across possible worlds means is precisely that: identity across such worlds, and criteria should be specified from one's theory of identity, if and when we have it. Again, if this is still felt to be unclear, we could follow Brody and abandon essentialistic-possible trans-world talk. The problem about the explication of identities will constantly plague us in this chapter.

3.0 Culturology, innateness and the human essence

We at least have some idea of the concept of nature now but hardly as much as we would wish. The issue of concern to us is, once more, the question of innate human nature and our tangent ended we return to Midgley's book. Midgley is the latest, and *philosophically* most respectable, of the authors who have given a philosophically important role for both socio-scientific and moral evaluative discourse to the idea of an innate human nature and thus her position must be criticized if my own position is to be granted credibility.

The first point which I make in response to Midgley's position is that the argument that socio-cultural explanations of aggression fail, because the socio-cultural sciences cannot explain the origin of aggression as a whole, for humanity, is a *non sequitur*. If Midgley is not making a mere arbitrary judgement about the explanatory powers of the socio-cultural sciences then it seems that she is appealing to a general methodological principle. The principle is, I believe, this:

(MMP) Necessarily for any Science S, if S explains phenomena P, then the origin of P is explicable by S.

(MMP) is, I maintain, false. It would follow if (MMP) is accepted, that the science of physics would not provide a satisfactory explanation of a phenomena such as atomic decay unless the origins of the atoms which decay are explicable. But, not only *is it possible*, but it could be argued that no generally satisfactory cosmology is *actually* taken as constituting a consensus amongst thinkers. According to (MMP) there will be few, if any satisfactory physical explanations. But this conclusion is absurd and there is no independent reason given for its acceptance. I maintain that we can understand the nature and operations of mechanisms in the world even if the world was inexplicably created by an indifferent God. To reject this claim must lead inevitably to a deep explanatory scepticism, since cosmological matters remain the most speculative and controversial areas in the natural sciences. Thus if (MMP) is rejected Midgley's argument collapses. We reject (MMP). Therefore Midgley's argument collapses.

To possess an "aggression instinct" is, Midgley tells us, to possess a natural tendency or predisposition "to attack other members of his species sometimes, without being taught to, without needing to as a means to another end, and without always having what seems to be adequate provocation" [Midgley, 1978, p. 59]. Let us recall that "aggression" is defined by Midgley as "a general tendency to attack members of one's species" [ibid., p. 54]. Now explicating the concept of aggression has proved especially difficult for Sociobiologists. Wilson [1975, p. 242] accepts that Sociobiologists can only improve upon the standard definition of aggression (i.e. an act which abridges the rights of another) by specifying that the really important losses to a victim are losses in inclusive fitness. Pleading, he states "we should not worry too much about terminology" [ibid., p. 242]. But one cannot worry enough about one's terminology. If one uses a term so utterly obscure as to refer to both moral acts based upon abridgements of rights and to biological abridgements of inclusive fitness contradictions are easily deducible. Raping a nun who, counterfactually would never have had a child, and getting her pregnant, is an act which increases the nun's inclusive fitness and hence is not an "aggressive" act – but surely it is an abridgement of the nun's rights, and hence is an "aggressive" act, and thus we have our contradiction. If we differentiate between two distinct senses of the term "aggression", then of course we have no contradiction. However, what must now be established by argument, is the existence of various bridging principles which enable the possibility of causal-explanatory accounts of moral-aggression. Otherwise instances of moral-aggression cannot be taken as phenomena within the explanatory domain of a theory of biological-aggression any more than "charm-properties" of micro-physical entities explain why some people are "charming". In short, we will lack a unified theory of "aggression" in the absence of such arguments. Sociobiologists have supplied no non-question begging arguments to support such a conclusion.

Midgley's definition is inadequate even on her own terms. If she wishes to use her term to refer to phenomena from marital disputes to nuclear warfare and acts of institutionalized "aggression", the term "attack" cannot be used to refer to reflex like actions which occur even without environmental cues of provocation, or without being a means to some end, even if the end is "irrational". Even the most stereotypic closed instinctual behaviour patterns of insects do not occur without the necessary environmental cues and this can readily be supported by selectionist arguments. She admits that human beings have few closed instincts and recognizes the importance of culture in shaping the open instincts which human beings have – therefore her account of human "aggression" is inconsistent with other theoretical claims which she makes in her book.

Midgley gives no analysis of power ascriptions, but we could plausibly analyse these as Harré [1970] has proposed. To say that human beings are aggressive by nature is to say that if they are subject to conditions of a specified kind, then they attack other human beings by virtue of their intrinsic biological nature. On this account, it is impossible to eliminate human aggression, for to do so, one must eliminate the power of agents to exert force upon others at all, and to do this, one must drastically alter their natures so that they could not act in ways which would constitute under some description, an "attack" upon others. But given the indefinite number of ways in which we can "attack" others, physically, psychologically and socially, this would involve eliminating the possibility of human beings acting at all. This is, to say the least, counter-intuitive, and trivializes the position, for an "aggression instinct" is nothing more than a specific capacity for human action.

In reply it may be argued that the power-analysis of aggression is inadequate precisely because of its triviality. It would be maintained that not only do human beings have a power to attack others, but there is a conditional probability in a specified environment that they will attack others. The major difficulty here now is to specify what such probabilities are and to produce a theory which in this situation does not involve one in circularities; for example if a propensity theory is adopted [Popper, 1957, p. 70] and talk of propensities is taken to be talk of dispositions or powers, then the circularity is evident. It is unclear to me how a frequency or subjectivist account of probability would succeed here, since our discourse is about the nature of a particular, not the degree of reasoned belief, and indeed about the action of people under possible unactualized environments, not merely those actually observed, or likely to occur as a matter of brute contingent fact in the "long run".

The final problem which I present for Midgley is this: if we allow "aggression" to be a basic "predisposition" of human beings, nothing prevents us from allowing any number of human qualities to be taken as part also of this fundamental human biological nature. Now the ranks of the instinctual swell to such a level that the thesis is revealed as being logically vacuous. What point is there at all in postulating an instinctual basis to a potentially infinite array of qualities such as love and hate, kindness and aggression, creativeness and destructiveness and so on? There seems to be no reason for this at all; the concept seems to be devoid of explanatory power. Midgley, in short, only tells us that human beings can act "aggressively", and this phenomena remains unexplained.[5]

This thesis will be generalized: the concepts of the "instinctual" and the "innate" are in general explanatorily useless in the causal analysis of both animal and human behaviour. They should be put on the conceptual scrap heap just like, as relativity supporters allege, the idea of absolute simultaneity has been put. With regard to the human domain consider the well known nativist theses of Noam Chomsky and Jerrold Katz [Chomsky, 1975] [Katz, 1975] (although Katz himself now has second thoughts [Katz, 1981]). This is surely a reasonable nativist thesis to take as a test case. Ignoring the utter diversity of phenomena to which innateness is predicated (structures, mental functions, mental capacities, principles of mental operation etc. [Kitcher, 1978, p. 1]),[6] let us restrict our discussion to the so-called "innateness-hypothesis" ("I.H."): that the human brain is "programmed" such that every linguistically competent language user innately possesses a set T of transformational grammars $T = \{G_1, G_2, \ldots G_n, \ldots\}$ such that all members of T share general features known as "linguistic universals" including an active-passive distinction, a transformational component, and linguistic categories such as concrete and abstract nouns, verbs taking on abstract subjects and so on [Putnam, 1975, pp. 133–134]. Further, part of the innate furniture of the brain is some neural mechanism which may be modelled by a function f which assigns "weights" to the elements of T with respect to experiential interaction.

Chomsky advances I.H. to provide an explanation of these alleged facts: (F_1) the linguistic "competence/knowledge" of an adult "extends far beyond anything that he has learned" [Chomsky, 1975, p. 123], so much so that the number of sentences which one can interpret; (F_2) the initial linguistic input to the child is extremely "degenerate" as *most* of the sentences the child hears are non-grammatical, hence refuting inductivist accounts of language learning; (F_3) mastery of language does not appear to be affected by motivation levels, intelligence and one's social environment; (F_4) despite differences at the level of the surface grammatical structure in human languages there are basic similarities ("universals") at the level of the deep structure such as the various transformation rules.

I.H. however fails to provide an adequate explanation of (F_1)–(F_4) and its coherence is open to question. (a) Putnam [1975, pp. 143–144], whilst open to some criticisms on other points in his interpretation of Chomsky's I.H. [Horner, 1976], is undoubtedly right, in my opinion, to point out that Chomsky has given no sketch of the neural generative mechanisms involved here. The point in this criticism of relevance to our concern is that "the innate" serves then as an intellectual storeroom to which pressing psycho-linguistic problems are pushed – but not resolved [Luria, 1975]. (b) There may well be good reasons for this failure: the concept of a rule presupposes social life insofar as criteria for discriminating between conforming and non-conforming behaviour is required [Margolis, 1978, p. 115] – and thus it remains obscure as to what such pre-cultural rules are. Margolis' point, can I think, be restated as follows: grammar, be this of the deep or surface variety, is a property of a language and does not exist independent of the language anymore than the structure of a building exists independently of the building itself. But if this is so, grammatical competence could not exist independently of linguistic competence because grammar is a property of language itself. Therefore, to ascribe grammatical competence to an individual is already to ascribe linguistic competence to that individual. But now our appeal to I.H. leads us into circularity, since the latter phenomena is the very one we sought initially to explain![7]

I.H. fails as an explanatory hypothesis in another way. Given facts (F_1)–(F_4) above it is unclear how in fact language acquisition is possible at all. Chomsky does not presumably deny the relevance of the social environment to language acquisitior such that if this was denied acquiring a language would be much like the action of any other "closed instinct". If this was proposed, the existence of feral children [Curtiss, 1977] would be impossible to explain. Note that we cannot without contradiction claim that it is merely the absence of a linguistic environment which explains the phenomena of feral children if the truth of the "closed instinct" hypothesis is assumed, by the definition of "closed instinct". But to use an equally unsatisfactory term, if language acquisition is an "open instinct" the causal influence of the social environment is a necessary condition for language acquisition. One can now readily explain the phenomenon of feral children. But, granted this, how can language acquisition be explained at all given (F_1) and (F_2)? The competence of a speaker far exceeds his learning, and his socio-linguistic environment is composed mainly of non-grammatical sentences. Now if either the socio-linguistic environment is not as degenerate and impoverished as Chomsky says it is, or the child could still, from the grammatically satisfactory speech acts performed, gain substantial linguistic competence, then Chomsky's case against Empiricist theories of language acquisition collapses. On the other hand, if his case against Empiricism is taken to stand the non-grammatical linguistic environment of the child must be taken as informational-noise – yet it is only by environmental interaction that language acquisition occurs at all and the child is diverted from becoming a feral child. Thus, if on this ground all Empiricist theories must fail, we conclude that a Rationalist theory of language acquisition is logically impossible – requiring both that the causal influence of the social environment is a necessary condition for language acquisition, and that it is not so. The latter horn of the dilemma takes language acquisition on the Rationalist position to be impossible, and the former to be simply unjustified, giving us no good reasons for rejecting Empiricism. In either case Chomsky's I.H. is unsatisfactory. Talk of innateness solves none of the difficulties of language acquisition, but merely transfers them to another level. Innateness-talk here, as elsewhere, serves as a gigantic storeroom for the classification of problems which are merely hidden rather than reasonably resolved.[8]

The developmental biologist Daniel S. Lehrman maintains that the "innate" and the "acquired", are an inadequate set of concepts for analyzing development [Lehrman, 1970, p. 33], and with this I agree. Here I would like to converge to Lehrman's biological conclusions from philosophical considerations.

The concept of the "innate" is usually explicated (if it is at all) in one of three ways. Firstly, it is equated with the "inherited". We have already considered this matter. Secondly, it is defined simply as the "unlearned". Thirdly, and infrequently, an independent analysis is given.

One brave attempt to give an independent account is made by Stephen P. Stich [1975] from whom we can take the following account of an "innate disease", but which is nevertheless, an interesting definition:

(SDI) A person has a disease innately at time t if, and only if, from the beginning of his life to t, it has been true of him that if he is, or were of the appropriate age (or at the appropriate stage of life) then he has, *or in the normal course of events*, would have the disease's symptoms.

SDI is open to both conceptual criticism and readily manufactured counter-examples. Stich himself recognizes that the definition fails to distinguish between "innate diseases" and "generalized susceptibilities", such as the effects of illness resulting from the accumulation of either chemicals or radiation from one's environment. One need only vary the counter-example such that the environmental agent is some necessary condition for life at all (such as oxygen, sunlight etc.) to produce instances where it is impossible to say whether the disease is "innate" or "environmentally" caused. This conclusion follows independently of our metaphysical image of biological development and the nature of the gene.

The second difficulty with SDI is to somehow explicate the idea of "the normal course of events". It is true that in some instances we have some very clear intuitions about this matter. But in other cases, which can be readily cited or invented, we have not. Stich faces this problem squarely in his attempt to explicate the idea of "innate belief". SDI modified by replacing "disease" and "disease's symptoms" with "belief" allows us to claim that beliefs such as "that night follows day", "that there are rivers" and so on are innate. The ranks of the innate swell to uncontrollable lengths. On the other hand, to take the normal course of events as "any physically possible course of events", makes the ranks swell *further*, not diminish to zero as Stich suggests [ibid., p. 9]. Here the belief that there are unicorns would be taken as an innate belief. After this point Stich abandons his attempt to analyse the phrase "the normal course of events". I suggest that the phrase is not sufficiently precise to analyse the concept of "innateness" whatever partial intuitive plausibility it possesses and despite its limited plausibility elsewhere.

Facing these difficulties one may then attempt to explicate the "innate" by reference to that which is not learned. The price which one pays now is that "acquired" or emergent developmental phenomena which are "non-learned" in an intuitive sense and also "non-innate" (i.e. phenomena arising from "developmental noise") are taken as innate. Indeed, if God implanted in us a soul at some stage of development (as John Eccles proposes), then this too would be "innate" by this definition, as it is certainly not learned! The "innate" now becomes, once more, an intellectual blackhole, a giant storeroom, analogous to an unsympathetically interpreted Freudian Unconsciousness into which one can place any number of inhomogeneous phenomena, which ensures its utter explanatory bankruptcy as a reasonable generative mechanism. In so classifying such a bewildering array of phenomena we are no closer to providing a causal explanation of the phenomena than when we began our quest. Therefore we could eliminate innateness-talk without loss of explanatory and predictive power. Indeed, we gain nothing by retaining it.

The Culturologist is not committed to affirming that all behaviour is learned for it is unclear what in fact this could mean. The position rather denies that the "innate-acquired" distinction is either semantically clear or methodologically useful and wishes the nature-nurture debate to be declared ill-defined and the traditional questions written off as cognitively meaningless. This is not to deny that human beings have no biological essence at all or to affirm further caricatures of the Culturologist position such as "human beings are infinitely plastic". The point to be made is that it is unclear as to what such slogans mean. The adjective "infinitely" is either an obscure metaphor or a literal absurdity. Such totally ill-defined debates must be brought to a close and positions such as Sociobiology, which despite the unconvincing lip service by its defenders depend upon such distinctions, should also be written off.

Discussion could perhaps be left neatly, cleanly and dishonestly at this point. However it seems necessary if my own position is not to be classified as obscure and as problematic as my opponents, to address the question of the nature of human essence. The question which we are addressing is "What is the essence of a human being?" The question is, I maintain, reasonably interpreted as a question about *general* rather than *individual essences* [Parsons, 1979, p. 73]. Keeping this distinction in mind, may help us avoid a sorites type problem,[9] where we take a human being and mentally dissect him/her, taking away first the limbs, and then other parts of the body, asking at each stage whether what we have left is a human being, hoping to drive the Essentialist into the dilemma of accepting either all properties of the human being as essential or none. Baruch Brody [1975] for example, has been led to conclude that there is only one property which is essential to humanity, "namely the possession of a brain that has not suffered an irreparable cessation of function" [ibid., p. 348]. However, whilst this may be an arguably plausible definition of death, it does not answer another question which Brody set himself to answer, namely, "what are those properties which constitute humanity as a natural kind and enable human beings as a kind to be differentiated from other kinds?" [ibid., pp. 342–343]. The property of "possessing a functional brain" does not do this because other *individual* organisms, at the very least other primates, undoubtedly would pass out of existence without a functioning brain.

Further, I believe virtually all of the traditional answers to the question of the nature of human essence can be demonstrated to be inadequate – including *Homo faber*, *Homo ridens*, *Homo civilis* and *Homo loquens*. The chief difficulty which has faced all of these accounts is that either some intuitively recognized human beings lack the specified qualities or else other animals, robots or supernatural beings possess these qualities.

My two part strategy to attempt to avoid this difficulty is as follows. First, we need, I believe, an ontology which admits six general categories of *concrete particulars* in the universe: *Things*, *Beings*, *Agents*, *Mentalities*, *Persons* and *Societies*. Our difficulties in understanding what the nature of human beings in fact is, is due to a failure to clearly distinguish between Things, Beings, Agents, Mentalities and Persons. For the moment I will not attempt to analyse each of these concepts because there are utterly enormous philosophical problems involved here which are best dealt with in the next section. We should remark at the onset, that this classification does not present a metaphysically-free account of the world. The classification would be rendered problematic if Things for example had mental properties as pan-psychists propose that they do. It is of course not possible within the scope of one work to rebut the critiques which could be made from the perspective of every conceivable metaphysical system. These can only be dealt with over an extended period of time in other works. Secondly, we need to obtain clarity about the nature of human nature. I address this latter question in the remainder of this section.

Becker [1975] has offered a number of insightful propositions about what it means to be human. Both the entry and exit from the class of human beings is a *process*, which in both cases is in part a biological one such that the biological part of the entry and exit processes must necessarily be completed before the general completion of the entry and exit processes can be said to have occurred. Human development is taken to be a process analogous to metamorphosis, as Becker puts it:

> . . . just as it makes good sense to speak of butterfly eggs, larvae, and pupae as distinct from the butterflies they become (to say that they are *not* butterflies) so too it makes sense to say that human eggs, embryos, and fetuses are distinct from the humans they become – that they are not human *beings*, only human becomings. [ibid., p. 337]

On this position, a particular is a human being rather than a human becoming "when the relatively undifferentiated mass of the fertilized human ovum has developed into the pattern of differentiation characteristic of the organism" [ibid., p. 337].

An alternative "life cycle conception" rejects the claim that entry into the human species coincides with the end of "metamorphosis". Here the egg, larva, pupa and butterfly and the conceptus/zygote, embryo, foetus, neonate, infant, child, adolescent and adult are all life-cycle stages in the development of one entity of one species.

Becker criticizes the "life-cycle" account on both logical and empirical grounds. If the life-cycle position takes being a human to begin at conception then monozygotic twinning refutes the position. Here conception is taken to be the moment when the sperm and ovum fuse. However, it can be argued in reply that in the case of monozygotic twinning that two moments of "conception" occur – first in fertilization in the sense of initial fusion and second in fission, in the sense of being the initial biological beginnings of two distinct organisms. The point to be made is that a plausible life cycle approach will not take "the process of becoming *a* human being [to end] at conception" [ibid., p. 340]. To be human it is necessary to occupy one part at least of a human life cycle (the circularity here will be dealt with shortly).

Becker's second argument [ibid., p. 339] is that the life-cycle supporter is fallaciously arguing as follows:

(1) The conceptus is a being (i.e. it is an entity and is alive)
(2) It is human (i.e. it is not of some other species)

Therefore

(3) It is a human being, an argument, which when symbolized, is like this: $BC \& HC \vdash (HB)C$.

This argument is fallacious as Becker points out, involving taking the term "human" in (2) to be an "adjective" (applicable to any living part of any member of the species), whilst in (3) it is treated as a noun meaning, "member of the species *Homo sapiens*". But this equivocation can be readily removed, by treating the term "human" in premise (3) also as an "adjective" (more precisely a predicate term):

(1) The conceptus is a Being (i.e. it is an entity and is alive)

(2) It is human (i.e. it is not of some other species)

Therefore

(3) It is a *human-being* (i.e. an entity and alive *and* not of some other species)

producing a sound, but not very informative argument, which when symbolized, is like this: BC & HC $\vdash$ BC & HC.

The life cycle position is thus defended against Becker's criticisms. But is it a more reasonable position? I would like now to press an objection against the "metamorphosis" metaphor. The difficulty with this metaphor is that metamorphosis is a rapid transformation from larval to adult form with frequently a destruction of larval tissues and a radical reconstruction of such material in the adult form. There is nothing quite like this in human development and hence the metamorphosis metaphor is inappropriate. Metamorphosis is a *special* type of developmental *life-cycle*.

The account at present establishes little. To be human it is necessary to occupy at least one part of a human life cycle. But the obvious circularity here must be eliminated. Stating what it is to be human is so difficult, and so many thinkers from Aristotle to contemporary writers have failed, precisely because we have not as yet adequate material (or have yet to organize it) to specify conditions of sufficiency and necessity. Sufficiency and necessity, are specifiable by a "Total Science" which itself is a part of a Metaphysical *Weltanschauung*. One Metaphysical assumption for example which is implicit in the life cycle account is that being a human being is a historical way of being, that historicity is part of the essence of humanity [Grene, 1978]. Consequently, insofar as a general evolutionary account of the origin of life is accepted, *part* of the human essence can be defined by an "evolutionary taxonomy". Let us state more precisely what this involves.

Human beings, belonging to the Kingdom *Animalia* possess essential properties such as the absence of (in general for *Animalia*) functional chloroplasts and as such must feed off of other bio-systems; belonging to the superclass *Tetrapodia*, human beings have under a range of environments, two pairs of *pentadactyl limbs*; belonging to the class *Mammalia*, human beings typically have hair and females secrete milk (which the young *could* be fed upon); they have three auditory ossicles in each middle ear, a lower jaw of a single pair of bones and a diaphragm which is used in respiration. One can continue like this, writing the *volumes* of the details of the steps which I have missed from this gloss distinguishing between *Hominidae*, *Pongidae* and *Hylobatide*, until one is forced to enter the debates of *Homo's* unique constitution and history. At this point one can specifically begin talking about *Homo sapiens* in specifically socio-historical terms: we are the species which has had history H_1, H_2, ... H_n, ... further, and most importantly for the perspective of this work, to have had such a history (*Geschichtlichkeit* [ibid., p. 16]) is to be a being with properties P_1, P_2, ... P_n, ... My answer to the latter question of what these properties are is this: such a being is (a) a social being and (b) a cultural being or a person, which has the properties associated with being an organism of the human species; properties ascribed by a "Total Science" of biology, history, anthropology, and philosophy – all of which are situated in the "best available" metaphysical *Weltanschauung*.

To "spell out" in precise details what the above description requires for completion, or at least to be relatively full, explicit and clear, will require the work of entire sciences – obviously! My critics are surely applying standards of criticism which are unreasonably strong, and which would indeed undermine much of human inquiry, if they required at such a point a detailed and relatively full account of human nature here: indeed I do not believe that we have one as yet. I should also point out that major metaphysical assumptions are made here which presuppose reference to other widely known arguments. The *Historicity assumption* is justified only insofar as positions such as Creationism (including Russell's hypothesis that the world was created five minutes ago) solipsism and ontological nihilism, among other positions, are argued to be unreasonable positions.[10] This can I believe be done but it cannot be done in the space available here, and so we accept as a hopefully temporary fact of our intellectual condition that many will remain unconvinced. Indeed, we note that what is being advanced here is a *research programme* based upon some central Root Metaphors, and this is all that is methodologically required to complete our rebuttal of the PSU metaphysical *Weltanschauung* – *it* after all *is* a research programme, not a completed *science*.

This is not to shy totally away from expressing a positive view on the nature of human biological nature and sociality. If there was no Sociobiological controversy, my position to be expressed here would rightly be taken to be banal and uninteresting, but this is not the case and it is informative to be clear about basic assumptions. As I have said, human beings are taken to have the essence which they have, discoverable by a "Total Science", by reference to the specific biological-bio-social attributes which they possess. Further, we note that it is not necessary to engage in the theoretical-reduction of the social sciences to produce a plausible "Total Science" of human nature. Rather, we would say, in line with a realist philosophy of science and account of causality [Harré, 1970] [Harré and Madden, 1975] [Bhaskar, 1978 (1)] simply this – human beings have their characteristic attributes and powers by virtue of being of a specific bio-social nature but the causal-explanatory and/or ontological reducibility of these properties and power to any "lower level" is neither methodologically necessary nor philosophically justified. I have argued throughout this work that the Sociobiological account of this essence is incorrect, especially through its acceptance of biologically untenable dichotomies, and is thus *excluded* from our "Total Science". But if our counter-arguments fail, then we have no choice but to accept the contributions of Sociobology towards our total picture of human nature, and the resultant image of ourselves.

4.0 Culturology defined and defended: Beings, agents, mentalities, persons and societies in the universe

We can hardly leave matters at this point, as poetically just as our conclusion was. The Culturologist position remains intolerably vague without further definition. Indeed, how could one adequately deal with the meta-scientific question of the nature of social-scientific explanation without regard to the ontological status of entities referred to in such explanations? I believe that one cannot. Once more I humbly remind my critics that what is to follow is an initial explanatory sketch of a research programme. Presentation of anything else is physically-intellectually impossible here once one appreciates the maze of philosophical problems which confronts one as the argument progresses.

Whilst ontological claims have been made aplenty in this work, the position is metaphilosophically inadequate insofar as no attention has been given to the basic semantical-metaphysical question of *ontological commitment*. This is especially important insofar as the programme advanced here involves a *critical* reworking a "socialization" and "enculturalization" of Popper's doctrine of the "Third World" [Popper and Eccles, 1977] and this position has been criticized by Keuth [1974] on the grounds that postulating the existence of entities only makes sense relative to a criterion of ontological commitment and Popper, for one, doesn't have this. Let us therefore turn to such matters.

Questions of ontological commitment are faced with an immediate and depressing paradox which has not been fully recognized: there can be no account of ontological commitment prior to the initial acceptance of an *ontology*. This is not to deny that we can say that some metaphysical position is absurd or false *nor* that we must necessarily accept an ontology before we can criticize some other ontology for the latter proposal begs the question badly against the philosophical sceptic who maintains that none of the ontological claims of any theory is justified.[11] Nor do I deny that there are various methodological, epistemological and conceptual conditions which a Philosophical theory must meet to be rationally considerable as being True. What I am proposing is that the category of "Beings" or "Existence" is *sui generis* and hence irreducible to some other philosophical categories. Thus, all criteria C of ontological commitment: "x's exist if and only if C" will be philosophically inadequate. This is a bold claim but one which can be very economically justified by a simple method of argument: for any such C we can empirically cite an alternative ontology which would be unjustified if C is accepted. Only if we could establish the rationality of the ontology legislated by C *already* could we reject alternative criteria of ontological commitment. But this presupposes advancing independent non-ontological arguments against competing positions accepting alternative criteria. Hence we are involved in a logical circle or else our criterion is useless as a guide to Being.

Consider for example Ron Harré's [1981] conditions for what it is to be a particular, which he believes social entities must also meet:

(i) it must be *continuous in time*;

(ii) it must occupy a *distinctive* and *continuous* region of *space* or a distinctive and continuous path through space;

(iii) it must have *causal powers*

[ibid., p. 141]

(i)–(iii) however, also presuppose a *particular* philosophy of space, time and causality which needs independent support (and which may be correct). Again, at least some of the entities which some philosophers have described as "existing" have neither existed within the spatio-temporal framework (God, Platonistic numbers), nor have possessed causal powers (epiphenomenal minds, Platonistic numbers).

A second set of attempts at explication fail because they are demonstrably circular. For example, Munitz [1974, p. 170] proposes that for x's to exist, necessarily x's must be a real part of the world. But what is the "World"? It is precisely the collective aggregate of that which exists [ibid., p. 141] and the claim is circular. The claim is applicable also to the Quinean claim that to exist is to be a value of bound existentially generalizable variables in some theory [Quine, 1961, p. 15]. The reason here is simple: to explain the difference between existentially loaded quantifiers and ontologically neutral quantifiers such as those which occur in *free logic* we must already presuppose the concept of existence.[12] If we take x's to exist, if they are the values of *any* bound variables, from the perspective of free logic we are making a contradictory statement in the metalanguage claiming that some non-existentially loaded quantifiers are, in actual fact, existentially loaded. So, in short, the Quinean criterion presupposes already our understanding of existence talk.

Again, if we say that x's exist if and only if x's are referred to in statements of a *true theory* (i.e. a theory which contains either as its theorems or as its "central theoretical claims" either all true or true and logically true statements), then this position is refuted by the occurrence of ontologically true theories or true theories about non-existent objects [Routley, 1980 (1)]. Further, when we come to give our account of truth, if we ban outrightly any position which does not accept the "ontological assumption" (OA)[Routley, 1980 (1)] and, if a correspondence theory of truth is accepted, then by virtue of accepting an (OA) account of reference, we implicitly presuppose the concept of "existence" as the criterion of correct reference is via (OA) to objects which *exist*. On a coherence account of truth, an alternative very well known problem exists: there are any number of coherent theories, which are mutually inconsistent, so at least some of the entities of some coherent theories do not exist.

These are in my opinion the most plausible candidates for criteria of ontological commitment and most other positions have either been successfully criticized elsewhere [Routley, 1980 (1), pp. 697–730] or else can be met by some variation upon the above type of arguments.[13] This leaves us with the possibility of either rejecting ontology as a coherent cognitive enterprise or else rejecting Keuth's claim [Keuth, 1974] that postulating the existence of entities makes sense only relative to a criterion of ontological commitment. But there is no reason to outlaw ontology and much good work is done in ontology which simply takes the concept of "existence" as an unanalysed primitive. Evaluation of ontological claims is made by an evaluation of a complex matrix of empirical, logical, semantical, methodological and epistemological arguments, as is done elsewhere in philosophy. There is reason, then, to reject the latter alternative of Keuth.

4.1 Culturology defined

What we shall attempt here is an explication of the Culturology position through an informal, yet precise, account of its central concepts. The main sources of influence here have been from philosophers rather than social theorists – especially Margolis [1978], Popper [Popper and Eccles, 1977] and Bunge [1979, 1981]. Margolis has convincingly argued that any philosophical theory, if it is not to be unsatisfactorily incomplete, must present an explicit account of the nature of *mind* and *persons* [Margolis, 1978, p. 9]. Margolis performs the needed task of directing us towards the peculiar and distinctive ontological properties and relations of *Cultural entities* but does not succeed, I maintain, in giving us an explicit or unproblematic account of such entities. On the other hand, Popper's doctrine of the "Third World" – of the objective contents of thought, of institutions, problems and theories, whilst subject to extensive criticisms, can be successfully defended if critically reinterpreted as the World of *Cultural Entities.* Throughout this we draw upon and modify freely material from Bunge as well as other cited sources. Our goal is to obtain an initially clear but incomplete account of concepts such as "person", "culture", "society", "agent" and "mentalite", thus rebutting the PSU objection that such concepts are intolerably vague, pre-scientific and in need of junking because their sense and reference are unclear.

4.1.1 Persons and culture

Margolis [1978] is quite right in maintaining that there does not exist a satisfactory philosophical theory of persons. The problem is distinguishable from the problem of the "Self" (or put in a semantic mode "what does the egocentric particular term 'I' refer to, if it refers at all?"), although both problems are not frequently distinguished. We could conclude that either "I" is non-referring or else refers to a "Cartesian Ego" [Anscombe, 1975], whilst still accepting that *persons* are *not* identical to Cartesian Egos. The question at stake is "what is the nature of the person?", and is not immediately solvable by solving the problem of Selfhood.

One attempt to deal with the problem of the nature of persons, which I believe is not satisfactory for our purposes is the Strawson-Puccetti account [Strawson, 1959] [Puccetti, 1968]. Strawson proposes that persons are an ontologically basic kind of entity which are in principle irreducible to either "minds" or "bodies". Strawson says, and is usually interpreted as maintaining, that the concept of a "person" is *primitive* relative to the concept of "mind" and "body", that persons are those entities to which it is meaningful to ascribe p-predicates (person predicates) and m-predicates (material/bodily predicates). However talk of *primitiveness* here seems to indicate primitiveness of definability (it is unclear how else to interpret such predicate-task), that the concepts of "mind" and "body" are definable by means of the concept of a "person". This does not seem to be possible since at least for the concept of a "body" as persons are *precisely* those entities to which it is meaningful to ascribe m-predicates and p-predicates and we are involved in circularity. However, we could eliminate this problem by taking the concept of a person to be *logically independent* (and this it must be, if not definable by, or in turn capable of defining, the other concepts) of that of the concepts of "mind" and "body". We shall explore this question in more detail shortly. Our remarks here are restricted towards Strawson's account.

However Strawson's position, and its modification by Puccetti, stands open to a further objection. If the concept of a person is taken as a sortal universal (i.e. it supplies a principle of distinguishing and counting individual particulars) as Strawson proposes that it does, we are involved in a further circularity as Puccetti has recognized: persons are individuated from other entities by being entities to which p-predicates and m-predicates are ascribable and "what is that to say except that persons are persons?" [Puccetti, 1968, p. 8]. Puccetti fails to see however that this indicates that a major problem exists in his and Strawson's methodology: the semantical criterion presupposes that we have already some idea about the ontology of persons, and it is the latter question which is of importance here [McMullin, 1980].

Strawson's claim that persons are *sui generis* entities, and are neither Cartesian egos nor simply complex bodies (e.g. [Williams, 1973]) has been receiving support from an increasing number of authors who have come to find the standard (usually unexplicated) concept of a "body" unsatisfactory [Long, 1964, 1974], [Cowley, 1971], [Bestor, 1976]. These arguments cannot be readily summarized for these papers generally attempt to show by eliminative arguments that philosophical body-talk fails to have a satisfactory referent at all. This claim is not the assertion that human persons have no bodies at all but rather that there is no non-circular criterion, or one not subjected to counter-examples which allows us to individuate living human bodies that either may or may not be the bodies of persons without implicit reference to person-concepts. This renders problematic reductionist accounts of persons which take them to be merely complex bodies. Further, insofar as bodily movement statements are taken to be problematic and action statements basic [Bestor, 1976, p. 23], one of the major problems of the philosophy of action – namely, what else is an action over and above a mere bodily movement must be reconsidered. We make this point knowing full well that the Culturologist can hardly outline an account of the nature of persons without a consideration of basic questions of agency. We shall comment upon that question in due course. Thus whilst Strawson's explication is inadequate, there remains merit in his general orientation.

My point of departure, I said, was with Margolis' *Persons and Minds,* so after an unfortunate false start, let us proceed. Persons are, Margolis tells us, "in a sense not natural entities: they exist only in cultural contexts and are identifiable as such only by reference to their mastery of language and of whatever further abilities presuppose such mastery" [Margolis, 1978, p. 245]. Persons are taken to be linguistically competent beings, culturally emergent entities which are physically embodied in either a physical body or a sentient organism.[14] Further, by virtue of their linguistic competence, persons are capable of making intensional distinctions and understanding and following rules. They are beings which can engage in reflexive judgements, impose normative constraints upon themselves (and others) and are capable of choice and responsibility [ibid., p. 16].

The coherence of this account of persons hinges around the coherence of Margolis' account of *Cultural entities*. To account for the ontological peculiarities of Cultural entities, Margolis introduces an explicit definition of the concept of "embodiment". The relationship is asymmetrical, irreflexive and intransitive. Further, embodied and embodying entities are (1) numerically distinct and (2) not both ontologically independent of one another. Embodied entities have at least some of the attributes of their embodying entities, but embodied particulars possess properties of a kind which embodying particulars do not possess. Cultural entities, including persons, are tokens-of-a-type: token-particulars can instantiate abstract type-particulars and it is because of this property that the relationship of embodiment is exhibited [ibid., p. 231]. The argument given for the thesis that it is possible to instantiate particulars of a certain type, as well as universals such as properties and relations is as follows. Universals cannot be created or destroyed. In creating a new kind of soup (or type of art work), something other than a universal must be created. But this, if it is a particular, must be an abstract particular. Yet if the soup is an abstract particular, how is it that it can be tasted – one can no more taste an abstract particular than one can kick a set-theoretical entity. This inconsistent triad is avoided by Margolis, by admitting that abstract particulars can be instantiated. The chef in making his/her new soup created a new "type-particular" through creating "a token-particular of that type" [ibid., p. 232]. The constitutive materials of the soup are not, however, tokens-of-a-type as "only objects that exist as culturally emergent entities can have the property" [ibid., p. 234].

Throughout his discussion Margolis has little to say about the nature of culture itself. He gives however at one point the following explication of the term "culture":

(MDC) By 'cultural', then is meant the property of any system in virtue of which certain entities emerge and exist in it, a system of *both* persons *and* what they characteristically produce or do as persons exist or once existed or occurred. A cultural system is a system of tokens-of-type. Such a system is at once rule-governed, in that what is typically produced is intelligible only in terms of rule like regularities, and rule following, in that persons are essentially capable of using language, of intentionally acting in accord with rules which they understand and are able to violate. [ibid., p. 237]

Disregarding here well known difficulties associated with the concept of "rule-following behaviour" [Giddens, 1976, pp. 44–51], (MDC) is no more informative as an account of culture than Strawsons' position is as an account of persons – despite the obvious merits of Margolis' account of persons – for as an explicit definition it is outrightly circular, explicating the concept of culture with respect to cultural entities such as languages. I will in turn suggest that no explicit definition of "culture" is needed, that an implicit definition as well as an informal outline of the sorts of things to which it refers will suffice.

Before attempting such an account, remarks should be made about Margolis' account of the ontology of cultural entities, especially with respect to the relationship of embodiment. Armstrong [1980] is quite right in pointing out that there is a dualistic tendency involved in postulating that particulars instantiate other particulars. This seems to be necessarily true insofar as Margolis maintains by virtue of the definition of the embodiment relationship, that persons are *numerically distinct* from their bodies, although not ontologically independent (i.e. cannot exist without a body). But if persons and their bodies are *numerically distinct*, substance dualism is inescapable where "substances" are taken to be particulars – for what else could the term "numerically distinct" possibly mean? This revives, rather than buries, the problem of the relationship between a person and his/her body, and is inconsistent with Margolis' stated scepticism of dualistic accounts of the mind-body problem [ibid., p. 60].

According to Margolis, it is because of the fact that persons and other cultural entities are token-particulars that instantiate abstract type-particulars, that the embodiment relationship is employed. Philosophers usually use the terms "token" and "type" as follows: a token is a particular element of a general class or kind (the type), each such element being a token of the same type. On this account Margolis' following comments are open to criticism:

. . . the particular printed word 'the' is a token of the type word THE,

> but the physical marks that convey the token word do not, on any familiar view, constitute as such tokens of any type. Only subjects have *such* intentional properties as that of 'being created' or of 'having meaning' or 'symbolic import' or the like can have the property of being a token of a type. [ibid., p. 234]

This claim is not correct. The physical printed words are (imperfect) tokens of types of shaped signs, the token "t" being one token of a class of "t-shaped signs". Margolis' claim that only objects having intentional properties can be tokens of a type is therefore false, and if it is for the reason of explaining the token-of-a-type phenomena that Margolis has introduced the concept of embodiment, then his reasons for introduction are poor ones.

4.2 Cultural being: Towards an alternative

My alternative proposal is to take the concept of culture as primitive with regards to explicit definability but as a concept which can be given an intuitive explication. The domain of "culture" is, as Popper rightly points out, composed of "embodied" objects such as books and journals, as well as "unembodied" objects such as numbers, mathematical structures and, in general, the domain of abstract reference [Popper and Eccles, 1977, p. 41]. Where I differ substantially from Popper is in taking "World Three objects" to include *Persons*, as well as the products of persons – including mundane things, which seem beneath the notice of all philosophers except Joseph Margolis – such as tins of paint, soups, popular records, the Beatles and so on. The domain of culture also includes phenomena which would not have existed *except through the productive activities of persons*, including some short-lived particles created in high energy experiments and all of the so-called, and rightly called, "man-made elements". Included must also be the organisms which would not have existed apart from human intervention in breeding and domestication.

This view faces what might seem to be some insuperable objections. One is that if the World was created by God, surely a being who is a person, then the World and all its contents, including electrons and unseen rocks, are all cultural entities, the cultural artifacts of God. Then the position collapses because it is utterly trivial. What this objection shows is that substantial socio-scientific hypotheses are not "metaphysically-theory-neutral", contrary to the stated beliefs of some social theorists [Berger and Luckmann, 1967]. This is true, if only because the social sciences must necessarily take some position on the ontological status of referred entities, even if this is a scientific instrumentalist position. Thus I align myself with the atheistic, agnostic and alternative theistic traditions which maintain that there is no such God and that there is no reason to consider the mere logical possibility of His existence as affecting philosophical theorization here.

Another objection can be made from the perspective of classical logic and rebutted from the perspective of noneism [Routley, 1980 (1)]. Keuth [1974, p. 410] is quite right in pointing out that Popper's Third World forms an inconsistent set: it must contain inconsistent objects such as inconsistent scientific theories. This is only problematic within the framework of classical logic and set theories and the problem is readily solvable from a noneist alternative.

Yet another objection: Mario Bunge [1981] considers Popper's Third World to be a Platonistic "fantasy". We consider now whether his objections are applicable to my position, and if so, if they are sound. Do I attribute ontological autonomy to the cultural domain apart from its producers? To answer this question, Bunge invites us to answer the question raised by the *Gedanken experimente*; what would remain after the actions of a Divine Annihilator [ibid., p. 147]? Clearly no *concrete* cultural entity would remain for they are constantly destroyed by human beings, not merely by Divine Annihilators. However the question of annihilation does not arise for *conceptual* cultural entities, since it presupposes existential assumptions which are rejected by the accepted logical orientation here. What is being proposed then, is that not all cultural entities exist outside of the matrix of social practices of their producers. A spark-plug which magically appeared in ancient Rome and was used as a gem cannot be said to be a "spark-plug" anymore. Spark-plugs have been traditionally taken as "*functional-entities*", entities whose essential properties are derived from internal relationships with other components in a *system* (even if they are temporarily on the shelf as spare parts: they are at least potentially usable in a mechanical system). On the other hand some cultural entities such as specially bred animals, would survive, if only for a time if the Annihilator disintegrated all human beings. Thus the ontology of cultural entities is quite free from the charge of being a "Platonistic fantasy" for concrete entities. But of course, well known problems about abstract reference remain outstanding. This is a general philosophical problem not one peculiar to my approach.

At this point, some additional concepts are needed if the Culturology position is to be further explicated and systematized with our previous account of human nature. Consequently, we shall introduce some definitions and postulates which are taken to characterize the Culturology position. These are summarized in the following table; we take the signs:

"$\sim$", "$\vee$", "&", "$(\forall)$", "$(\exists)$", "$\Box$" and "$\Diamond$"

to have their usual classical meanings, and

"$\rightarrow$" and "$\equiv$"

to be signs of (satisfactory) implication, and mutual implication respectively. We introduce, where convenient, other signs.

Table 8.1. Definitions and postulates of the culturology position

(CD1) Reality $\triangle(p) = df \{(\forall F)(\forall R^{\infty}_{n \geqslant 2})(\forall p_1)(\forall p_2)$
$\ldots (\forall p_n) \ldots [\text{Concrete } (p_1, p_2, \ldots$
$p_n, \ldots) \;\&\; F(p_1) \;\&\; F(p_2) \;\&\; \ldots \;\&\; F(p_n)$
$\&\; \ldots \;\&\; R^{\infty}_{n \geqslant 2}(p_1, p_2, \ldots p_n, \ldots)]$
$V[(\forall F^*)(\forall R^{*\infty}_{n \geqslant 2})(\forall p_1)(\forall p_2) \ldots$
$(\forall p_n) \ldots)] [\text{Conceptual } (p_1, p_2, \ldots$
$p_n, \ldots) \;\&\; F^*(p_1) \;\&\; F^*(p_2) \;\&\; \ldots \;\&$
$F^*(p_n) \;\&\; \ldots \;\&$
$R^{\infty}_{n \geqslant 2}(p_1, p_2, \ldots p_n, \ldots)]\}$

where "F", "F*", are property variables, "$R^{\infty}_{n \geqslant 2}$" and "$R^{\infty}_{n \geqslant 2}$" and n-placed, and possibly infinitely placed relations and where "$\triangle$" is the relation "is an aspect of", such that "Reality $\triangle$(B)" reads "B is an aspect of Reality".

(CD2) Universe $\triangle(p) = df \{(\forall F)(\forall R^{\infty}_{n \geqslant 2})(\forall p_1)(\forall p_2)$
$\ldots (\forall p_n) \ldots [\text{Concrete } (p_1, p_2, \ldots$
$p_n, \ldots) \;\&\; F(p_1) \;\&\; F(p_2) \;\&\; \ldots \;\&\; F(p_n)$
$\&\; \ldots \;\&\; R^{\infty}_{n \geqslant 2}(p_1, p_2, \ldots p_n, \ldots)]$

(CP1) $\Box$ [Universe $\triangle(p) \equiv \sim$ Universe (p)]

(CD3) Universe $\triangle$(p) & Particular (p) = df [Thing (p) V
V Being (p) V Agent (p) V Mentalite (p)
V Person (p) V Cultural entity (p) V
System $\binom{n}{1}$(p) V Mixed System $\binom{n}{1}$(Things)
Beings, Agents, Mentalities, Persons,
Cultural entities)]

(CD4) System $\binom{n}{1}$(p) = df
(1) System $\binom{1}{1}$(p) = System (p);
(2) System $\binom{2}{1}$(p) = System (System (p));
(3) System $\binom{k}{1}$(p) = $\underbrace{\text{System (System(... (System) ...)}}_{k \text{ times}}$

(CP5) $\Box$ [System $\binom{n}{1}$(p) $\equiv$ (System $\binom{n}{1}$(Things) V System
$\binom{n}{1}$(Beings) V system $\binom{n}{1}$(Agents) V System $\binom{n}{1}$
(Mentalities) V System $\binom{n}{1}$ (Persons) V System $\binom{n}{1}$

Table 8.1 (Continued)

(Cultural entities) V Mixed System $\binom{n}{1}$(Things, Beings, Agents, Mentalities, Persons, Cultural, entities])

(CP6) $\Box$ [Thing (p) $\equiv$ ~ {Being (p) V Agent (p) V Mentalite (p) V Person (p) V Cultural entity (p) V System $\binom{n}{1}$ (Beings) V System $\binom{n}{1}$(Agents) V System $\binom{n}{1}$ (Mentalities) V System $\binom{n}{1}$(Persons) V system $\binom{n}{1}$ (Cultural entities) V Mixed System $\binom{n}{1}$(Beings, Agents, Mentalities, Persons, Cultural entities)}]

(CP7) $\Box$ [Being (p) $\equiv$ ~ {Thing (p) V Agent (p) V Mentalite (p) V Person (p) V Cultural entity (p) V System $\binom{n}{1}$ (Things) V System $\binom{n}{1}$(Beings) V System $\binom{n}{1}$ (Agents) V System $\binom{n}{1}$(Mentalites) V System $\binom{n}{1}$ (Persons) V System $\binom{n}{1}$(Cultural entities) V Mixed System $\binom{n}{1}$(Things, Beings, Agents, Mentalities, Persons, Cultural entities)) & [Capacity-to-Become (p, (Agent (p) V Mentalite (p) V Person (p)] }

where "Œ" symbolizes "the essential property of".

(CP8) $\Box$ [Agent (p) $\equiv$ ~ (Thing (p) V Being (p) V System $\binom{n}{1}$(Things) V System $\binom{n}{1}$(Beings) V System $\binom{n}{1}$ (Agents) System $\binom{n}{1}$(Mentalities) V System $\binom{n}{1}$ (Persons) V System $\binom{n}{1}$(Cultural entities) V Mixed System $\binom{n}{1}$(Things, Beings, Agents, Mentalities, Persons, Cultural entities)) & ~ (Cultural-entity-not-a-Person (q) & (q = p)) & ~ $\Box$ (Mentalite (p)) & ~ $\Box$ (Person (p)) & (Capacity-for-Action (p))]

(CP9) $\Box$ [Mentalite (p) $\equiv$ (Agent (p) & ~ $\Box$ (Person (p)) & Œ Mind (p))]

(CP10) $\Box$ [Person (p) $\equiv$ (Mentalite (p) & Cultural entity (p) & (Œ (Power-Engage-Cultural Practices (p))))]

(CP11) (1) $\Diamond(\forall p)(\exists t)$(Concrete Cultural Entity$_{t_n}$ (p) $\rightarrow$ ~ Concrete Cultural entity $_{\sim t_n}$ (p))

(2) $\Diamond(\forall p)(\exists t)$(Person$_{t_n}$ (p) $\rightarrow$ ~ Person$_{\sim t_n}$ (p))

(3) $\Diamond(\forall p)(\exists t)$(Mentalite$_{t_n}$ (p) $\rightarrow$ ~ Mentalite$_{\sim t_n}$(p))

(4) $\Diamond(\forall p)(\exists t)$(Agent$_{t_n}$ (p) $\rightarrow$ ~ Agent$_{\sim t_n}$ (p))

where "t" is a time variable, and "$\sim t_n$" is any time not t_n.

Table 8.1 (Continued)

(CD5)	Society (p) = df [Social System (p) & $(\forall q)(q \sqsubset$ Composition (p) $\rightarrow$ Social Being (q))] & (Social Structure (p))]
(CP12)	$\Box$ (Social System (p) $\equiv$ [System $\binom{n}{1}$(Social Agents) V System $\binom{n}{1}$(Social Mentalities) V System $\binom{n}{1}$(Persons) V Mixed System $\binom{n}{1}$(Social Agents, Social Mentalities, Persons)])
(CD6)	Social Being (p) = df $(\exists q)(\forall p)$(Mutual Co-operative Actions (p, q))
(CD7)	Mutual Co-operative Action (p, q) = df $(\exists R)$(Project (R) & Acts (R, p & q))
(CD8)	Project (R) = df [Object of Action (R, p) & (Agent (p) V Mentalite (p) V Person (p))]
(CD9)	Social Structure (Society (p)) = df Nature (p)

4.3 Mind and Culturology

So much then by way of initial semi-formal exposition and we turn immediately to comment upon this material. I should state at the outset that the table does not attempt to provide satisfactory non-circular accounts of a number of concepts, nor does it strive for economy of presentation attempting definition through as small a number of concepts as possible. However I have outlined concepts which were defined or accepted from the work of other authors earlier, from which such an account may be developed.

Our central task is to understand the nature of persons and systems of such persons: societies. Persons are cultural entities, but a special type of cultural entity: one which has a *specific type of mind*, and engages in cultural practices. Having such a *mind* and having the capacity to engage in cultural practices – these being any actions involving cultural entities such as languages, logics, symbols, scientific theories, art works, tools – are not taken to be merely accidentally related properties but rather it is by virtue of the peculiar mental properties of persons that they can engage in such practices. But in making this qualification we need to specify more precisely what "mind" is. This dreaded task is unavoidable and I certainly cannot present here anymore than a very general account, the supplementation of which is an enormous task on its own.

Supposing we reject the idea of the mind championed by Descartes and Eccles which arguably takes the mind as a numerically distinct, but interacting particular with the body. And supposing that we are also sceptical of current materialist accounts of the mind, we are faced then with the prospects of presenting our alternative. The problem from the perspective of a *Systemic Emergent Realist* perspective is no longer causal-ontological insofar as one must explain how two *things* so unlike such as the mind and the body interact, but rather conceptual and theoretico-explanatory. Let us work our way into these difficulties.

It should be noted at the outset that there can be no non-trivial metaphysically-neutral account of mind. Consider for example the mutual conflicting claims which any such account would need to accommodate: many social scientists still remain reluctant to attribute a full range of mental capacities to at least other animals, whilst pan-psychists would attribute some range of mental qualities (proto-minds) to inorganic particulars. This situation, fortunately, supports one general conclusion which this work has arrived at: the need for placing one's position within the framework of a systematic *Weltanschauung*. But lacking this here, we cannot avoid begging-the-question against some positions from the outset. Thus our aims here must be very modest indeed hoping merely to present some initial clarification of presupposed concepts consistent with the broad theoretical commitments of this work.

We present the following account of mind summarized in Table 8.2.

On the account summarized below, mentality is not a unified phenomenon which some particulars have in total, and other particulars lack totally. True, there are particulars without mentality at all. We maintain against the pan-psychist that rocks are one such particular, but there is also a qualitative difference in the mental capacities which we can reasonably attribute to other animals and machines. The position which strikes me as most plausible is to recognize three types of mentality and to attempt to classify both animals and machines (and possibly even supernatural entities) on this basis. Hence so that the principle categories of Tables 8.1 and 8.2 do not conflict, we must maintain that some particulars have a capacity for action even though we would not ascribe mental properties and capacities to them. This is, to say the least, a thesis which most philosophers of action will find utterly absurd. Indeed, even on the question of mind my proposals stated merely as bare definitions would still be taken by many philosophers and social theorists to be faced with insuperable philosophical problems. For I must then attribute sentient-mentality to all Mentalites. But then I must admit that non-persons can be attributed properties of perception, belief, intentionality and thought. In any case, many of the properties and capacities which I must then allegedly attribute to non-persons are propositional (and highly intensional) in nature. For example in presenting an account of perceptual experience, it has been plausibly argued that to have perceptual experience is (in many cases) to acquire true or false beliefs about the objects of perceptual [Pitcher, 1971] [Margolis, 1973 (2), 1978] [Armstrong, 1973]. However this claim conflicts with widely accepted views that belief, intentionality and thought, requires a linguistic capacity [Vendler, 1972] [Hartnack, 1972] [Malcolm, 1972–73] [Davidson, 1975].

We concede at the outset that on the assumption that animals, being non-persons, lack linguistic ability and are thus incapable of beliefs or thoughts

Table 8.2. The nature of mind: some explications

(CD10)	Mind (p) = df {Sentient (p) V Sapient (p) V Enculturalmind (p)}
(CD11)	Sentient (p) = df {Experiences (p) → ~ [Propositional-Perceptual Experiences (p) V Thought-Conceptual Experiences (p) V Beliefs (p) V Intensionality (p) V Knowledge (p) V Mental Images (p) V Dream-Hallucinatory Experiences (p) Memories (p)] & ~ □ Conscious (p) & ~ (Self-Conscious (p))}
(CD12)	Sapient (p) = df (∃ Properties)[Sensations (p) & Propositional-Perceptual Experiences (p) & Thought-Conceptual Experiences (p) & Beliefs (p) & Intentionality (p) & Knowledge (p) & Mental Images (p) & Dream-Hallucinatory Experiences (p) & Memories (p) & Conscious (p) & Concrete Rationality (p)] & ◇ [Self-Conscious (p)]
(CD13)	Consciousness (p) = df Aware of its Experiences (p)
(CD14)	Self-Consciousness (p) = df Aware that it Experiences (p) = df Consciousness (Consciousness (p))
(CD15)	Multiple-Self-Consciousness $\binom{n}{2}$(p) $= \text{df} \underbrace{\text{Consciousness} (\ldots (\text{Consciousness}}_{n \text{ times},\ n \geqslant 2} (p) \ldots)$
(CD16)	Concrete Rationality (p) = df ~ (Abstract-Encultured Rationality (p) & Rationality (p)
(CD17)	Abstract-Encultured Rationality (p) = df Rational Practices involving Cultural Entities (p) (e.g. mathematical practices, scientific, philosophical, linguistic, ethical practices etc.)
(CD18)	Enculturalmind (p) = df {(Sapient (p)) & □ (Self-Consciousness (p) & (◇ Multiple-Self-Consciousness $\binom{n}{2}$(p)) & Œ (Abstracts-Encultured Rationality (p))}
(CP13)	□ [Person (p) ≡ Enculturalmind (p)]
(CP14)	□ [Mentalite (p) → Sapient-mind (p)]

about language, linguistic structures and propositional entities. We do this of course under the death penalty of (vicious) contradiction, for otherwise the distinction between mentalities and persons, and between sentient-mind and enculturmind collapses. Further we must grant that even if one accepts a neutral account of propositions, taking them to be *sui generis* inexistential objects, we must use sentences or some functional elements of speech to identify the individuate propositions. However whilst beliefs are relations between a believer s and an object believed, or some state of affairs, granting the thesis that propositions may only be identified and individuate linguistically, it does not follow that a capacity for belief presupposes a linguistic ability. The reason for this is that "Believes (s, p)" does not imply that "Believes (s, True (p))". I could believe that a certain fruit is good to eat, without believing that the proposition "Joseph Wayne Smith likes Bjuras" is true, for I may not know what the name of this tasteful fruit is. Further, the object of my belief then is about a certain complex psycho-physical state of affairs, not about a proposition's truth-value. The state of affairs in satisfying the sentence which in turn designates the proposition, would justify us in claiming that the cited proposition is true, but it does not follow from this, that the object of my fruit-belief is a proposition. Indeed it is not.

I maintain that at least some non-linguistic animals may be reasonably attributed at least some beliefs – as well as intentions and thoughts. In stating this I would also recognize that I know of no satisfactory reductive and/or extensional account of beliefs, intentionality and thought.[16] These concepts then may well be incapable of non-circular explicit definition, open only to explication by their relationship with other concepts in an intensional circle. It is impossible to fully analyse these concepts here, but I believe that the very general points which I wish to make can be made without this.

There are some well known objections which can be advanced against such a position, even if it is granted that such a position is not outright nonsense. It may be alleged that we are utterly incapable of saying what the content of animal beliefs are, even though we frequently do so, and seem to be able to predict, moderately well, the behaviour of some animals. Stich [1979] maintains that it is a necessary condition for an animal to be justifiably attributed beliefs that we can in turn attribute concepts to the animal. But, he asks with regard to the dog Fido, what would count as evidence for the claim that Fido could distinguish between a real bone and a fake one?

The answer to Stich's question is simply this: consistent identification and discrimination. "Concept", in one sense of the term is synonymous with the term "idea". Ideas are much like propositions, quite abstract epistemic phenomena. I agree that no mentalites have ideational powers. However in a more general sense we are justified to attribure conceptualization powers to any entity which has the capacity to correctly identify objects and do this not by mere accident but by its nature. I have observed that not only can dogs differentiate between real and fake bones, but they can readily become so pampered that they come to prefer some meats to others if given a choice, and consistently identify their favourite meats by smell, even if we go to the trouble of attempting to trick the dog. Therefore Stich's criticism is unsound, since we do in fact have (fallible) evidence that the required discriminations are made.

A stronger argument however has been presented by Davidson [1975, pp. 22–23]. The argument this time hinges upon what Davidson takes to be a necessary condition for having beliefs at all: namely having the concept of belief. Having the concept of a belief does in fact seem to presuppose linguistic ability if we are to take a restricted account of conceptualization which equates conceptualization with the possession of propositional knowledge, especially of the truth conditions of belief sentences satisfying Tarski's convention T. But taking such a restricted account of conceptualization leads to problems. It is unclear whether even we philosophers have as yet a satisfactory account of belief (and certainly of knowledge, as the debate began by Gettier demonstrates [Gettier, 1963]). Consequently it seems unreasonable to deny that animals have beliefs merely because they do not have the concept of a belief. Davidson's requirement thus seems to be far too strong.

We have concentrated here on critiques of the ascription of beliefs to animals but the arguments against attributing thoughts to animals can be met by similar types of arguments. At least my general line of response should be clear by now. I wish to state that we have done nothing here to further clarify the nature of thought, belief, intentionality and so on. This task is of course the general research programme of the Culturologist orientation, constrained by the metaphysical parameters of avoiding reductionist materialism on the one hand, and equally popular reductionist forms of (idealistic) social holism and structuralism. This is merely the philosophical task and its formulation and development remain – let alone the task to supply a comprehensive socio-evolutionary account of mind [Griffin, (ed.), 1982]. My very modest goal here has been to sketch and defend some basic categories and we shall immediately continue with this task.

4.4 Agency and Culturology

We have also made a controversial assumption which requires some elaboration. "Agents", as defined here, refer to those particulars without minds, but which are not merely things such as rocks, or even self-organizing systems found in physics and chemistry [Haken, 1978], or "powerful particulars" [Harré and Madden, 1975] such as coiled springs and acids. "Agents" here, must be distinguished from "Agents" in the more widely accepted philosophy of action sense – thus we distinguish between S-agents (the former) and PA-agents (the latter). PA-agents would on my classification be either mentalities or persons; S-agents are those particulars which are not in a developmental process of becoming either a mentalite or a person; an uncontroversial example of an S-agent is a multicellular animalicule. But more controversially, I would wish to ascribe some capacity for action even to non-microscopic fixed soil plants. Here, if panpsychism is false we have an instance of phenomena which lack mentality but are not merely, if Emergentism is correct, powerful particulars such as acids, or near mechanic self-organizing systems. The plants and bacteria, whilst not reasonably attributed purposes are undeniably involved in changing their surrounding environment in accordance with their *Bauplän* [Gould & Lewontin, 1979]. Plants in turn radically alter the environment: recall from evolutionary cosmology that without the oxygen producing plants, there would not have been a suitable environment for the animal kingdom.[17]

Yet, how can we meaningfully extend action categories in such a way that S-agency is not merely an entirely new concept from PA-agency which bears no philosophically significant relationship to it? And to further complicate matters, I cannot under pains of inconsistency deny that a significant distinction exists here. Unless this dilemma is solved, then it would be established that the category of (S)-Agency in Table 8.1 is redundant. It seems then that the Culturology position is faced by a major conceptual problem.

The problem may be tackled by reflecting upon a currently outstanding conceptual problem existing in the philosophy of action, posed in its most quotable form by Wittgenstein in section 621 of *Philosophical Investigations* [1953] : what is left over if I subtract the fact that my arm goes up from the fact that I raise my arm? This is to say, what is the difference between an action and a mere bodily event? As I stated earlier, this question is problematic once we reject the idea that bodily events are basic relative to actions and counter-instances can be produced once we reject this assumption. A result of this is that it is possible to support the so-called "Agent Causality Thesis" [Taylor, 1966, 1974] [Chisholm, 1966, 1971, 1976 (1), 1976 (2)], which I maintain appears to resolve my own dilemma.

I will assume at the onset that any adequate account of action must be a *causal account*, that the concurrency theory (action is a behaviour b, and there is a mental event M concurrent with b, and both b and M occur) and various contra-causal accounts of action have been refuted [Davidson, 1963]. We have not the space here for repetition of well-known arguments, and the reader is referred to the secondary literature. To support my thesis, I will argue against the accounts of two popular causal-mentalistic accounts of action, maintaining that they do not supply intuitively adequate accounts of action being subject to a range of counter-instances – or being outrightly circular. On this basis, by recourse to an Agent-causality account of action, I hope to show that there are no existing good reasons for excluding agency categories to plant life, although of course we may only justifiably ascribe an S-agency category, not, under further threat of contradiction PA-agency categories. We do not, and cannot, address here further major philosophical problems arising from the use of PA-agency categories.

Frankfurt [1978, p. 158] has maintained, correctly I believe, that an adequate causal (mentalistic) theory of action does not at present exist and it is unlikely that one could be constructed to avoid well known counter-examples, as the causal theory locates "the distinctively essential features of action exclusively in states of affairs which may be past by the time the action is supposed to occur" [ibid., p. 1958]. In this gap it seems possible that some involuntary event which occurs *accidentally*, resulting in the occurrence of a particular event, which would have occurred if an action was performed, but cannot be considered to be anything more than a mere bodily movement.

Donald Davidson [1963] has been taken by a number of authors (e.g.

[McCullagh, 1975]) as advancing the following account of action: behaviour b_1 of agent A is an action a_1 if and only if (1) there is at least one (true) description of it which corresponds to what the agent intended, and (2) the behaviour was caused by beliefs and wants which rationalize the behaviour. Insofar as we accept this account as characterizing Davidson's position (cf [Blumenfield, 1981, p. 314]) it, along with the more popular idea that an action is a behavioural event caused by a reason, stands open to a general criticism. Such theories of action merely require that bodily events occur as the effect of prior mental causal antecedents (these may well be physical under alternative descriptions), failing to specify that the agency-relationship must also apply to the intermediate periods of time in which actions are performed. One can then produce counter-examples where in this gap, regardless of the causal antecendents, the effect cannot be regarded as a mere bodily movement.

Recent defenders of the *volition theory* [Ripley, 1974] [McCann, 1974, 1975] may well seem to be able to avoid this problem – but only at the expense of failing outright to solve Wittgenstein's problem. For both McCann and Ripley, actions are to be distinguished from mere bodily behaviour insofar as volitions are the causes of behaviour which are classifiable as actions. I am not concerned with whether or not McCann and Ripley have avoided well known regress arguments which have faced past volition theories. Rather, I will argue that volition accounts merely transfer Wittgenstein's problem to another level.

Volitions for Ripley, are taken to be any "conscious or mental act which initiates and guides the physical change that is brought about deliberately in a physical act" [Ripley, 1974, p. 143]. For McCann, volition are thoughts which are causally basic: "there is no difference between an act of volition occurring on my part and my performing such an act, any more than there is a difference between an act of arm-raising occurring on my part and my raising my arm" [McCann, 1974, p. 466]. The difficulty facing both accounts as general theories of action is that they merely push Wittgenstein's problem to the level of mental acts. Now we need to know what in fact mental acts are, perhaps by contrast to externally caused mental events.

The agent causality position of Taylor [1966, 1974] and Chisholm [1966, 1971, 1976 (1), 1976 (2)] is in general, a claim that a person may bring about movements of his/her body without doing anything else or having some prior volition. Alongside causation by events, there is taken to be a second view of causality, agent-causality, this being the power of an agent to produce certain results. In any action description, essential reference to an agent must occur to convey this as an act; there must be a reference to myself as an "active being" [Taylor, 1966, p. 11] [Chisholm, 1976 (1), p. 69–72].

Taylor [1966, p. 120] admits that the statements:

(1) This man started that forest fire

and

(2) This match started that forest fire

are grammatically identical in every way except for their grammatical subject and truth content. The match cannot literally start a fire as a man might, rather, all that (2) asserts is that certain conditions were causally sufficient for the starting of the forest fire and the match figured among those conditions. These conditions are such as to provide no ultimate explanation of the fire [ibid., p. 121]. But in the case of the man starting the fire, quite unlike the event causality illustrated by the match example, the man is, Taylor maintains, not a *passive agent* undergoing changes in response to external conditions, but the initiator of change himself. Likewise for Chisholm, in cases of event causality every event results from some other event; in agency-causality, there is some set of events caused, not by other events, but by the agent. The answer to Wittgenstein's question is, then, that what is left over after "subtracting" the arm going up, from my lifting of it, is simply *the Agent*. Actions are then nothing more than the doings of agents, that is, their causal intervention in the world.

It is difficult to maintain the distinction between agency-causality and event-causality given recent developments in the philosophy of causality which have taken all causal relationships to be relationships between "agents" (= "powerful particulars") [Harré, 1970], [Harré and Madden, 1975] [Bhaskar, 1978 (1)]. However Taylor's point remains that there is a fundamental distinction between S-agents and mere powerful particulars, even if these powerful particulars are not passive in Taylor's sense. Perhaps the spontaneous disintegration of some element is a change which is not merely a passive change. The difference between S-agents and radio-active nuclei, is that, despite the uncertainty effect, the reaction between the nuclei and its environment is "stereotypic" – no change in morphology occurs in the nuclei in response to the environmental conditions: either decay occurs or it does not. Further there is no novelty in response between the decay products and the environment, any degree which could significantly compare to the extremely complex and novel interactions between plants or animalicules and their environments. The environment is not changed in accordance to any *Bauplän*. I would thus maintain that both human beings and plants are S-agents in this sense – they are capable of dynamic interactions with their environments, which may result in mutual morphological changes. Therefore I maintain that my "agency" category is not a logically vacuous one.

It has been argued by Thalberg [1972, 1978] that the agent-causality position is incoherent even for human action. In the case of a person walking, there is no other event except the agent causing the leg movements. The agent therefore causes the motion of her body. But, as Thalberg argues:

> Earlier, to avoid giving a circular analysis of action, namely that her act of walking consists in her agent-causing her act of walking, the agent causationist denied that she caused her action. But if we continue assuming that it is events that are agent caused . . . , then we realize that the agent causationist is both asserting and denying that the housewife agent-causes the same event. [1972, p. 41]

Thalberg is right insofar as the agent-causation asserts that an act of walking consists of a further act of agent-causing the act of walking: this seems to generate an infinite regression. There is only one act *in this respect* (which is not to beg questions of further act identification): this is the agent walking. Now insofar as causes are thought of as explaining the effects they bring about, appealing to the causal interventions of the agent *does* explain why certain movements occurred rather than not at all, as only certain sorts of movements can be taken to constitute the act of walking. Getting to the shop by falling down the hill is not "walking to the shop". On the other hand it is not necessary to deny that nothing further can be said either physiologically or socially about this activity. Walking is not a miracle, and has *necessary* physical causal and social causal antecedents. None of these claims are demonstrably inconsistent with the agent-causationist position.

I should stress once again that my aim here has been to defend merely the coherence of my "agency" category. This is the bare starting point of a *general* philosophy of action, not its conclusion. Questions of act identification and individuation have not been considered. Insofar as I am maintaining that general agency, S-agency may be attributed to non-mentalities, a new area of research is outlined for action theorists. It may well be that the conjectures here are ultimately demonstrated to be incoherent, but this will only be seen by further exploration of the position. Thus I advance as a conjecture, one which seems to me at present plausible and logically coherent, that our use of non-mentalistic/non-intentional action categories (S-agency) is a legitimate one as a general metaphysical classification. However, once more I stress that such a classification in itself does not take us very far at all methodologically; logical reconstruction must begin somewhere and the initial stages seem always to be plagued by the problem of extreme generality.

4.5 Societies and social being

Just as many philosophers have considered animals to have a "degenerate" or minimal mental life, so have social theorists narrowly restricted their attention to human societies. Sociology has not therefore been the "science of society". Whilst we make social ascriptions to animals constantly, understanding animal behaviour *sociologically* has been a task left to ethologists and physiologists, usually of mechanistic orientations. A satisfactory alternative cannot be presented here, if I even had the rough outline of what this might be. The task is of monstrous philosophical proportions because relative to the foundational work done in the philosophy of physics, the philosophy of Sociology stands relatively undeveloped – and hence the "crisis of Sociology" tradition.

Unfortunately nothing more than a relatively minor altering of this situation has been undertaken here. Whilst having responded to the criticisms of various reductionists little more than a first step development of an alternative position has been undertaken here, and if there is one point which must be communicated above all others, it is of the work to be done. This and further points will be taken up in the concluding section. Thus the limitations of these remarks as well are acknowledged from the outset.

Societies have been taken to be systems of social beings. Like all particulars for the Systemic Emergent Realist, social beings are not unrelated atoms, but possess essential relationships with other social beings – be these Agents, Mentalities or Persons who have the capacity to engage in mutually co-operative practices. Thus, given that I am prepared to accept that plants may well be taken to be "agents" from an emergentist, anti-reductive anti-vitalist perspective, talk of "mutually-co-operative practices" between plants is not outrightly incoherent if our goal is to produce a definition broad enough to fit all of nature. The "practices" or "actions" here of course are not explicable by reference to any ideas, reasons or mental aspects of the agents involved. Nevertheless, relationships such as that of mutual cross-fertilization, seem to me to involve mutual co-operation, and thus a specifiable system of such plants, constitutes a plant society or plant community. Most frequently this community will consist of a multitude of organisms involved in various biological relationships. That most likely biochemical explanations can be given of many of these relationships, hardly seems a sufficiently strong reason to deny sociality/communiality to such organisms.

Societies most frequently, and uncontroversially recognized are those of mentalities and persons. To simplify the enormous difficulties of this area somewhat, my remarks will in general be restricted to societies of persons. Those social theorists who have expressed an "ontological commitment" (i.e. that they exist) to societies, especially of a functionalist and structuralist orientation, have tended to take societies to be "structures". This tendency is replicated in Bhaskar's work, where he takes societies to be "an articulated ensemble of . . . relatively independent and enduring (social) structures" [1978 (2), p. 14]. This tendency leads, as Anthony Giddens [1981 (1)] has argued, to disastrous consequences for the philosophy of "Sociology". Let us review such consequences, as well as the difficulties facing Gidden's own social ontology. Whilst my following discussion of Bhaskar and Giddens is critical, my remarks should not be interpreted as suggesting that I do not regard these works of high importance; a contemporary starting point for research in this area.

For Bhaskar [1979], society is "both the ever-present *condition* (material cause) and the continually reproduced *outcome* of human agency" [ibid., p. 43]. His social ontology, called "the transformational model of society", views the individual/society connection as follows: both the society and the individual are necessary conditions for each other, but neither can be explained or reconstructed in terms of the other. Social structures, unlike natural structures are only relatively enduring, and are therefore, he maintains, possible objects of transformation. Social structures neither exist independently of the activities they govern nor independently of the agent's conceptions of what they are doing in their activity. Further, societies are taken to be necessarily imperceptible and not only cannot be identified independently of their effects, but do not exist independently of their effects either [ibid., p. 57].

Bhaskar's view of society bears fruitful comparison with some aspects of Giddens' sociological project [Giddens, 1976, 1977, 1979, 1981 (1), 1981 (2)]. Here I wish to concentrate only upon a fragment of that project: his social ontology. Social systems are taken to exist only insofar as they are reproduced in social encounters by social agents, but such reproduction of social systems does not take place completely under conditions intended by agents. The process of reproduction depends first upon the processes of material reproduction in which human beings labour to produce food, clothing and shelter. But practical activity must be co-ordinated, and language serves as the medium by which this is done. The production of society through social interaction is a skilled accomplishment of actors, a factor stressed, he has argued, by "interpretative sociologies" at the expense of considerations of structure. Structuralism has on the other hand tended, Giddens believes, to regard the reproduction of society as a mechanical process without a subject. Giddens attempts to dissolve this dilemma by reworking some basic sociological categories. The actionist/structure dilemma, is of course a principal concern of Bhaskar. Undoubtedly if this could be done, it would be a major sociological accomplishment.

Social systems for Giddens are reproduced relationships between either individuals or collectives or between both, throughout "time-space". Social systems then are taken not to be structures, but rather to have structural properties and relationships. Structures are recursively organized generative rules and resources. The precise details about such roles and resources need not concern us here.[18] The models by which social systems are reproduced by agents through the use of generative rules and resources constitute the *structuration* of a social system. This concept of structuration commits Giddens to accepting two propositions: "(a) that social systems are structured only in and through their continual and contingent reproduction in day-to-day social life; and (b) that the capability/knowledgeability of social actors is always *bounded* (although in historically mutable ways)" [Giddens, 1981 (1), p. 172].

A central concept connects the positions of Giddens and Bhaskar: *the duality of structure*: "social structures are both constituted by human agency, and yet at the same time are the very *medium* of this constitution" [Giddens, 1976, p. 121]. Language illustrates the duality of structure: language exists as syntactical and semantical structures only insofar as there are syntactical and semantical linguistic rules. Now the rules of syntax for example, Giddens proposes, "refer" to the reproduction of "like elements", and yet at the same time, the rules generate the totality of speech acts which thereby constitute the language.

We will now briefly review the problems which the concept of the duality of structure generates, beginning with Bhaskar's position. The major allegation is this: it is not clear what it means to say that societies exist only in their effects, given Bhaskar's own realist theory of causality [Bhaskar, 1978 (1)]. If it is "powerful particulars" which are causally related, "causal-effects" or pure "effects in-themselves", without any powerful particular as a cause are impossible as, since one of the relata of the causal-relationship is missing, the concept remains ill-defined [Smith, 1982 (2)]. Furthermore his transformation theory of society renders any evolutionary account of the origins of society impossible: a regress into the past is generated by the claim that society *always* pre-exists the individual. The problem is not altered by pointing out that our evolutionary ancestors were social, but merely shifted one stage back. If there was a time in the evolutionary past where a transmutation from non-social individuals to social individuals occurred, then Bhaskar contradicts himself. But if not, it is difficult on his assumptions to see, given basic evolutionary assumptions (by contrast to Creationism) *how there could be societies at all.*

The duality of structure according to Giddens, results in social structures being constituted by human agency as well as being simultaneously, the medium of such constitution. But this *cannot* be so. If something requires a medium for its production, such as sound requires a substance medium, then that something cannot simultaneously produce that medium. We seem here to be taking action as a (prior) necessary condition for structure and structure as a (prior) necessary condition for action, which leads us immediately into a vicious infinite regression. The example of language production does not support Giddens' position. The rules of syntax do not "generate/produce" language. Rather it is the agents in speech enacting situations who do this. That the sounds they produce constitute language rather than mere unstructured babble, depends *in part* upon syntactical relationships between sounds. It is true that in speaking I draw upon various syntactical rules constitutive of English, and thus reproduce them as structural properties of the English language [Giddens, 1981 (1), pp. 171–172], but it remains unclear as to what is the "medium" of such constitution. Is it the token speech act, the English language itself, or what? To admit that the medium is the *syntax* of English itself, leads us into the vicious infinite regress outlined above.

Societies for the Systemic Emergent Realists are systems among others in the universe possessing their own characteristic emergent properties. This in part explains "Sociology's" pre-occupation with organic analogies for what, in fact, "Sociology" and biology share is a common source of metaphysical categories. This is not to deny that both fields possess their own unique set of concepts which are seemingly irreducible to each other. Social structures are natures or essences of societies, described by the best available socio-scientific theory. This is a concept which seems to fit quite well with Bhaskar's [1978 (1)] realist philosophy of science, a position which in fact takes the job of science to be the exploration of the nature of particulars. Thus it is impossible to say without empirical inquiry what such natures will be. In general, a society may have a social structure by virtue of being part of a general super-system of societies (a *general essence*) such as for example the world capitalist system. On the other hand a specific society such as British society, will, insofar as it is a particular which can be individuated, have an individual essence which distinguishes it from other societies of the kind, most likely based upon the characteristic historical, political and economic properties and relations of Britain.

A sufficiently general social ontology will avoid, as accounts such as Bunge's [1979] does not, attributing properties to societies in general, which only known societies have had, but could well be lost in some imaginable society. Thus whilst economic relationships are present in all known societies, even if such relationships are internally related to various cultural practices such as religion, we can imagine, without theoretical violence to our imaginative faculties, "Logan's Run" type, technocratic, fully automatic societies, where production is carried out by machines and the citizens wallow in hedonistic pleasures. Beyond stating that such societies will still contain people who will be engaging in perhaps at present unimaginable forms of cultural practices, very little can be said about such societies. However if we take the set of human societies to be *all* human societies – past, present and future as Bunge does [ibid., p. 190], our account is open to counter-examples from the theoretically possible future. Thus we should be extremely careful in the generalizations made and incorporated into our definitions. My account, I believe, is not subject to this difficulty.

Problems of individuation and reidentification throughout time, are especially acute in social ontology. It would seem that with regards to the former problems, accounts which distinguish societies from other socio-systems by reference to the concept of self-sufficiency, [Bunge, 1979, p. 177] are hardly cogent in a world where, according to neo-Marxism, the world monopoly capitalist system could hardly survive as a socio-system in its own right without imperialist adventures and where according to neo-classical economics, trade is the laxative of the economy. On my account, any socio-system may be treated as a society insofar as the concept of social structure can be defined. There seems to me to be good philosophical reasons for accepting such a broad account. First the term "society" is used in a wide number of senses which more restricted explications must ignore. Second, the account here can readily deal with sorites style paradoxes which would, if found, establish that no societies at all existed: take one individual from a society and one has still a society, take two individuals from a society, and one still has a society, and so by the principle of mathematical induction one may take away all individuals from a society, and still have a society! But this conclusion is absurd for we can only recognize the (virtual) existence of social groups without members, such as a vacant university dart club, insofar as there are other membered non-virtual social groups which maintain the club on the books. It seems to be the case, given the tacit assumptions of this argument, that no societies exist at all. (Methodological individualists should not rejoice since the same argument has been used against the existence of persons [Unger, 1979].) However, if we are prepared to admit that whilst *one* person can never constitute a society, (although he/she can be regarded as the tragic last survivor of what was once a society), two persons engaging in mutually co-operative social practices can, for some purposes. Two survivors can still communicate, build huts (or fall out shelters) and other cultural entities. Beyond this, it is up to the working Sociologist to "conceptually carve up" the social world in ways which are theoretically satisfactory for presupposed tasks. Isolated cities and prisons may well be regarded as "societies" for some theoretical purposes.

Problems of the reidentification of social entities through time raised some extremely difficult philosophical problems. As particulars, societies may have parts (that is mutually co-operating members) which are, for some temporal periods, not on the planet Earth at all. This occurs for example, in space flights. Astronauts, whilst leaving the Earth seem still to remain a part of human society insofar as their activities are rigidly governed by the commands of home base and to lose such communication is regarded as a major disaster. Hence human societies as particulars have some rather ontologically unique spatial properties.

Further, it seems that whilst a society may over time replace all its members, or at least many members, without a significant change in nature, some social entities such as rock groups like the Beatles, disintegrate precisely when certain key individuals lose interest in engaging in mutually co-operative song writing practices and produce solo albums (even prior to legal dissolvement of the partnership). Further, without reincarnation, it seems now (post 8/10/80) logically impossible for the Beatles to reform, even though only one member is dead ("it would not be the same"). On the other hand, some rock groups seemingly remain the "same" even though they may change most of their members over a number of years or radically change their musical styles. Mundane facts like this, which occupy our everyday lives (the reader can substitute his/her own examples) certainly produce difficulties for philosophical theories of individuation and reidentification, in addition to all the current general difficulties raised by this topic. Here, I can do no more than recommend the subject for their research. This is I believe, no trivial recommendation, given the relative uninterest ontologists have shown towards the domain of Cultural being.

5.0 Conclusion: The state of the argument

Without laboriously repeating the arguments made in this chapter, a fair and general summary of what has been attempted, would be that a very general, abstract and *certainly not* fully detailed, account of human nature and of the particulars of the universe has been sketched. The account fails to be, I concede from the outset, even a satisfactory ontology, since to understate the issue, a large number of important questions have been avoided. Indeed the account of culture presented here, to be given adequate exposition, rather than mere initiation, will need an intellectual synthesis of logic, semantics, metamathematics, meta-science, and so on – in short a world system. Even small fragments of this programme are brisk with difficulties. How, for example, do cultural beings such as persons manipulate abstract cultural entites such as numbers, that is to say, how is the knowledge of and practice of mathematics possible?

I, in general must maintain, under pains of having my entire argument collapse, that none of the pressing difficulties *of detail* raised *and demanded* by the "crisis of Sociology" argument are met by the Culturologist position in its present superficial and schematic form except to rebut very basic sceptical arguments such as the utter vagueness of sociological concerns. Further, whilst my case against Sociobiology and Physicalist Scientific Unificationism stands or falls on its own merits, only the bare abstract shell of an alternative is hinted at here. To expect more in one work is to make utopian demands.

In particular nothing specific has been said as yet about some of the severe explanatory and methodological problems of "Sociology/Culturology". I do not believe that these can be solved without first achieving considerably clear understanding of what for example "reasons" are, and what in fact "acting on reasons" involves. We need detailed accounts of the mind – what is it to "think", what is involved in believing and conceptualization, and especially, if a rationality other than abstract encultured rationality (itself a landmine of problems as current debates in meta-science demonstrates) is admitted, what precisely is involved in "judgements" of such a rationality? I have no ready answers to now offer to such cognitive "black holes".

Nothing has been said about the role of critique in the social sciences, and nor can any *specific answer* be given here in closing. *Schematically though,* the answer is rather simple: if the narrow specialization and departmentalism present in our educational institutions is rejected, in the case of "Sociology" being reduced to absurdity by the "crisis of Sociology" argument, insofar as ethics are taken to be a cognitively meaningful intellectual activity, systems of ethics are merely a part of one's world view, and just as one judges some proposals via methodological standards, so can one ethically evaluate various programmes. Critique in the social sciences then consists of the evaluation of the truth and rationality of positions, as well as their ethical goodness, and rightness. If the idea of the specialized, departmentalized social theorist, who can satisfactorily understand social reality and the domain of culture from a metaphysically neutral position is rejected, questions of not merely understanding the world, but changing it in the interests of all *Beings* cannot be ruled out as an out-of-bounds-question, without prior demonstration that such questions are cognitively meaningless (a demonstration non-cognitivist ethics has failed to give us [Allen, 1975]).

The *good society* is literally one which conforms to the prescriptions of *the* satisfactory ethical theory. The "satisfactory ethical theory" is just that one which produces the right answers to all the pressing questions of metaethics and normative ethics. It is no wonder that all the "utopias" of past social theorists have been so vague, for the resolution of all these questions in such a way that one can obtain a coherent system is a task of utterly enormous difficulty. The same of course applies to the general systems metaphysical proposals made throughout this work. If the critic could show that no condition of "systematicity" [Rescher, 1979] can be satisfied, especially coherence conditions (even non-standard noneist coherence conditions) then my position here will be refuted.

Further significant progress cannot be made here. Therefore this work concludes, and our major intellectual labours continue.

CONCLUSION

The state of the overall argument of the work

It is maintained that the Principle Thesis (P.T.) of this work has been rationally supported. That is to say, there are no good reasons to believe that a "reduction" of Culturology/Sociology to Sociobiology, in total or in part is theoretically possible. It has been argued in this work that the PSU idea of the unification of sciences by theoretical reduction to physics is not merely unjustified, but is a claim which upon analysis is quite obscure. Further the system of metaphysics of PSU has been critically examined here and found lacking. This work has in general attempted to answer many of the questions of its central theoretical concern by arguing that these questions themselves are far too obscure for rational contemplation and cogent philosophical defence. This work may be characterized from other works of its genre by extending this profane philosophical scepticism into the sacred realms of the domain of the natural sciences, especially biology.

It was argued that Sociobiology has a major role to play in PSU unification programme. The reasoning here was that since PSU universally assume that much of non-bogus biology, especially evolutionary genetics, is in principle reducible to chemistry via biochemistry, the seemingly strong explanatory power of Sociobiology enables much of human and animal behaviour to be understood in purely physico-chemical terms. It has been argued here that Sociobiology is theoretically continuous with the leading view of evolution, the neo-Darwinist synthesis. Critics of Sociobiology have frequently been unaware or simply unprepared to recognize that their epistemological and methodological arguments damage neo-Darwinism. I have given reasons, in what I hope was a strong and sustained argument, for rejecting the neo-Darwinist position as an adequate theory of evolution. Further in the course of the argument many other aspects of received mechanistic biology have been criticized, from genetics to sex role research and ecological Sociobiology. The reductionist thesis collapses immediately in such a situation of general epistemological crisis and immediately leaves us with the difficulty of offering a rational reconstruction.

The solution which I gave to the "Crisis of Sociology" argument was that "Sociology" needs a detailed *Metaphysical Weltanschauung*, and the same must be said for the fields of biology considered here. Such a *Weltanschauung* would guide theory construction and enable a choice between mutually conflicting orientations. This is my conjectured solution and only initial thoughts have been given here about these issues. This limitation is legitimate, although regrettable, given both the radical nature of my proposals and the human intellectual condition. Looking at each field of knowledge from symbolic logic to "Sociology", one finds foundational difficulties of an acute magnitude. Such difficulties must of course in part be explicable by internal factors to the respective disciplines, yet for the epistemological holist this cannot be the complete explanation, so that such widespread difficulties must indicate a major defect in current conceptions of knowledge. Part of this difficulty is, I conjecture, the widespread lack of metaphysically satisfactory "world pictures". Indeed, even in advancing this last statement further difficulties should be evident since it remains to clarify the nature of metaphysics and metaphysical arguments and to explicate precisely the relationship between metaphysics and science.

This work has not produced a satisfactory resolution to the problems raised from establishing (P.T.) by its adopted methodology. Human beings, especially academics, long for neat, parsimonious solutions to difficult problems, and are generally irritated by critical and sceptical treatises. Given the present state of human knowledge, there is I believe, a historical place for irritating works such as this one, as cognitive cathartics. The resultant state of irritation and disorientation should not result in *ataraxia*, but rather in the need to construct a new world view.

APPENDIX 1

Sociobiology and ideology

1.0 Introductory statement of the argument

The text gave no explicit consideration to the question of the ideological nature of Sociobiology. I shall attempt to correct this omission to some degree in this appendix. The Sahlins-Science for the People position, that Sociobiology is ideological, will be defended here. The strength of this position can be demonstrated by distinguishing between a number of questions which the defenders of Sociobiology simply collapse; these include the questions of the truth-value or theoretical reasonableness of a position on the one hand, and the mutual relationships between scientific theories, political ramifications of a position and the ethical consequences, as well as the ethical relevance of such a position on the other.

In defending the thesis that Sociobiology is an ideology, I must give a brief account of what I consider to be an adequate theory of ideology. I will maintain that the equation of the domain of the ideological with the domain of the false is a mistake, which suffers from a severe paradox, first adequately formulated by Karl Mannheim. To avoid this paradox, it is necessary to distinguish between the domain of ideology, and the domain of truth. Hence it is logically possible that *true theories* can be ideologically laden. By such a strategy I hope to show that most of the rejoinders made by defenders of Sociobiology are quite irrelevant to the central issues of the debate, formulated by critics such as Sahlins and Science for the People. Consequently Sociobiology, even if true, is indeed a rationalization of reactionary politics and the hungers of virulent Western Capitalism.

2.0 The Sociobiology-as-ideology argument

Two of the most well known critiques of Sociobiology are the Sociobiology Study Group of Science for the People [1977] and Marshall Sahlins [1977].

I too, wish to build up as strong a case as possible against Sociobiology on "ideological" grounds. We use this term as a primitive for the moment: we will need to examine various accounts of ideology and opt for a position of our own. First however, I hope to clarify and sharpen the debate by criticizing some objections which have been made against Sociobiology. Then I consider the arguments which I think succeed.

2.1 A Criticism of some of the weaker arguments

There are two arguments – one from the Sociobiology Study Group, and one from Sahlins, which I will consider here. Both arguments are unsound and unsatisfactory in my opinion.

One argument from the Sociobiology Study Group, Ruse [1979, pp. 93–96, 1981, pp. 232–234] has gone to great lengths to rebut, although there are some dilemmas facing Ruse himself. The argument is that both the language of the Sociobiologists and the pictures in their texts are "sexist" in nature. Let us accept the term "sexist" as unproblematic, as both parties at the dispute do. The mere existence of the use of universal masculine pronouns [Sociobiology Study Group, 1977, p. 138] is not restricted to Sociobiologists, nor sadly is the use of terms and examples with sexist connotations. The former can be found in Marx and Engels themselves, and the latter is widespread in a sexist society such as ours (i.e. at the *least*, the Capitalist West). So this argument could only at best show that Sociobiology reflects patriarchical prejudices. Yet the language could be altered, and the pictures which offend, be removed if necessary. Indeed even the worst case of sexism, ironically the Pallas paperback edition of Michael Ruse's *Sociobiology: Sense or Nonsense?* (1980), with its semi-pornographic-surrealist drawing of two machomen fighting for a "shapely" naked female prize, would probably be defended by Ruse on the grounds that Pallas did this, not him. Yet in rejecting this line of attack it is difficult to restrain oneself from asking: why do book sellers and publishers do such naughty things to the profound and sincere writings of such a respectable philosopher? The answer obviously is to sell books, oblivious of ethical standards, and it seems that even academic booksellers and publishers are not above the tactics employed by publishers of pulp novels.

A second argument which I believe is very problematic is found in Sahlins' text [Sahlins, 1977, p. 101]. Sahlins' argument is that Sociobiology is the logical and historical outcome of an utilitarian social ontology, the essential elements of which "from differential success to the competitive struggle to reproduce one's stock and the transfer of powers" [ibid., p. 101], is torn straight from the pages of Hobbe's *Leviathan* and he cites Talcott Parsons [1964] as one important source. Now implicit in this position are two arguments. The first is the shallow claim that Sociobiology sees human kind as being in a Hobbesian state of nature: a claim which the Sociobiologists would rebut by pointing out that one of their central concerns is *altruism* (if only a genetically selfish one at that), and in any case the Hobbesian world is maladaptive and if it ever existed, would be most open to invasion by alternative Evolutionary Stable Strategies. Sociobiology cannot be defeated that easily, unfortunately. The second argument is more interesting and will be considered in depth.

The argument from Parsons' utilitarian dilemma is that (1) The Utilitarian dilemma (Ud) is epistemologically problematic; (2) Any social theory committed to Ud is epistemologically problematic; (3) Human Sociobiology is committed to Ud; (4) Therefore human Sociobiology is epistemologically problematic. What is wrong with this argument is premise (1).

In *The Structure of Social Action*, Parsons took Utilitarianism (U) to be a *Philosophy of action* comprising four theses: (1) Atomism; (2) Rationality; (3) Empiricism; (4) The randomness of ends. Atomism takes the unit of analysis of social inquiry to be conceptually isolated act units. Individuals are conceptualized as being "discrete" with regard to ends ("ends" being the future states of affairs to which intentional action is orientated) except insofar as these ends are indispensable to the social system qua social system [ibid., p. 52]. Hence there is in this approach a principal conceptual concern with the acts of individuals and an ignorance of the integration of ends into systems, or if this is given, then laid stress is given upon the lack of *integration* of such ends into social systems. The *Rationality premise* (U2) is the proposition that action is rational insofar as the selection procedure for ends is supplied by the canons of "positive empirical science" [ibid., p. 51]. The *Empiricist premise* (U3) on the other hand, is the proposition that the categories of a given theoretical system (for the "positivist theory of action", positive empirical science) are adequate to explain all significant aspects of the phenomena in the domain of explanation of that theory. Finally Utilitarianism has according to Parsons, no theory of the relationship of ends to each other, but only of the means–ends relationship. If taken as a complete theory, this would lead inevitably, Parsons believes, to the conclusion that there can be no significant relationship between ends at all, that they are random. This is taken by Parsons as a *reductio ad absurdum* of the Utilitarian theory of action.

If (U1), (U2), (U3) and (U4) constitute an inconsistent 4-ad, one or more premises must be denied. Parsons considers the modification of (U4) and (U2). *Radical Rationalistic Positivism* denies (U4), that is that ends are random, and alleges that the basis of non-randomness is through a scientific knowledge of reality. This Parsons believes, results in the elimination of the "analytic" independence of ends, and a collapsing of the distinction between the *conditions* of a *situation* (that which the actor has no control over) and the *means* (that which the actor has control over)[ibid., p. 64]. Consequently the agent is reduced to rationally understanding and adapting to his/her situation, although Parsons rightly puzzles over the causal importance of this reflective process, given the collapse of means into conditions.

Radical Anti-Intellectualistic Positivism modifies (U2): the actor's knowledge of the situation may not be fully adequate. Causally, this may arise from (1) error or ignorance, or (2) elements intrinsically outside the domain of "positive empirical science", both elements of which are in the conditions of the situation, and according to Parsons' definition, are therefore outside the bounds of intervention of the agent.

The significance of all this for the Sociobiology debate, which Sahlins has directed us to, seems to be as follows: both departures from the Utilitarian theory of action, lead to a position where the explanation of human action is made by reference to non-subjective conditions and environment, with the difference being that *Radical Rationalistic Positivism* allows a mediation of heredity and environment through scientific rationality, whereas in *Radical Anti-Intellectualistic Positivism*,

> . . . this medium is dispensed with and it is by means of an "automatic" process which, if it is subjectively manifest to the actor at all, is so only in terms of which render effective adaptation and control impossible, positively only as error. [ibid., p. 67]

Now, insofar as hereditarian and physico-environmentalist categories are scientifically adequate, this is, insofar as the empiricist premise (U3) is justified, the rational legitimacy of the action frame of reference is called into question. It can be no more than a pre-analytic scaffolding to build a "non-subjectivist" theory, which is dismantible once the theory is built, which Parsons believes Social Darwinism does, and which Sahlins believes Sociobiology must do as well. Here, ends are not only random, but they are irrelevant insofar as human beings are determined by impersonal forces outside of their control. In addition the Hobbesian problem of order is readily solved: it is "meaningless" [ibid., p. 113].

Such an argument is only as good as Parsons' premises, and these are not good. A strong case can be made for Sociobiology being a form of "genetic atomism" involving molecularly false myths of a "central directing agency" [Webster and Goodwin, 1982], but this thesis is distinct from the definition of atomism presented to us by Sahlins through Parsons. Second, it is quite problematic as to whether Sociobiology is a form of empiricism, given its reference to underlying genetic mechanisms, but it is by far more problematic, as to whether Parsons' account of empiricism is even coherent. If we reflect upon Parsons' definition of empiricism a paradox arises: surely Parsons would wish to allege that the categories of social theory themselves are scientifically adequate to account for all social facts, but *why* then regard the solutions to the Utilitarian dilemma as problematic? But if this is so, then any wide ranging social theory – especially grand scale Parsonsian social theories, would be rendered problematic. So Parsons refutes himself. Second, the Utilitarian dilemma argument is quite questionable: the concept of *randomness* as defined, does not support Parsons' argument, as an actor basing his/her goals on "scientific knowledge" does not therefore reduce the means of a situation to their conditions. Ethical naturalism does not imply determinism. Third, even for Bentham and J.S. Mill, surely the paradigm cases of Utilitarians, ends are simply not random in Parsons' sense. The principle of utility, that the good is that resulting in the greatest happiness for the greatest number, must by *definition* consider the relationships between ends, insofar as it is a universalizable moral principle. Utilitarianism may be an objectionable position, but it is not on the grounds that Parsons and Sahlins think it is.

This completes my criticism of two arguments that have been advanced against Sociobiology from a normative position. I turn now to the constructive part of my argument.

3.0 The idea of ideology

In this section, some clarification must be given to the concept of ideology. I would like to reject the equation between ideology and pseudo-science/science distinction. To further complicate our task, this question needs to be distinguished from the various ethical ramifications of a scientific theory. We shall discuss, briefly, both questions in the following section.

3.1 Pseudo-science and the is–ought distinction

The idea of a "pseudo-science" is one plagued by considerable vagueness, and innumerable conceptual difficulties. Let us take a recent account of this distinction by J.M. Blum [1978] and review its problems. For Blum, a pseudo science involves the following four elements:

> . . . (1) essentially incorrect results are generated: (2) these are successfully and persuasively disseminated to a substantial audience; (3) dissemination occurs by convincing the audience that the results are bona fide scientific conclusions; (4) the normal processes of error correction in science are retarded or prevented from functioning, so that incorrect beliefs are sustained over time. [ibid., pp. 145–146]

What is wrong with such a characterization, is that any "degenerating research programme" or paradigm in crisis would satisfy these conditions. (1) is satisfied by virtue of the fact that the research programme is *actually* degenerating, and that the paradigm is *actually* in crisis. It would seem from Kuhn's work [Kuhn, 1970] that all "paradigms" have the means to collect recruits, and (3) also is satisfied as most of these champions of particular positions, would surely argue that *their* position is rationally justified. (4) also would seem to be satisfied, insofar as the paradigm/research programme-tradition one opts for, does involve one in accepting certain "hardcore" assumptions, which may not be examined satisfactorily and if incorrect, will be perpetuated over time by social processes.

An alternative account of the nature of pseudo-sciences has been presented by Thagard [1978 (2)]. For Thagard a pseudo-science is any theory or discipline which (1) has been less progressive over an extended time period than competing theories and (2) practitioners make little effort to resolve outstanding difficulties which the position faces. Thagard's account is quite intuitively plausible. Nevertheless condition (2) raises problems for this account. All that is necessary to satisfy (2) is that *some* effort to resolve outstanding difficulties be made. This is clearly met by astrology, Thagard's paradigm case of a pseudo-science, as Watson [1981 (1), pp. 61–76] shows. Recall, that it is not necessary for such efforts to be ultimately rationally satisfactory (a requirement so strong as to rule out much of theoretical physics). It is difficult to see how we can tighten Thagard's principle, without thereby excluding some intuitively acceptable science in a state of theoretical difficulty, or at the other extreme, including too much.

Yet we need not accept cognitive relativism in admitting that there is no clear distinction between science and pseudo-science. It is still in principle possible to philosophically and empirically examine one's "hard core" assumptions, even if the myth of a logically conclusive falsification/critique is rejected. A popular argument by conventionalists has been that one can in fact prevent any critique of a research-programme from devastating it, by the addition of *ad hoc* hypotheses or by rejecting some auxiliary hypotheses. Indeed one can. But as was stated in the text, whether it is *reasonable* to do this depends upon the difficult dialectical process of balancing reasons both for and against either the hard core assumptions or the auxiliary ones. In general there is not, nor is it necessary to have, any general decision procedure for this. We have to dirty our hands and work through each of the arguments as they are made. Perhaps we may find that our very intuitions about the scientificity of a position are challenged in the light of such a critical review of arguments and counter-arguments, respectively for and against the position in question.

To sum up: the pseudo-science/science distinction is quite difficult to explicate, and I am sceptical whether this can be satisfactorily done, so that pseudo-sciences can clearly and distinctly be distinguished from false or technically flawed science.

The science–ideology distinction is of course one which I wish to retain. With special regard to the topic of our concern, Sociobiology, we must clearly distinguish this issue from the question of the ethical ramifications of Sociobiology, for such a distinction is, I maintain, both real and reasonable.

The ethical question of direct relevance to the "Sociology-as-Ideology" thesis, is the reasonableness of the appeal to the factual-evaluative distinction made by authors right throughout this tradition in rebutting claims of racism, sexism, ageism and so on. The argument is that moral conclusions cannot be deduced from non-moral premises. From Hume onwards, this has usually been advanced more as a challenge to the ethical naturalist, than as an independent argument. The issues are admittedly very complex. For each naturalistic account critics look at the premises and find them to some degree value-laden, and the argument is rejected. One wonders what the anti-naturalist is looking for, and what, if it is found, its relevance to moral reasoning is. Let us assume as all Sociobiologists do, that not only this distinction, but the debate itself is coherent. Now it is usually argued that if it is a fact that certain positions are racist or sexist, it does not mean that the proposers are saying that this is *good*. It is logically possible that there could be an author, call him Hensen, who believed that Negroes were utterly inferior in mental powers to whites. Yet Hensen despaired about this and raised his fists to heaven, cursing God for this terrible state of affairs.

So it is logically possible for one to accept scientific racism and yet reject certain amoral racist conclusions. But so what? It is logically possible also to accept both positions. Yet, even so, this issue is largely irrelevant. Even if a purely factual statement is made, it is still an open matter as to whether the *speech act* or *act* of publishing the statement is morally right or wrong. Hence the act of making certain factually true statements may be morally wrong. Possibly the best examples of this are works in the fields of IQ and race, IQ and social class, sex difference research and patriarchy. In a sexist, racist society doing such research is wrong, for it readily supplies intellectual weapons for racists and sexists. We know from the extensively documented history of such research that it has produced only misery and further crippling inequalities and exploitation, not the liberation of human beings *per se* [Chase, 1977] [Hubbard, Henifin and Fried (eds.), 1979]. In the cases of genetic engineering and nuclear energy there are at least *prima facie* proposable arguments as to why such research should continue, although it is another matter again whether these arguments are in fact sound. No so for much of human Sociobiology. In a racist society, Jensen's research, for example, can be reasonably regarded as morally wrong; in a non-racist society, it is at best irrelevant, such as the study of the length of one's eyelashes and the colour of one's eyes.

We have discussed this argument based upon the factual-evaluative distinction because it has been constantly used as counter-argument by the defenders of the ideological purity of Sociobiology. More of course, must be said on this issue. However, the next step in our argument is to gain some clarity about the nature of ideology. Presenting a Sociologically adequate theory of ideology is an enormous intellectual task. So also is the task of formally explicating the concept. It is the latter task, which must now be considered as the next logical step in our argument.

3.2 Theories of ideology

The term "ideology" is vague. It has been used in a number of senses, some not as relevant to the sociological tradition as others. In one sense of the term, an ideology is any set of articulated beliefs which constitute a *Weltanschauung*, that is, gives a comprehensive picture of human beings in the world. A less general use of the term restricts its reference to the value, interest and action orientated positions of social groups within a society and between societies themselves and with respect to other societies in the world system. Yet such a position is virtually too general to be of much sociological importance as it stands, without some theory as to why various groups have the values and interests that they have, and why they act as they do, and what in fact these groups are. Our concerns here must remain with the purely philosophical task of initial clarification and explication.

We owe the term itself to the French liberal philosopher Destutte de Tracy, who had used it to denote the Science concerned with the study of the genesis and reproduction of ideas, and it was assumed that an ideological study was a necessary condition for a scientific understanding of society. Yet such a science was not divorced from specific values and a political orientation: namely to provide the theoretical and conceptual machinery necessary for the reconstruction of French society, on the basis of laissez-faire economics and political individualism on the ruins of the old regime. So the term was born true to form, with a front garden of scientific neutrality, and with the main structure embodying a specific programme of social reform. This is how the critics of the early theories of ideology viewed it.

With Marx we find perhaps the most important development of the theory of ideology in nineteenth century social thought. In Marx's writings, at least two senses of the term "ideology" can be identified. One tendency is to draw a distinction between ideology on the one hand as involving a false and distorted view of the world, and science on the other, as a giver of truth and clarity. Marx's criticism of the classical political economics of Smith, Say and Ricardo includes the claim that they falsely regarded certain historically specific social relationships of capitalist social orders, as of trans-historical universality, and consequently as beyond social change. Marx's attack on the young Hegelians was on the grounds that their theories of the conditions of successful radical social change were flawed. Note, Marx did not, according to this interpretation, dismiss all of the work of classical political economy or neo-Hegelianism. Rather some of this work was false, some of their categories "reified" the social relationships from which they in turn were derived.

In the second sense of the term, Marx added a precise theoretical interpretation to the sectional groups mentioned at the beginning of this section. Here ideology is analyzed with reference to class categories. In this second sense he referred to the "ideology" of the bourgeois class, and to that of the revolutionary working class. In this sense the term refers to the value, interest and action orientations of these two main classes.

This story has been told in more detail thousands of times. What is of interest to me here is the response of Karl Mannheim to Marx. Marx believed that his theory of capitalist society whilst obviously value-laden in the support which it gave to the working class, was not ideological in the first sense of the term. It did not distort reality. Yet how can the distinction between appearance and reality be drawn in an objective fashion which is not merely dependent upon one theory-laden description, such as Marx's, and why ultimately should Marx's theory be accepted in contrast to that of say Weber or an élite theorist such as Pareto?

Mannheim [1960] pointed out that Marx discovered that criticizing one's opponents as being ideological (in Marx's first sense) was a way of deflating and discrediting their views. This strategy he argued, could be played upon Marx himself. Mannheim held that the sociohistorical and material ambitions of a group inevitably produced value-laden, distorted and consequently false positions. His own sociology of knowledge was directed towards an empirical analysis of the social causes of thought, and an evaluation of these positions with reference to their ideological distortion factors. Immediately now it seemed that the sociology of knowledge programme faced an argument from self-referential inconsistency, precisely to the conclusion of its own inconsistency. Mannheim's position itself is or is not value-laden and distorted, and hence false. If it is not, then his position is inconsistent, since such claims equating social causation of belief and distortion is precisely his position. But if it is, then Mannheim refutes himself since his own perspective can claim no more objective validity than any other.

He sought to escape this paradox with his doctrine of *relationalism*. A familiar strategy to cope with the self-referential argument against relativism, is to deny that there is any absolute concept of truth. Rather there is only truth relative to a framework. This is the strategy Mannheim opted for, in maintaining that knowledge is necessarily perspectivistic. Social groups allegedly were capable of grasping one perspective of history. Yet these positions remain opposed to each other, and illuminate inadequacies in various aspects of the other's position. A synthesis of the good pieces of fragmented picture of social reality, could presumably be made by the relatively classless intelligentsia. They are united in the quest for the truth, since they are not fragmented by class biases.

At this point the argument turns full circle. Given a relationalist solution to the relativist dilemma, it is not clear how one can speak of a position illuminating inadequacies in another. It can only be by the internal standards of a position. According to the allegedly inadequate position, they are not inadequate at all. So what then is there to synthesize? Nothing. It is difficult to see how such inadequacies can be isolated, given relationalist assumptions in the first place. Second, in admitting that the intellectuals have a God's eye view of society, Mannheim must modify his relationalism. Further, from at least some perspectives (e.g. [Gouldner, 1979]) it seems that the intellectuals are hardly classless seekers of truth. If inconsistency is not to arise again, one needs to know what are the correct perspectivistic views of each position. One cannot compare perspectives with things in themselves from an aperspectivistic position. Mannheim did not say how one proceeded to justify one's synthesis from this point on. Consequently relationalism collapses back into relativism.

What is wrong with this argument, is Mannheim's assumption that human thought is completely "determined" by one's social condition, that is, the assumption that sociological determinism is true, and the assumption intimately related to this, that such thought is therefore distorted and false. Part of the problem hinges around interpreting the term "determined" and explaining how these "systematic illusions" of understanding are produced on a quite extensive scale, given the universal distortion of belief. In short, how do societies "function" under such conditions at all? In particular what is needed is some non-questioning argument giving us reason to believe why all group positions should produce distortion in the first place. Indeed what has group positions got to do with truth anyway? What has the theory of ideology got to do with truth anyway?

I contend that the most satisfactory response to Mannheim's dilemma is to abandon the equation of Science with "that free of ideological aspects", and Ideology, as that which is distorted and false. Consequently I maintain that true positions are logically possible candidates for being classified as ideological in some or perhaps all aspects.

The attempts of Althusser to distinguish between science and ideology are inadequate [Glucksmann, 1972] [Glucksmann, 1974] [Hindess, 1977], and the theory of ideology advanced by Habermas [1971] does not radically depart from the science-knowledge/ideology-distortion distinction. Stated briefly, Habermas maintains that ideology is distorted communication, a notion which he illustrates by means of the example of psychoanalysis. Ideology is intrinsically related to the critique of ideology, since to claim that such and such a position is ideological is to expose various non-rational processes distorting the thought, just as the object of psychoanalytic theory is to enable the patient to gain a rational understanding of his/her repressions, and thereby overcome them – or at least so he believes. Yet psychoanalysis is from the perspective of anti-psychiatry, anything but the hermeneutic weapon which Habermas characterizes it to be [Torrey, 1974]. In any case, even if this objection is met, Habermas' work does not lend much weight to the required distinction. Ideology, he tells us, is distorted communication, and distorted communication is any act of social communication which does not sustain the validity claims of the mutual intelligibility of that communicated, the propositional truth of that communicated, the sincerity of the communicator, and finally the rights of the communicators to speak as they do. Hence any symbolic system of communication – from logic, mathematics, physics to literature could be ideological. Such a characterization, independent of its merits, seems of little use to critics of Sociobiology such as Sahlins and the Sociobiology Study Group. Habermas has given a *rational* concept of ideology: Sahlins and the Sociobiology Study Group want to classify not the communications *between* Sociobiologists and critics as ideological, but the theory itself, and this is not permitted by such an account.

Roy Bhaskar's attempt to reconstruct Marx's theory of ideology also accepts the knowledge-science/ideology-distortion distinction [Bhaskar, 1979, pp. 86–87]. Systems of beliefs and theories are ideological if and only if *critical*, *explanatory* and *categorical* criteria are met. Hence to claim that Sociobiology is ideological, one must possess a theory S or set of theories Σ, which explain most of the phenomena explained by Sociobiology under its own descriptions, or those of Sociobiology's if incommensurable, as well as explaining an additional set of phenomena. Second, for the *explanatory* criterion to be met, S or Σ must explain the conditions for the continual acceptance of Sociobiology and specify its limits and conditions of transformation in the social structure, and S or Σ must be able to explain the conditions for its own acceptance and so specify its limitations. Finally, according to the *categorical* criteria, Sociobiology must be unable to meet the necessary conditions for scientificity or the minimum necessary conditions for the theory to sustain the nature of the subject matter, and S or Σ must satisfy both of these conditions.

These conditions represent mutually necessary and jointly sufficient conditions for Sociobiology to be classified as ideological. Note that Bhaskar's conditions are *strong*, that for any theories I and S, if the critical, explanatory and categorical criteria are met, I and S are ideological. So religions, inadequate "scientific" theories and inadequate epistemologies are all so classified. However, human acts are not classified as ideological, since the term is defined by reference to theories and beliefs. So the definition is both too strong and too weak. It also fails to classify Sociobiology as ideological. Possibly both the explanatory and categorical criteria can be met. The critical criterion is not. Sociobiology would seem to have a *prima facie* great explanatory power, so great in fact, that many critics have taken the theory to be flawed precisely for that reason. Yet even if from a philosophical point of view Sociobiology is flawed, it is still logically possible that it can offer explanations for a wide range of phenomena, albeit flawed ones. The critical criterion would fail even if according to S or Σ, given that human Sociobiology might be untenable, S and Σ are both incapable of accounting for phenomena in the non-human world. Thus Bhaskar's position renders it impossible to classify most flawed biological reductionist theories – even if they are quite blatant in their legitimization of the status quo.

Thus, most importantly, Bhaskar's position does not explicate the important insight which is found in Marx: that ideologies legitimize the interests of the dominant classes. Bhaskar's position does not capture this, although the explanatory criterion would seem at first sight to do so. But one can explain the conditions for the acceptance and transformation of a theory without necessarily investigating the relationships between the power structure of a society and the dominant ideas and theories about economic, political and social phenomena, in short questions of *hegemony* and *domination*. For ideological ascriptions can still be made of theories, on Bhaskar's account, even in non-hegemonic social orders.

I would propose that a theory of ideology must investigate the ways in which actions, practices, theories, institutions – in short the domain of culture itself, legitimates the interests of hegemonic groups in society, and the influences of those groups upon other groups in another society (or possibly the entire society). The concepts of hegemony and domination cannot be understood independently of the concept of *social power*. I cannot consider all alternative explications of this concept here. I will however state my account. The concept is definable with respect to the ideas of actual and possible *wants*. The *interests* of an individual are those wants which S has already or may come to have in the future. The *real interests* of S are the set of possible wants which it is possible for S to have, which if satisfied promotes the development, flourishing and well being of S. The *interests* of a social group SG are simply those wants which members of SG already have or may have in the future. The *real interests* of SG are the set of possible wants for members of SG, which if satisfied promotes the development, flourishing and well being of members of SG by virtue of their membership of SG. To say that power relationships exist between individuals A and B is to say that either A has the capacity to use B and/or B's labours to satisfy A's interests, or vice versa. For social groups, SG1 has power over SG2 if and only if, at least some of the individuals of SG1 can use SG2 individuals and/or their labours to satisfy their interests. Mutual relations of power of course exist. Note the use of the modal "can". Powers can exist, even when not manifested in actions. The teacher has the power to keep John in after school, even if John is a good boy, and companies to stand down workers, even if they do not.

Note also that it is not necessary that agents be actually aware of the precise nature of the power relationships existing, for them to be truly ascribed as powerful. Indeed ideology manifested in the satisfaction of empirically given interests is only one form of power relations. Yet there are deeper *implicit levels* of ideological penetration. An example is the phallus-oriented sexual practices of Western societies, where "real sex" is equated with phallocentric acts of some sort [Rotkin, 1976]. Such a relationship, it can be argued, is a subtle expression of the power which men have over women in this society, and indeed others throughout history: it defines the legitimate and satisfactory modes of sexual conduct. It is not by itself conclusive evidence of this of course, but we would expect further power relationships to interlock and mutually support this one. This example does serve to illustrate the common feature of ideology: particularistic interests hiding behind claims of general applicability and universality. I suspect very deep ideological relations would be found in many other areas of human social relationships, such as the experiences of time, place and order, although the sociology of knowledge programme has yet to supply such deep hermeneutics. But at least we can now return to the Sociobiology debate with a rough idea of what we mean by the term "ideology".

4.0 Sociobiology as ideology

I will outline and defend two arguments against the criticism of Ruse and the Sociobiologists. The first is that Sociobiology is sexist. The second is a more general claim that Sociobiology is reactionary and a support to virulent capitalism, even though it is logically possible that a politically radical and biological reductionist Sociobiological theory might be constructed.

4.1 Sociobiology as sexist ideology

What August Strindberg put on the stage, Sociobiologists put in biology books; Dawkins [1976] entitles the chapter in his book discussing the relationship between the sexes as "The Battle of the Sexes". The title well summarizes the conflict which allegedly exists of biological necessity between the sexes. Women stand at a genetic disadvantage to men: not only do they have a capacity to produce per individual (although in population terms the general relationship must be one-one) but the female human is left to care for the baby (even if she is also working in the public sphere as well). This is not a contingent part of the theory like the clumsy and often silly language of the Sociobiologists, but is a deduction from the theory itself, by virtue of Trivers' notion of parental investment. Males therefore have a tendency towards "promiscuity", females towards "monogamy". Men are the *dominant creatures*, women are the *domestic creatures* employing different strategies to trap males and make them support their share of their genetic investment. Wilson [1979, pp. 129–130] sums up this image neatly:

> It pays males to be aggressive, hasty, fickle and undiscriminating. In theory it is more profitable for females to be coy, to hold back until they can identify males with the best genes. In species that rear young, it is also important for the females to select males who are likely to stay with them after insemination.
>
> Human beings obey this biological principle faithfully.

This position under one interpretation is logically incoherent. It is not possible for men as a class to be "promiscuous" whilst women as a class are "monogamous", for to be a "promiscuous" male is to precisely ensure that some females are not "monogamous". It is true that an arbitrary individual male can in fact inseminate and fertilize many females, and Sociobiologists well recognize that in doing this, there is one less female for another male: hence sexual competition. But if the unit of selection controversy has taught us anything, it is that one cannot infer from the sociobiological properties of individuals, that groups of individuals will have certain sociobiological properties. We cannot infer from the fecundate powers of individual males, that males in general are "promiscuous" and that females are "non-promiscuous", for we know that the conclusion is not merely false, but necessarily false. Hence Wilson's claim is merely a (perhaps deliberately) misleading way of stating that an arbitrarily chosen individual could fertilize many females. The Panglossian optimization assumption providing a theoretical justification for this claim has already been criticized in the text. Nor need we ask the Sociobiologists for some explanation and evidence that females do identify males with optimally excellent genes (rather than genes which score a good pass), how they in fact do it, and what "excellent genes" are; we also ignore the logical jumps from reproductive categories, to sexual categories, to marriage categories.

Now the Sociobiological view of the sexes provides an excellent argument, if it is sound, for patriarchy. What better legitimization of dominance can be given than that it is "natural", the optimally excellent state of affairs.

Michael Ruse [1979, 1981] would strongly disagree with this. Ruse's first counter-argument goes as follows: Sociobiologists have used imprecise metaphors, and yet given a technical restatement by the philosopher of science, any alleged ideological implications can be removed from the theory. Ruse then shows us how to do this [1979, p. 98, italics added]:

> . . . because some genes are passed on via sperm and some via ova, adaptively speaking they *must* give rise to different behavioural characteristics.

Unless a low redefinition of the term "behaviour" is given, the above statement is false. We grant immediately that physiological differences exist between "males" and "females" (ignoring epistemological difficulties raised by sex research with regard to these terms) and that certain behavioural differences might exist solely because of these physiological differences. Men for example will not have babies. But so what? This is trivial, and Sociobiologists want to say more than mere banalities. They would wish to predict a wide range of social behaviour from certain biological parameters. Ruse has committed himself to quite a hopeless task. The terms "sperm" and "ova" are names applied to the type of haploid reproductive cells which are respectively usually smaller and in a larger number in a given space-time slice, by contrast to the larger cells in a smaller number. There are real philosophical difficulties in these definitions, since they lead to the ancient "paradox of the heap" or sorites. But we need not consider this question here. The point to be made is that there is no necessary connection between gamete size and number on the one hand, and behaviour on the other in the animal kingdom, which could provide a theoretical grounding for Ruse's counter-argument. Certainly the claims of the Sociobiologists do not follow solely from Ruse's proposition.

Ruse's second argument is that not all the behaviour which genes give rise to is typical male chauvinist behaviour, as at least some behaviour of males indicates that they have "substantial domestic virtues" [ibid., p. 98]. But Ruse is immediately uncertain about the strength of this argument – as well he should be, since it begs the question at issue by merely multiplying the stereotypes – and he drops this argument and presents another one. The story this time is that even if men are conceived as being more dominant and "polygamous" than women (note the abuse of terminology confusing sexual and marital categories), this is not a normative claim at all, as it does not entail any evaluative assessment of the behaviours in question in terms of goodness and/or badness. Yet this objection is quite irrelevant, since we can conclude from our previous ethical inquiries and illustrations, a misunderstanding by Ruse of his opponents. Whilst most would want to argue that terms such as "dominant" and "polygamous" are value-laden, even if this was not the case, such a position could still be, and indeed is, ideological as defined here. All that is necessary, and sufficient for this assessment is if the legitimization of hegemonic groups occurs. The act of making such statements about the conflict, superiority and dominance of males does not escape ideological ascription. Nor in fact does the act of studying and arguing for such positions, for one is by the very process of inquiry accepting that such phenomena are of value as objects of scientific inquiry out of the potentially infinite number of questions which could be investigated, and putting one's arguments on the market.

The final argument given by Ruse [ibid., p. 100] is that even if all the other objections are met, at least there is nothing in Sociobiology to support "genetic fatalism", so we are not absolute slaves of our biology. I agree. This however does not affect our argument, since the ideological ascriptions are not made on the assumption that Sociobiology is a genetic fatalism.

So it is concluded that the defenders of Sociobiology have yet to meet the objection that Sociobiology is sexist. The rejoinders illustrate a basic lack of understanding of the significance of the Sahlins-Science for the People argument.

4.2 Sociobiology as the ideology of virulent capitalism

There has been a tendency in the critical literature to equate Sociobiology with a revamped form of social Darwinism. Yet it is clear that the Sociobiological theory itself is firmly based within the neo-Darwinist synthesis: Hamilton and Maynard-Smith first used kin selection theory and genetic based games theory to understand animal behaviour. If there is something wrong with human Sociobiology, it must be because of a radical ontological difference between humans and the rest of nature or else that there is something wrong with the position itself. I have argued that the latter position is correct. Here I would like to briefly explore the socio-metaphysical foundations of neo-Darwinism and its extension, Sociobiology.

First, I believe that Derek Freeman [1974] has established that Darwin was not significantly influenced by Herbert Spencer at all, and that it is problematic to the degree, if any, to which an aprioristic thinker like Spencer was influenced by Darwin. Darwin, Freeman points out, in his *Autobiography* denies that Spencer's work influenced him, and confessed difficulties in understanding it. Indeed Freeman goes on to argue, that there are extreme metaphysical differences between Spencer and Darwin, especially insofar as Spencer accepted that there was a force that transcends human knowledge and conception responsible for the drive to perfection. Evolutionary themes were characteristic of the Victorian era, so Darwin lived in a social environment which gave him exposure to a way of looking at the world. Yet evolutionary themes are not exclusive to the Victorian age: even some of the early pre-Socratic thinkers toyed with these ideas.

I believe that we can also accept that the idea of natural selection came to Darwin from observations of animal breeding, and that Darwin formulated the specific mechanism of evolution from reflection on the data collected on the *Beagle* voyage of 1831–36. Even so, Darwinism was influenced by specific metaphysical assumptions about the nature of organisms, assumptions which were themselves socially grounded.

As Webster and Goodwin [1981, 1982] have argued, Darwin was influenced, and indeed accepted unquestionably, the ideal of a historical science produced by the Enlightenment, which contrasted with the secondary place which earlier "Rational morphologists" gave it. This is a question logically distinct from the one which Freeman has discussed, and it is also one to which one could reply "well so what?" However Darwin accepted more metaphysics than just this; from the tradition of Protestant Natural Theology he accepted the theoretical importance of "content" over organic "form" and from a Humean empiricist and Utilitarian position he accepted an epistemology [Webster and Goodwin, 1982, p. 22]. Darwin himself acknowledged his debt to Natural Theology [Darwin, 1889] and it is hardly surprising that Darwinism conflicted with traditional Christian thought, especially Natural Theology, since it is essentially a secularized form of it [Cannon, 1961]. Natural Theology whilst accepting the designing powers of God, was concerned mainly with the empirical study of the appearance of the world, rather than with questions of form and type. Indeed organisms were conceived both as mechanical and functional entities, in particular to concentrate attention upon external relationships between organisms and the environment in demonstrating the utility of the parts of the organisms. In Paly's famous design argument, this was taken to illustrate the designing providence of God. There was a secondary role given to internal structural relations of the parts themselves of organisms, and a preoccupation with the individuality of organisms and the differences between them, rather than with common relationships of form. Consequently Natural Theology embodied an atomistic approach to the biological domain, with a concern with questions of adaptation and a lack of concern with questions of structure. In this sense Natural Theology fitted well with empricism with its equation of the real with the empirically observable. Darwin and the neo-Darwinist, and especially the Sociobiologists, have tended to conceive of organisms as functional entities, and restrict their attentions to observable facts at the level of the populations, or more recently the gene, and strive for the development of statistical theories. Yet all this is mere description, not science, unless there is some theory of how in fact organisms come to adapt to their environment once God has made an exit. In particular Darwin was faced with the problem of giving a generative mechanism for evolution, since natural selection is impossible without the existence of variations. Further Darwin lacked a satisfactory theory of inheritance and ontogenesis.

From Mendel, neo-Darwinists would gain their idea of the gene as an atomistic entity which acted as the "central directing agency" of the cell, and of the ontogenesis of the organism itself. The position is significantly modified to account for the interaction between non-allelic genes (epistasis) and to allow pleiotrophic effects, but in general genetic atomism is accepted, and to be dominant especially in Sociobiology.

From August Weismann, the position is taken that the increase in complexity in development occurs through the unfolding of a pattern that was already possessed by the germ cells, entities which remain protected from environmental influences. It is however generally conceded, that the unfolding of the pattern is only possible through an interaction with the environment. Weismann's dogma was gladly translated into the central dogma of molecular biology: DNA $\rightarrow$ RNA $\rightarrow$ Protein during the molecular revolution of the 1950s. Mendel's units of heredity were identified with segments of DNA, and variations in phenotypes alleged to be caused by mutations of the DNA either as genes in point mutations or through chromosome mutations, which occurred at random. Associated with the discovery of the nature of Mendel's mechanism, has been, we have seen, a trend to identify the "atom" or unit of natural selection, not as the group, or the individual, but the gene itself, and to deny the operation of natural selection at any other level. This reaches its logical conclusion in Sociobiology, with its mythology of "selfish genes" and its belief that genes are not only "central directing agencies" as Weismann thought, but "central controlling agencies" as well. Consequently Sociobiologists opt for a causally epiphenomenal view of organisms, where non-genetic phenomena such as reasons are of secondary importance. Hence, for example, parental manipulation may well occur even if it is contrary to the beliefs and reasons of human agents. Even where cultural forces are given some relevance, as in the Wilson-Lumsden gene-culture co-evolutionary theory [Wilson and Lumsden, 1981], any form of pure cultural transmission must be rejected as an evolutionary unstable strategy, and culture that wicked beast, placed back upon its "genetic leash".

If one views Sociobiology as the logical extension of the neo-Darwinist synthesis, then clearly if Sociobiology faces major foundational problems, these transfer immediately to neo-Darwinism in general. Whilst rejecting earlier Sahlins' Parsonsian account of Utilitarian atomism, I nevertheless agree that the atomistic individualism of capitalist society is projected upon the biological domain. Of course, neo-Darwinists and Sociobiologists are hardly alone in their pre-occupation with atomism: we witness the same preoccupation in the leading orientation in high energy physics [Shrader-Frechette, 1977], and atomism and

and varieties of mereological reductionism have occupied favoured positions in philosophy, logic [Routley, 1980 (1)] [Meyer and Routley, 1977] and as is well known, in social theory. Now such a general thesis illustrates but a vague and weak ideological component to a wide range of theories, and unless general and abstract considerations are presented the position may be rightly taken as an unchallenging one.

At a more concrete level Sociobiology has viewed animal behaviour through bourgeoisified concepts and has especially assumed that specific historical institutions such as the nuclear family are biological givens. Michael T. Ghiselin [1974], (certainly no mainstream Sociobiologist), has taken this thesis even further [Ghiselin, 1978] : biology and mainstream economics should be organized into a new branch of knowledge called "general economy". This discipline would provide laws and principles applicable to "everything in the universe having the requisite organizational properties" [Ghiselin, 1981, p. 309]. Needless to say such principles are in short supply, and may suffer radical deflation given the present foundational crisis facing mainstream economics [Hunt and Schwartz (eds.), 1973] [Schwartz (ed.), 1977].

Wilson [1975] has taken a large number of terms from natural languages and used them to describe animal behaviour: among these are terms such as "slavery", "altruism", "homosexuality", "division of labour", "transvestism", "investments", "exchanges", "barterings" and so on. Both Sahlins and the Sociobiology Study Group have given extensive discussion of this issue, and thus I shall not dwell upon the initial exposition and documentation of a well known argument. Now one is of course free to use terms as one wishes given one gives precise technical definitions of these terms, and consistently uses them. This has occurred in physics for example with respect to terms such as "charm", "strangeness" and the term "work", this latter term having some relationship to the ordinary language term. Sociobiologists have not done this, indeed part of the programme would immediately disintegrate if this was done, and the ideological ramifications of the position made more evident.

The ill-definition and misuse of terminology is further evident in Wilson's [1975] discussion of barter and reciprocal altruism. It is simply assumed without argument that barter is a form of reciprocal altruism [ibid., p. 551]. This mistake of definition, I argued in the text, permeates the Sociobiological tradition. Further Panglossian assumptions of the optimal excellence of the selfish Genotype, complete the picture of the bourgeoisizing of the natural world.

Ruse [1979, p. 75] has argued that the use of models is a legitimate part of the scientific enterprise itself, such as for example Kerkule's theory of the benzine ring based upon his model of a snake swallowing its tail. But the point in such modelling, he rightly points out, is that the source of the model and the modelled entity differ. And Ruse points out that physics has drawn its models from the human world as terms such as "force", "repulsion" and "energy" illustrate.

In reply to this argument it is pointed out that the cited terms of physics are given precise definitions by a physical theory. This, as I have argued, is not the case for Sociobiology, and it is especially evident in discussions of ethics where the commonsense and genetic concepts of altruism are used interchangeably. It has not been shown how the latter explains tokens of the denotation of the former, and this in my opinion is impossible for many instances. Secondly, Ruse and the Sociobiologists seem to believe that Sociobiological descriptions can be shorn of all normative content. I will believe this when they abandon the ambiguous and often contradictory use of commonsense for technical terms. Clearly the commonsense terms are value-laden, and if they are so analytically careless as to not even define them, and use them correctly, incoherence is their deserved fate.

Ruse protrays his misunderstanding of the sociological theory of ideology in maintaining that even if the critics such as myself are right in their claims that Sociobiology is infected with Capitalist ideology, one can still wave the is-ought distinction and rebut the charge [Ruse, 1979, p. 82]. The critics would be content in accepting the antecedental claim, which is the very point of their argument. This seems to be the solely relevant arguments which Ruse has given in rebuttal of our position. The arguments are rejected as being either unsound, or irrelevant to the point of debate. On this basis, the central thesis of this paper is alleged to be established, and the case made by Sahlins and the Sociobiology Study Group against Sociobiology therefore stands.

APPENDIX 2

A critique of Alexander Rosenberg's *Sociobiology and the preemption of social science*

This appendix presents a critical response to Rosenberg [1980 (2)]. The text has dealt, I believe, with the general issues upon which Rosenberg has attempted to construct his argument for the success of Sociobiology over Culturology, but since Rosenberg's text has gained considerable recent citation, a separate consideration, and a repetition of some material is justified. Further, this popularity is well deserved by Rosenberg's book, for it is a rich harvest of controversial arguments, brilliant insights and cogent summaries of the current foundational debates in economics and the social sciences. Such scholarly merits are, I maintain, consistent with a general failure by Rosenberg to justify his own central presuppositions of his orientation. It is at this level of debate, that the Culturologist should make his/her critical response to Rosenberg. I shall attempt such a response, referring to points argued for my position, and against PSU and "metaphysical materialism" in the course of this work, finally considering likely rejoinders.

Rosenberg accepts the metascientific thesis of "methodological naturalism", that is, the social sciences can adopt the methods of the natural sciences and provide explanations and predictions approaching the same degree of precision as the natural sciences [ibid., p. 3]. However there is an immediate problem facing those who believe that human action is subjected to invariant laws:

> The absence of such laws, or even of successively improved approximations to them, remains a continuing embarrassment to those empiricists who agree with Mill that methods, and the sorts of knowledge which the applications of such methods are to eventuate in, must be broadly the same in social and natural science. [ibid., p. 2]

Thus the methodological naturalist is faced with the trilemma of having either to reject his/her metascience, or admit that there are substantial methodological differences between the social and natural sciences, or else explain how in fact, given the truth of methodological naturalism, the human sciences have failed to produce the laws and experimental results which approach the natural sciences in their extent. Rosenberg takes the latter branch of the trilemma, claiming that to embrace "either of these [first] two alternatives is too sweeping a price to pay for an explanation of the failures of social science" [ibid., p. 3]. Unfortunately the text does not respond to the immediate reply that could be made to this claim: that such a "sweeping price" is one which many are prepared to make given these propositions:

(P)(1) The trilemma facing methodological naturalism, is conceded by all parties to the dispute to be a major philosophical problem, challenging the cogency of the assumption that human actions are governed by strict deterministic and invariable laws, or perhaps even by inductive-statistical laws of a general non-trivial nature. There is a strong presumption against methodological naturalism, especially given the present general concern with questions of parsimony: the failure of methodological naturalism, seems a simple and reasonable proposal.

(P)(2) Methodological naturalism is *prima facie* inconsistent, since it must maintain that if naturalism is presently true both that the social sciences are methodologically unified with the natural sciences, and also that they have failed to conform to such methodological tenets, *or* it is maintained that methodological naturalism is an ideal which any reasonable rationally reconstructed social science must conform to, and that no such unification exists.

The problem facing Rosenberg is that no non-question begging reasons are given for accepting the basic metaphysical assumptions of "epistemological empiricism" and "metaphysical naturalism", which would be required to shift the burden of proof back against the anti-physicalist orientation such as that adopted by myself. Rosenberg is explicit about this neglect:

> "I will not in this work devote much attention to expounding or defending any particular version of empiricism and physicalism, but shall simply assume that some particular version, perhaps one not yet even expounded or defended, is correct. The general lines of these two "isms" are clear enough, and the precise and defensible expression of these views is clearly among the most important tasks in philosophy. The failure of so many attempts to coherently expound these philosophical theories, let alone defend them, in the face of counter-examples and counter-arguments, reflects the enormous difficulty of such exposition and defense". [ibid., p. 22]

In the following passage he is explicit about the position which his opponents must therefore take:

> . . . the justification for my assumptions [i.e., of empiricism and physicalism] is nothing more or less than the claim that if my explanation for the failure of social science is a valid argument, anyone who disagrees with its conclusions and its imperatives for future work in the sciences of man is *ipso facto* committed to rationalism in epistemology, and to anti-materialism in metaphysics. [ibid., p. 9]

Again, this is a conclusion which one may willingly embrace given Rosenberg's own self-confessed acceptance that counter-examples and counter-arguments face empiricism and physicalism. Given that Rosenberg accepts that differing metaphysical and epistemological assumptions "must decisively determine the character of strategies of research in the natural and social sciences" [ibid., p. ix], the central arguments of the book constitute little more than a *reductio ad absurdum* for those who reject Rosenberg's central metaphysical and epistemological assumptions. Reason has been given in the course of this present work to reject such assumptions, and we shall find at the conclusion of our discussion here, no reason to alter our position.

So much then for a counter-argument to the basic assumptions of Rosenberg's book, from the perspective of my own philosophical position. We have not yet summarized the course of the argument which has led Rosenberg to accept Sociobiology as preempting Culturology. This shall be now done, with the warning that Rosenberg's writing style makes short precise summaries difficult (this is quite reasonable: one writes to express one's vision and opinions, not to make matters easy for one's critics), and that the following material is selected, biased by my own issues of concern in this work.

Central questions of the philosophy of social sciences are taken to "depend either for their motivation or for their answers on how we account for the difference between the natural and the social sciences" [ibid., p. 19]. Further, it is a long standing tradition amongst those who have dealt with the "crisis of Sociology" argument, to recommend a radical revision of the discipline along the lines of their favoured science. For Rosenberg, this model is supplied by the biological sciences: "except in the works of Herbert Spencer and some sociological functionalists, biological thinking has hardly been appealed to at all as the model on which a social science should be constructed" [ibid., p. 21]. Rosenberg's position would be severely questioned if this proposition was in fact false, for then, the biological model is merely one among the many theoretical orientations.

Rosenberg is historically wrong in claiming that the biological model of "Sociology" has been given only minor consideration. Certainly Social Darwinism was more widespread in its acceptance than being merely restricted to the works of Herbert Spencer [Harris, 1968] and is a *Weltanschauung* which has had profound social consequences. Further evidence of this claim is readily at hand: (1) biological explanations of sex-related cognitive differences by far outweigh in quantity sociological explanations [Sherman, 1978, p. 136]; (2) the existence of extensive biological explanations in the general study of women and society, especially with regards to the existence of patriarchy, is well acknowledged [Goldberg, 1977] and criticized [Hubbard and Lowe (eds.), 1979]; (3) the social history of IQ and mental testing, and the concern to supply biological explanations in this field, is also well documented [Evans and Waites, 1981]. Further, Rosenberg underestimates the impact of functionalism in his claim, given the prevalence of organic and physical metaphors within "Society". There is no one source which contains a detailed argument to readily support this last statement, nor is there an extensive historical discussion justified at this point, merely to argue through this point. Neither, however, can this statement be dismissed as a mere arbitrary conjecture. The organic analogy, found from Spencer and Durkheim, throughout functionalist thought in this century, and the physicalist language and metaphors of "Sociology" ("social forces", "living pressures", "social interactions" etc.) are phenomena in need of explanation. Thus in conclusion, it is clear that not only is Rosenberg historically wrong, but points (2) and (3) indicate that the failure of *contemporary biology*, especially molecular biology and neuroendocrinology, to resolve the "crisis of Sociology", a thesis which has already been argued for. Therefore, the second part of Rosenberg's argument is based upon questionable premises.

Let us press on now after having presented counter-arguments to two major assumptions of Rosenberg's position.

The next step in Rosenberg's argument is to argue that those who have denied a principal explanatory role of desires, beliefs and reasons in the explanation of human behaviour, including Durkheim and the French Structuralists, are in actual fact philosophically inconsistent if they also embrace empiricism. The argument is interesting and disputable, but to contest this step of the argument and do justice to Rosenberg's argument, is not possible in the space available here. But it is not a conclusion which in any way challenges my own position, so such a treatment is not necessary. The implication which is drawn immediately from this proposition about the determinants of human behaviour by Rosenberg is of interest: empiricism plus the assumption that the determinants of human behaviour are reasons leads inevitably to the position that any metascientifically adequate social science must seek "nomological regularities" [Rosenberg, 1980 (2), p. 49].

The connection between actions and the reasons for actions (or more accurately action-descriptions and reasons-for-action descriptions) is for the empiricist contingent rather than necessary, and non-accidental. But non-accidental contingent connections can only, given the position of empiricism, be distinguished from accidental ones by the subsumption of such non-accidental contingent connections under general laws. Thus the connection between actions and the reasons-for-actions must be law-like, short of denying that actions exist (by contrast to mere bodily movements) or empiricism is true [ibid., pp. 52–53]. Yet accounts of *Homo economicus* based upon utilitarian assumptions of maximization, have in all their various forms, despite an impressive mathematical formalism, failed [ibid., pp. 53–66]. The failure is that the utilitarians' claim that actions are a result of the maximization of preferences under constraints, requires a specification of the relevant beliefs and desires, and many commonsense counter-examples are constructable. Nor is this problem solved by the alternative reconstruction offered of rational economic man by

Becker [1976]. Thus one is presented with a choice of rejecting one or more of these assumptions:

(R1) Human action is a form of maximizing behaviour;
(R2) Human action is determined by beliefs and desires;
(R3) There are laws connecting reasons and actions of the sort desired by empiricism.

Rosenberg [1980 (2)] attempts to meet this challenge by arguing that mental predicates do not describe natural kinds:

> . . . that the events, states, and conditions that involve human beings, and that we describe by use of these terms, do not constitute causally homogeneous classes – sets of events, states and conditions that can be subsumed under a manageably small number of stateable general laws. [ibid., p. 111]

"True causal claims" made in daily life about particular reasons and actions, are made even though such predicates do not figure in any general laws, because "the events we bring together on the strength of false general beliefs, or no general beliefs at all, are in fact related by nomological generalizations that we are ignorant of" [ibid., p. 113]. That is to say, we are correct in such matters seemingly by accident, just as referential statements made about the products of combustion, from the perspective of a phlogiston theory, may be true, even though phlogiston doesn't exist. No laws have been produced by the social sciences because their typology does not express "natural kinds". This is a proposition to which I would assent, given Rosenberg's definition of a "natural kind". In doing so, one is not inconsistent in accepting the position of essentialism, for one merely must deny that all empirical/natural kinds can in fact be subsumed under a parsimonious set of general laws.

Whilst action and mentation are not natural kinds, Rosenberg finds consolation in the laws of biology, a science which we are led to believe possesses the needed natural kinds [ibid., p. 151] – although in the course of this argument, Rosenberg argues [ibid., pp. 118–124] that species are spatio-temporally bounded things, rather than natural kinds, and that terms like "gene" do designate natural kinds (although no explicit arguments are presented for this presupposition). It is then quite difficult to see how Rosenberg can justify the following conclusion, given his own assumptions about what many would take to be the paradigm case of a natural kind in biology, species: "it is incompatibility with biological laws *or their conceptual presuppositions about species which vitiates the candidates offered by social sciences*" [ibid., p. 151, italics added]. Yet if species are by hypothesis spatio-temporally bound things, rather than natural kinds, it is difficult to see how biology is immuned from the criticisms already advanced against the social sciences.

Indeed, whilst Rosenberg finds the concept of causality quite vague [ibid., pp. 52–53], the idea of a "law of nature" cannot be taken, after David Hume, as being intuitively clear and philosophically unproblematic. Yet it is precisely the fundamentality of laws of nature which constitute "the sciences". Brentano's "science of intentions" "will be a science without laws, and therefore no science at all" [ibid., p. 147]. Such a strong claim will require not only a detailed theory of laws, providing reasoned principles which distinguish them from accidental generalizations, but also an argument that all sciences must be nomological. The latter claim, of course, is a part of the empiricist *Weltanschauung* which Rosenberg assumes, ignoring twenty years of debate in the philosophy of science. But without some explication of the former claims, a sceptic would see no reason why "laws of nature" should not go to the same scrap heap as the concept of causality, for the main philosophical problem facing empiricist accounts of causality, is that they haven't answered Hume [Harré, 1970] [Bhaskar, 1978 (1)]. If this is so, Rosenberg has returned us to "the Crisis".

My response to Rosenberg's thesis that biology, especially Sociobiology, can resolve the "crisis of Sociology" should be evident from the textual material. Further, my general assessment of Rosenberg's position should also be evident. The cogency of the positions rests upon explicitly undefended, and contested theses. The Culturologist rejects these assumptions, and rebuts Rosenberg's challenge. Rosenberg may well state that he does not regard the sorts of criticisms which I have made as an objection to the argument of his book [Rosenberg, 1980 (2), p. 210], that he was explicitly attempting to follow through the implications of physicalism and empiricism. If this is so, then he does not address adequately his initially stated objective of not merely identifying the false beliefs of the social scientists, but to provide a "prognosis" and "prescription of improvement" [ibid., p. ix]. Improvements are not obtained by adding to "the Crisis", one more ultimately unjustified position.

NOTES

Chapter 1

1. I spoke in the text of Wilson's "scientific materialism". Wilson whilst quite prepared to write of Marxism as "mythology", shows recognition of the worth of "Liberal Theology", enough to criticize "scientific materialism by asking questions about the nature of human mind [Wilson, 1980, p. 433]. This work, will not attempt any detailed textual study of the stated philosophical implications of the works of the leading Sociobiologists themselves. By such a study the critic could demonstrate the inconsistencies between their statements. This is insufficient to produce a challenging critique of Sociobiology.
2. It does not follow from this that Sociobiology has no ideological implications. The issues of Sociobiology and ideology are discussed in Appendix 1. This appendix supplements the textual material, but its importance as an issue in itself make this material neither tangential nor irrelevant.

Chapter 2

1. Some relevant papers on the eliminative materialism debate include [Bernstein, 1968]; [Cornman, 1968]; [Guleserian, 1971 (1)], [1972 (2)]; [Lycan and Pappas, 1972]; [Edge, 1974]; [Shirley, 1974]; [Bush, 1974]; [Savitt, 1975]; [Carter, 1975]; [Berger, 1976]; [Shope, 1979] and [Churchland, 1981].
2. Edgar Wilson [1979] is another physicalist who sees the "orthodox-animistic" model of common sense being undermined by a "physical-objective model" much in the same way as the Copernican cosmology undermined the Ptolemaic. In particular from the perspective of the "physical-objective model" our ideas of freewill and guilt and responsibility and punishment are debunked. Crime in particular is seen as a "maladjustment" or a "disorder" rather than merely the bad actions of an individual.

This model has some quite progressive implications with regard to the understanding of crime, namely in debunking the view that crime is primarily the actions of an autonomous individual. However the view that crime is a "maladjustment" implies that the objects of such crimes (in most cases private property) are morally good and worth preserving in any case, so that criminals must be adjusted rather than having a society which is based upon private ownership of the means of production and produced goods abolished, and an arguably more distributively just social order established.

3. Unless it was claimed that there was not in fact "one world" but many distinct worlds. This point will be taken up below.
4. Laudan is in my opinion wrong; this can be shown by pointing out two counterintuitive consequences of the position. If our intuitive idea of theoretical satisfactoriness is to be judged by progressiveness relative to competitors, it follows that if this condition is taken to state necessary and sufficient conditions of theoretical satisfactoriness, one is committed to claiming that if there is only one existing theory in a field, it is logically impossible to assess its progressiveness. Progressiveness is a *relationship* between theories on this account, not a *property* of a theory. A tradition of epistemological scepticism seems to be ruled out of the debate by definition, for some traditions of thought would wish to claim that even if there was only one theory existing in a field, it could still be judged to be inadequate (defeating the necessary condition claim), and if in fact theory T_1 was more progressive than theories T_2, T_3, T_4, ... T_n, T_1 may still in fact be inadequate since the entire set of theories T_1, T_2, ... T_n are untenable (defeating the sufficiency claim). Laudan's position is thus meta-methodologically inadequate since it arbitrarily rules out the research tradition of scepticism.
5. Sztompka [1979] has given a breakdown of six fundamental philosophical problems which have divided the various schools. One can add to the list, but this is not necessary to make our point here.

   a. *The Naturalism versus Anti-Naturalism Question*: the methodological question as to whether there is a methodological unity between the social and natural sciences.

   b. *The Reductionism versus Anti-Reductionism Question*: the meta-scientific question as to whether "Sociology" is a theoretically autonomous discipline or whether it is in either practice or principle theoretically reducible to some other discipline (psychology, biology).

   c. *The Cognitivism versus Activism Question*: the epistemological question as to whether only factual propositions are permitted within the domain of socio-scientific theories, whether normative statements are as well, and whether the primary function of theory is to supply knowledge (cognitivism) or offer normative-political advice for action (activism).

   d. *The Neutralism versus Axiologism Question*: the axiological question of whether value judgements should play a role within the research process (axiologism) or should not (neutralism).

   e. *The Passivism/Determinism versus Autonomism/Freedom of Will Question*:

the metaphysical question of whether the actions of a person are determined primarily by external social or biological forces, or whether people are active shapers of their society, not at the mercy of such forces.
f. *The Collectivism versus Individualism Question*: the ontological question of whether Society is a *sui generis* real phenomena having trans-individual existence and/or properties (collectivism), or whether Society is simply the sum of all individuals and their activities (individualism).

It seems to be quite true, that which philosophy of "Sociology" one accepts will alter the way research is done, the formulation and recognition of problems, the selection of areas of initial research and the research strategies adopted. If one accepts a metaphysics whereby social agents are not taken to be knowledgeable agents, but are rather manipulated, erroneous, ignorant and passive things, some methods of empirical research such as questionnaires are immediately problematic. This is hardly an original insight, and has become increasingly recognized since the contemporary critique of empiricism gained momentum.

6. Relativism is itself a doctrine which does not escape peculiar problems of its own. *Cognitive Relativism* is the doctrine that a proposition p may be true relative to conceptual scheme C, yet false relative to conceptual scheme C^*. For Cognitive Relativism to be tenable it must be possible for p to be expressed in both C and C^*; if not, the thesis collapses. Now a proposition p_1:

   (p_1) material objects exist independently of any mental activities of human subjects

is true within the *Weltanschauung* of materialism and false within the *Weltanschauung* of subjective idealism. Reznek [1979, p. 71] argues that p_1 must violate the law of non-contradiction, that no proposition is both true and false. This will of course follow if both materialism and subjective idealism were to be "true", which is to say that the World is such that material objects exist independently of mind, and do not exist independently of mind. To maintain that the World is consistent, the Relativist is then driven to claim that the materialist and the subjective idealist inhabit different "Worlds". If this is claimed we are now at a loss to explain how in fact it seems to be the case that the materialist and subjective idealist causally interact during debate, if "Worlds" are taken to be ontological rather than merely conceptual entities. If this was also the case, the debate which does exist between the materialist and subjective idealist would be dissolved. They are not talking about the same phenomena in the same World at all. Because this doctrine is so utterly counter-intuitive, the absolute burden of proof must be put upon the shoulders of the Cognitive Relativist to show how in fact changes in theories cause changes in world-ontologies. I know of no non-question begging argument for this conclusion.

An alternative strategy, now becoming popular with the development of para consistent logics and set theories is to claim that the world may well be inconsistent [Routley and Meyer, 1976] [Rescher and Brandom, 1979]. By this it is meant that there is a possible world w such that $(\exists p)(\mathrm{True}_w(p) \,\&\, \mathrm{True}_w(\sim p))$. As far as I know no one has denied the minimal principle of contradiction $\sim(\forall p)(\mathrm{True}_w(p)\ \&\ \mathrm{True}_w(\sim p))$. One can easily present interpretations using "impossible objects" to the strong principle of non-contradiction. Consequently there is at least one interpretation of all possible worlds for which $(\exists p)(\mathrm{True}_w(p)\ \&\ \mathrm{True}_w(\sim p))$. It does not follow that w is in fact the actual world. Indeed if w was the actual world we would be unable to draw the distinction between possible and impossible entities (for if the actual world is not an actual entity, what is?). Since the main motivation of para-consistent logic is not irrationalism – but to allow us to deal with inconsistency without reducing discourse to nonsense, such a distinction must be preserved, under pains of having their entire position collapse.

It is also a lot to ask of the present author to present a sketch of a strategy to deal in a unified fashion with arguments for the consistency of the actual world; I maintain however that all these arguments are extremely weak. One approach which one might be inclined to put faith in, because of the near irrational faith which our culture has in the speculations of physicists, is the Everett-Wheeler Many-World Interpretation of the reduction of the wave function [Jammer, 1974, pp. 507–521] [Rescher, 1973 (2) pp. 88–106]. Interpreting each superposition of vectors as representing, not merely an unactualized possibility, but an actuality, leads to the claim there are many "branching worlds". An argument which can be developed against this claim, which can only be sketched here, is that if an "explosion of worlds" does occur, then talk of conflict is unjustified. What the para-consistent logician is after is contradictory aspects within one world, and this example fails to supply it.

7. Theoretical idealism is, as Milton Fisk [1976] has pointed out, a popular position today – even with eliminative materialists such as Rorty [ibid, p. 373] [Edge, 1974, pp. 201–202]! Rorty [1972] himself seems to take the position, that "experience" is a function of the language used and what exists depends upon "experience", so that what actually exists depends upon language. This is Edge's interpretation of Rorty, and leads to the conclusion that Rorty the eliminative materialist living in such a culture does not have sensations, whilst we at present do merely by virtue of both of our linguistic practices! This seems to be a *reductio* of the position itself, and it is especially difficult to reconcile with Rorty's claim that most of our present beliefs are true [Rorty, 1972, p. 662] [Aune, 1972, p. 667]. I would point out that if Sellar's criticism of the "myth of the given", an argument cherished by many, entails the conclusions that Rorty thinks it does, then Sellar's position is also problematic. The case that *all* perception is "theory-laden" has yet to be made in my opinion, even if we accept that Sellars has demolished phenomenalism [Shimony, 1978 (1)] [Routley, 1980 (1), pp. 813–831]. But problems will not end here. Rorty can also be interpreted (as Fisk does), as maintaining that the whole idea of a correspondence theory of truth is incoherent and that anything distinct ontologically from the social practice of discourse is utterly incomprehensive. If one adopted this interpretation of Rorty's work, there are obvious difficulties now in squaring this position with that of Sellars, to whose work Rorty

owes a great debt. This occurs insofar as Sellars is opposed to the very ideal of "Givenness" [Sellars, 1963 (2), pp. 127–128]. Rorty seems to be replacing one given "awareness", with that of another: "language". This is problematic because it is contrary to the stated implications of an epistemological position (Sellars'), which Rorty uses to criticize those who argue against eliminative materialism, that the reporting function of sensation-talk cannot be eliminated. This seems to indicate that Rorty's epistemology is incoherent.

8. In concluding the discussion in this way, I do not wish to be taken to underestimate the difficulties of these issues. One would wish most intensely that one had an adequate "theory of evidence" at hand (which in turn would seem to presuppose the existence of a fairly well developed epistemology) but this is merely a wish, for this does not exist.
9. Carlo Giannoni [1976, p. 164] argues that Hempel [1965, pp. 101–119] has demonstrated the Duhem-Quine thesis, by a purely logical analysis of scientific theories. Giannoni's reconstructed argument is this: (1) theoretical terms are implicitly defined by scientific theories; (2) statements are deducible which do not contain theoretical terms, *therefore* (3) theories as a whole face experience. Such an argument is in my assessment invalid. It remains totally unclear as to why a premise about holisitic testability, follows merely from a premise about the implicit definability of theoretical terms; indeed one would have thought that premise (2) would have in fact conflicted with premise (3), for here, seemingly, we have at least some statements outside the circle of implicit definition, and it is this which is in turn responsible for holism of structure.
10. By use of the "Strawmen" predicate, I am stating an explicit caricaturization of a position which may or may not be the position of any actual philosopher. Questions of what Popper and Lakatos actually say are not of importance in *this* context.
11. The order is unimportant because either strategy would seem to prevent the existence of the conflict between theory and evidence with a minimum of labour. Koertge's [1978] metaphor of comparing the Duhem-Quine problem to the search for sources of trouble in one's motor cycle is quite informative – "motorcycles", after all, "have no hard cores" [ibid, p. 253]. Putting in either a new engine or an entirely new theory is needless work if all that is at fault is a loose spark plug or one stray inference. Moreover if one made major adjustments first, then one ignores the possibility of the fault lying in incorrect inferences. It may well be that even if such simple methodological rules are violated by actual scientists [Feyerabend, 1975] this would only lead to methodological anarchism if we had reason to believe that such rules are never followed, *and* scientific progress occurred. But the ideal of scientific progress as an increasingly closer approximation to the Truth seems to be also utterly abandoned by Feyerabend [ibid, p. 230]. This means that one of Feyerabend's premises in a counterargument is false, and hence that his argument is unsound.
12. For Routley [1980 (2)] a general optimization model of theory choice takes the following form: there are n general factors $a_1, a_2, \ldots a_k, \ldots$

$a_n, \ldots$ (n not necessarily finite) which must be considered for the selection of some goal Σ, where there exist functions P_r, V and W such that for an arbitrarily chosen factor a_i, $p_r(a_i)$ is a probability function, $V(a_i)$ a value assessment function and $W(a_i)$ a weight. One must then maximise the function:

$$\Sigma(W(a_1).\ P_r(a_1).\ V(a_1)\ \ldots\ W(a_i).$$

$$Pr(a_i).\ V(a_i), \ldots\ W(a_n).\ P_r(a_n).\ V(a_n))$$

subject to a set of restraints $R = \{R_1(a_{i1}, \ldots a_{im1}), R_j(a_{ij}, \ldots a_{imj})\}$ such that $a_{i1}, \ldots a_{im}, \ldots a_{ij}, \ldots a_{imj},\} \subseteq \{a_i, \ldots a_n.\}$. A best choice maximises Σ subject to constraints R.

13. This is not to say that the use of this mode of reasoning constitutes a necessary condition in a demarcation of philosophy from science – it does not, for the technique of argument is also used in logic and metamathematics. I have no need, as Popper and the logical empiricists had, to demarcate metaphysics and science. Indeed sciences are no more than "cognitive systematizations" of knowledge about some object of inquiry [Rescher, 1979], and metaphysics then could be regarded as a science in this sense as would musicology insofar as it has (at least) justified true propositions. There is hence no such thing as a "Scientific methodology", which stands apart from a general epistemological methodology, which can give a shining verisimilitudinous stamp of approval.
14. In my opinion the dominant style of philosophical argumentation is an "inference to the best explanation [Harman, 1965] [Thagard, 1978 (1)]. Philosophical theories are judged as to whether they provide the best explanation of metaphysical facts. The case of external world scepticism illustrates this: does common sense realism "better" account for my sense experiences than the Cartesian demon hypothesis. The mode of argument employed here is an inference from the premise that a certain theory provides a better/more satisfactory account of the world, to the conclusion that the given theory is true. One must be able to reject all competing theories in this process, including the sceptics' position which serves as a null hypothesis. I differ from Thagard in not restricting this assessment process to criteria of consilence, simplicity, and analogy. Philosophers have wished to criticize the cogency of entire research programmes regardless of these factors. It is in such a tradition that I locate myself.
15. This is not to enter the Taylor-Dreyfus-Rorty debate [Taylor, 1980], [Dreyfus, 1980], [Rorty, 1980], [Rorty, Taylor and Dreyfus, 1980] and to claim that there is no methodological distinction between the social and natural sciences. To restate my position, I allege that Rorty's critique of systematic philosophy must, if it is to be cognitively meaningful, assume that systematic philosophy is non-hermeneutic, and non-edifying. This assumption is in my opinion false, since I know of no systematic philosopher who (officially) wishes to end "the conversation of mankind" (including Feyerabend). Systematic philosophy should both incorporate and transcend the insights of hermeneutics.

16. Ellis [1977] has given the following argument which might be taken as a reply to this criticism: A falsifiability claim which relies upon information given to us by other subjects "is not a method which ever permits the testing of *hypotheses* about purposes, goals, or wishes themselves" [ibid, pp. 197–198], because *all teleological hypotheses are unfalsifiable*. He states that if a

    > Sociologist can identify one inherently teleological hypothesis that has *ever* been scientifically disproven (within conventional limits of statistical probability), I will retract every word of my article . . . . [ibid, p. 198]

    However, we have good reason to believe that neo-Darwinism, and hence Sociobiology itself, is inherently a "teleological" science [Ruse, 1980, pp. 85–101]. Consequently Ellis' criticism undermines itself. Further, we should note that no explicit argument has been presented here on Ellis' part for the unfalsifiability claim, and it is surely problematic and thus requires it. We are merely threatened with the repetition of a contested thesis.

Chapter 3

1. At least three senses could be distinguished here: (1) T is rationally preferable to T* insofar as T is the best available theory (including the "null hypothesis") and consequently T provides a "better" explanation of T*; (2) T is rationally preferable to T* insofar as T is more fruitful in suggesting modes of future research and (3) T is rationally preferable to T* as a basis for social action, in dealing with problems of life. The remaining problem is to develop a coherent explication of each of these aspects in turn. For some further relevant reflections see [Radnitzky, 1980].
2. See for example the previous chapter for a discussion of Ellis [1977] and the introductory chapter for further supportive arguments.
3. The textual exposition draws upon [Radnitzky, 1970] for initial inspiration.
4. The account of physicalism at this point is extremely broad and extremely vague, nevertheless it conforms fairly precisely to the account present in the works of the cited leading physicalists. Later in the chapter we shall have more to say on this issue: namely that the physicalist's account of the "physical" or "material" is hopelessly vague or else outrightly circular.
5. For a discussion of Eddington's two tables problem see [Stebbing, 1937] [Ryle, 1949, 1954] who put the case for the common sense table, and [Sellars, 1961] [Feyerabend, 1962, 1966], who put the case for scientific realism. For a further discussion of these issues see [Mellor, 1969] [Noren, 1973, 1975] [Girill, 1976] [Cornman, 1974] [Holman, 1979].
6. On this account, laws of nature must be taken to be universal in space and time, meaning, as Earman [1978, p. 173] points out, that (1) Laws do not make reference to specific spatio-temporal locations; (2) Laws have an unrestricted range in space and time; (3) Laws do not contain explicitly, space and time co-ordinates and (4) Laws are invariant under spatio-temporal translations. Smart [1963 (1), pp. 50–61], has argued that there

can be no laws of biology, so biology should properly be considered a discipline like engineering presumably merely applied physics (cf [Ruse, 1970] [Munson, 1975]). Smart would no doubt wish to say the same thing about "Sociology". In this work, I will attempt to turn the tables on Universal Mechanists: "Sociology" is not applied biology-chemistry-physics, and insofar as "Sociology" lacks a comprehensive set of non-trival laws, accounts of theoretical reduction depending upon conditions of deducibility, isomorphism and various other formal or pragmatic relationships between the set of laws of the two disciplines, fail outrightly – so much the worse for universal mechanism.

7. For an interesting discussion of an encounter between Wilfred Sellars and phenomenology see [Bernstein, 1976, p. 117–135].
8. This is an idiosyncratic (but I believe justified) interpretation of Sellars. An alternative position, adopted by Margolis [1978] for example interprets Sellars' argument straight forwardly as an argument for the irreducibility of the intentional and the sensational. However, given Sellars' view that there are no *non-dissective* emergent properties of systems (terms which will be explicated later in the text), and that consistently ascribable properties of systems must be properties of parts of the system [Sellars, 1963 (1), p. 26], we seem to fall into a regress which involves us attributing sensa-properties for example first to neurons, then to bio-molecules and finally to sub-atomic particles, taking sensa to be additions to the Scientific image and Sellar's former assumption. Such a conclusion would seem to reduce this interpretation of Sellars' position to adsurdity (cf [Wilson, 1979, p. 82].
9. I have not classified PSU as being committed to the doctrine of *Naturalism* here. By "Naturalism", we mean the doctrine that "reality consists of nothing but a single all-embracing spatio-temporal system" [Armstrong, 1978, p. 261]. Naturalism is, in this sense (as Feigl [1953, p. 382] recognized), consistent with an emergentist anti-reductionist position such as my own. To introduce this sense of the term "Naturalism" into the argument of the text seems little more than self-defeating. But on the same token I have no wish to defend naturalism and shall suggest later that it is most probably false anyway.
10. Smart [1978 (1)] seems at first glance to raise a problem for this classification. He addresses the problem that clearly faces those, who in wishing to make physicalism a falsifiable doctrine, relativize *the Physical* to particulars, properties and relations designated by individual constants and predicates (respectively) of current physics [Leven, 1979, p. 61]. If this strategy is adopted is physicalism then an empty doctrine as a theory about the nature of mind? Smart's answer is "no", because (1) "the physical" is relativized to the physics of the day [Smart, 1978 (1), p. 340] and (2) future changes in physics will, he maintains following Feinberg [1966], be only in the micro-physical domain.

With regards to the second point, it seems to me that the Smart-Feinberg claim cannot as a matter of logic and epistemology, be justified. Any argument produced will necessarily contain empirical premises making substantial assumptions about the nature of physical reality which are not (1) necessarily true; (2) *synthetic a priori*. Hence it is logically possible that they are false. Indeed, since these assumptions are empirical, it is "empirically possible" that they could be false. It is still possible, it might be thought, to argue that in fact, it is improbable that these assumptions are false. But how is such an argument established? We seem at this point to face in full force Hume's problem of induction, and the argument in question is revealed as being question begging: we would seem to require knowledge about the future which we *cannot* have, that no general crisis awaits physics.

Let us take up the first point now: if physics is relativized to that dictated by the physics of the day, what metaphysical implications does this have with regards to the ontology of physical objects? In my opinion, it seems to make physicalism into a contentless doctrine. Over the history of physics, some physicists have not behaved as good "*physicalists*", and have for example appealed to God in matters cosmological, either to start up the clock of the world, or to close some theoretical gap which present naturalistic mechanisms did not. Today Creationists seem to be taking a similar strategy. Unless we make the question-begging claim that such discourse cannot be *a priori* physics-discourse, this seems to leave us no choice but to accept that it might be reasonable to regard God as a physical object. This is a philosophically absurd conjecture, since whatever Theists want to say about God, He is certainly not a "material object". We shall have more to say about this issue in section 5. In conclusion, Smart's exit strategy does not seem tenable.

11. Indeed, the development of Sociobiology and its extention into explanatory domains once exclusively occupied by "Sociology", is a way in which fulfilment of the PSU programme may occur, despite protests by any practising Sociobiologist.
12. The claim that the unification of science by theoretical reduction is necessary to achieve an intelligible and plausible world view [Hooker, 1979, p. 83] will be contested in section 5. Ontological and theoretico-explanatory economy are only two "cognitive values" [Rescher, 1979]. To exaggerate the status of these cognitive values to the degree physicalists do (constituting a near fetish for "simplicity") is to ignore the evaluation of other parameters of systematicity such as "wholeness", "completeness", "metaphysical plausibility", "self-sufficiency", "cohesiveness", "consonance", "functional regularity" and so on.
13. A further difficulty, necessitating the textual methodology adopted here, is that traditional survey sources such as [Schaffner, 1967] are in many ways inaccurate, cf [Hooker, part I, 1981, pp. 52–53].
14. An example of a homogeneous reduction, cited by Nagel [1961, pp. 338–339] is the extension of mechanics from a theory of the motion of point masses, to include the motion of rigid and deformable bodies.
15. The form of this theorem of relevance to Quine's concerns is this:
    (LS) If Σ is a set of formulas of a first order language L_1, and Σ has a model in a non-empty domain D_1, then Σ has a model in the domain of positive integers.

The problem which Quine grapples with is that of *neo-Pythagoreanism*, the thesis that the only ontological commitment one needs to make is to positive intergers. Insofar as reduction involves *merely* modelling, this thesis seems inevitable [Bonevac, 1980]. Since we have good reason to believe that there is more in the world (indeed for the Platonist, positive integers are not *in* the world), we have good reason to regard this implication as a *reductio ad absurdum* of the reduction-as-modelling view. We give another argument for this conclusion in the text, with particular reference to Quine's solution.

16. There does not exist, as Suppes [1957, p. 262] points out, a completely general characterization of isomorphism. However, an account suitable for the Adams–Suppes account of a scientific theory seems to be this:

    Let $\langle X_1, R_1\rangle$ and $\langle X_2, R_2\rangle$ be two models of a theory, where X_1 and X_2 are sets, and R_1 and R_2 relations. Then $\langle X_1, R_1\rangle$ is isomorphic to $\langle X_2, R_2\rangle$ = df there is a function f such that (a) f establishes a 1–1 correspondence between X_1 and X_2; (b) if x, y ϵ X_1 and x R y, then f(x)R f(y); (c) if x, y, ϵ X_2 and x R_2 y then $f^{-1}(x)R_1f^{-1}(y)$.

17. Oppenheim and Putnam's [1958] claim to *unify science by their own form* of micro-reductions is subject to further severe epistemological problems. For a micro-reduction to occur, for each branch of science, there must be a universe of discourse U(B) and a part-whole relation Pt(x, y) such that Pt(x, y) = df x is spatio-temporally contained in y. They offer this definition:

    (OP/D4) B_2 microreduces to B_1 = df B_2 is reduced to B_1 and the objects in $U(B_2)$ are wholes which possess a decomposition into proper parts, all of which belong to $U(B_1)$.

    It becomes here virtually impossible to deal with "reductions" (*and we must remember OP's unification claims*) such as that of Newtonian mechanics to Special Relativity, and of "Sociology" to Sociobiology, as in neither case is (OP/D4) satisfied. The universe of discourse of "Sociology" and (human) Sociobiology, both contain social entities such as human persons – and no such entity is a *proper part* of itself.

18. Yoshida's [1977] less than clear criticism of the "incompatibility argument" is seemingly this: even though T_1 & $AH_1 \rightarrow p$ (where "AH_1" is a set of auxiliary hypotheses) and $T_2 \rightarrow \sim p$, it is still logically possible that T_1 & $AH_1 \rightarrow T_2$. However, if this is granted then:

    (1) $T_1 \ \& \ AH_1 \rightarrow (T_2 \rightarrow \sim p)$

    and

    (2) $T_1 \ \& \ AH_1 \rightarrow p$, thus $T_1 \ \& \ AH_1 \rightarrow (T_2 \rightarrow \sim p) \ \& \ p$.

    But in the classical propositional calculus:

    (3) $(T_2 \rightarrow \sim p) \equiv (\sim T_2 \vee \sim p)$

    Then:

    (4) $T_1 \ \& \ AH_1 \rightarrow p \ \& \ (\sim T_2 \vee \sim p)$

which is equivalent to:

$$(5)\ T_1\ \&\ AH_1 \rightarrow [(p\ \&\sim T_1) \vee (p\ \&\sim p)]$$

which is an outright classical absurdity, whichever disjunct one chooses.

19. One counter-argument which a critic might make here, is that there is a sense in which the Special Theory of Relativity "reduces" to Newtonian mechanics [Nickles, 1973], so my claims about asymmetricity violation are misplaced. It is true that for velocities much lower than the speed of light as $v/c \rightarrow 0$, relativistic laws tend towards classical laws. However all this indicates is that one sense of the term "reduces" is "yields approximate values to". But even *logically competing theories* may do this over at least some range (or they would not be competing). In Nickles' case we could easily eliminate reduction-talk here without any loss of explanatory and comprehensive power. Consequently this counter-argument fails.
20. Causey [1977, pp. 26–27] in fact recognizes that there are some compelling counter-examples to the D–N mode of explanation. He attempts to meet these objections by stating that we can in a sense speak of the "cause of a law" [ibid., p. 26]. But note now the inconsistency of his claims. On the one hand he accepts the D–N account of explanation. On the other, he seems to accept that there is a more basic notion of explanation via causal relationships as argued by critics of the D–N mode such as Harré [1970] and Bhaskar [1978 (1)]; a claim inconsistent with the former allegation of *accepting* the D–N account that *all* reasonable scientific explanations should follow this position (what else could it mean to accept a position?) Obviously they cannot be reconciliated if the D–N account of explanation is accepted.
21. To achieve such micro-reductions, Causey places a number of conditions upon both a theory T_1 and L_1, the language of T_1. First, there must be a subset of the predicates t_1 of T_1 which denote various kinds of elements of Bas_1 [ibid., p. 55]. Second, no basic thing-predicate is to be definable in terms of other basic thing-predicates, with the exception of analytic co-extensionalities (e.g. $(\forall x)(P_1 x \equiv P_2 x)$, where "$P_1$" and "$P_2$" are co-extensional predicates). Third, each element of $Comp_1$ is designated by some thing-predicate of L_1, defined by thing and attribute predicates of Bas_1 [ibid., p. 56].
22. For a criticism of Causey, developed along lines different from those here cf [Achinstein, 1981].
23. For commentaries on the Structuralist View of Theories cf [Przełecki, 1974] [Tuomela, 1978] [Rantala, 1978, 1980] [Feyerabend, 1977].
24. Explicit argument for the claim that the position function is T-non-theoretical is not given by Sneed. His argument that mass in CPM is a T-theoretical function is simply that all mass-ratio determining techniques presuppose that CPM is applicable to the physical systems which make the comparisons. Sneed [1971, p. 32] recognizes that "position" could in fact be determined in a T-theoretical way by using a telescope with an automatic tracking device, the application of which involves mechanical principles, but he maintains (without argument) that values of the position function

can be determined independent of such an apparatus. True, but it cannot be determined independently of all devices – even if the only device used is the human eye. Thus, even if one determined, in a rough manner, the position of an object by unaided eye power, we are using a visual mechanism which Sneed would surely grant, obeys (at least approximately) mechanical laws. If an evil Demon began performing miracles, such that laws of classical mechanics failed to hold in the eye, locating objects would be a precarious business. Hence even the position function is T-theoretical. The distinction which the Structuralists depend so much upon collapses for their self-appointed paradigm case.

25. Does Ramsey-eliminability allow the Structuralist to avoid the problem of theoretical terms? The central criticism of the formulation of Ramsey sentences to solve the empiricist's problem of theoretical terms is that it did not! As Hempel [1965] noted, Ramsey sentences avoid the problem of theoretical terms only by refusing to explicitly mention them – but it still asserts the existence of certain entities by virtue of existential quantification. Much the same can be said with respect to the Structuralist's use of Ramsey sentences: the difficulty about the vicious circularity arising from the definition of 'T-theoretical' is not removed by Ramsey-sentences. If some *correct application* is necessary *before measurements can be made*, the impossibility of establishing the correctness of initial applications is not affected by the question of Ramsey-eliminability, for it is an *independent* epistemological question (resembling the "criterion problem"), one not resolved by the Structuralists, and reducing their position to absurdity.
26. The Structuralists unreflected use of naive set theory is yet another difficulty. Rantalla [1980] has for example shown how the derivation of the Burali–Forti paradox can be readily obtained [ibid., p. 271].
27. A further difficulty for Armstrong is this: given his thesis of naturalism and a causal criterion of existence, it seems that the naturalists own spatio-temporal framework conceived in either a relationalist or absolutist fashion is of problematic ontological status. For space and time have no causal powers [Bunge, 1977, pp. 328–329] and so by Armstrong's causal criterion of existence, do not exist. Hence either space and time do not exist (stated absolutely) or there are no spatially related particulars and temporally related events (stated relationally). But Armstrong also regards attempts to deny the existence of the spatio-temporal frame as misconceived [Armstrong, 1978, p. 262] – but this is precisely what one would conclude from a causal criterion of existence.
28. An attempt to analyze the "material" presented by Mario Bunge [1981, pp. 21–22] is subject to similar difficulties. Bunge's definition is this:

    (BDP) An object x is a material object = df for every reference frame y, if Sy(x) is a state space for x, then Sy(x) contains at least two elements.

    (For the definition of a "state space" cf [Bunge, 1977, p. 134].) The difficulty with this definition, is that some imaginable accounts of dualistic interactionism would meet the definition. Bunge rejects this claim on the grounds that there are no Cartesian minds in his materialist ontology, and in any case he asks, how could they be mathematically modelled? [Bunge, 1981, p. 21]. With regard to the last question, they can be modelled as easily as anything else (recipe for modelling: treat the mind as yet another particular, differing from others insofar as it is not spatially locatable). With regard to the first criticism, Bunge's remarks are question begging, and inconsistent with his own criticism of a previous attempted explication, where he himself used the example of "immaterial" particulars to undermine it [ibid., p. 20]. True, Bunge may believe that there are (contingently) no "changeable" Cartesian minds – *but if* there were, they would be "material" by his definition, which is contradictory. Therefore the account is unsatisfactory as a general explication.

29. Davidson's argument for the identity thesis is criticized by Katz [1977], who also argues for the implausibility of Davidson's account of the mental and the physical. Elgin [1980] has successfully criticized Davidson's appeal to Quine's indeterminancy of translation thesis to support anomalist monism. Other papers of relevance are [Goldberg, 1977] [Nickles, 1977] and [Honderich, 1982]. For an argument for the inconsistency of functionalism see [Bealer, 1978], and for further discussions of difficulties see [Kalke, 1969] [Gendron, 1971], [Block, 1978], [Doore, 1981]. For specific critical discussions of Fodor and related issues see [Causey, 1977, pp. 142–151], [Godbey, 1978], [Cartwright, 1979] [Richardson, 1979, 1982].
30. A contemporary Sextus Empiricus who wished to humble enthusiastic physicalists might outline the following foundational problems of physics and argue against all existing solution-candidates: (1) the problem of interpreting non-relativistic quantum mechanics [Jammer, 1974] – especially from a realist perspective [D'Espagnat, 1971] [Mehra, 1974] and especially in the light of quantum paradoxes such as Schrödinger's cat paradox and the EPR paradox; (2) the antinomy between quantum mechanics and classical logic [Finkelstein, 1972]; (3) the problem of the "mathematical infinites" in relativistic quantum field theory; [Bjorken and Bell, 1964, 1965] [Noyles, 1975], (4) the possible incoherence of special relativity [Essen, 1972]; (5) the current problems facing the micro-physical high energy paradigm [Shrader-Frechette, 1977].
31. Space limitations preclude any detailed discussion of the problem of universals here. My position closely follows the "no-nonsense" approach of Michael Loux [1978]: talk about particulars is just that, talk about particulars, and talk about attributes, just that, talk about attributes. I follow Armstrong [1978 (2), 1978 (3)] in taking these attributes to exist independently of the classifying mind (with qualifications to be given to various psycho-social particulars and attributes). Monadic attributes exist as the proper ties of particulars and of properties; polyadic attributes as relations between particulars or properties. Needless to say, careful qualification of what properties of properties and relations between relations is, and is needed to avoid well known paradoxes. It is of course the job of a detailed theory of universals to deal with such issues. Further, I follow Armstrong [1978 (2)] in taking it to be the job of a "total science" or Metaphysics to determine what universals there are, but reject his em-

piricist account of science and recognize the importance of *a priori* reasoning in this determination. Needless to say, these remarks will only give the reader the roughest of guides, and this temporary inadequacy is fully conceded.

32. In this respect it is interesting to review the phenomena of *resonance hybridization* in quantum chemistry. It seems, to state it simply, that for at least some molecules and ions, such as nitrous oxide for example, there are a number of ways to satisfy quantum valency requirements. Resonance hybridization involves a very complex process of "weighing" and averaging out bond orders in the resonance structure with respect to experimental data. But quantum mechanics itself, does not predict the bond length of nitrous oxide. The reason for this is, as O'Dwyer, Kent and Brown [1978, p. 130] state:

    > . . . there is no physical reality to the concept of resonance. It is merely a crutch we introduce to conceal the inadequacies of our very crude valency theory of representing bonds by dashes.

33. It is interesting to note that a leading physicalist Smart [1963 (1), pp. 44–45] in attempting to avoid the quantum causal paradoxes, considers the possibility of space-time being only a macro-micro property, so that truly basic particles are much like Kantian-things-in-themselves. But this seems to commit Smart to either (1) yet another Eddington's dilemma or (2) to a thesis of emergence. In addition, physicalism would be inconsistent with naturalism, as these neo-Kantian particulars are not part of the spatio-temporal framework, although like Cartesian minds, they interact with it.
34. It is recognized that many of the traditional arguments against emergentism have not been considered in the text. The reason for this is that others have considered these objections and criticized them in more detail than is possible here, cf [Nelson, 1969] [McShane, 1970] [Smith, 1983 (3)].

Chapter 4

1. A "reducing atmosphere" is, loosely speaking, one in which hydrogen based gases are found rather than oxygen; more precisely it is an atmosphere characterized by the predominance of *reduction reactions* consisting in either the removal of oxygen from substances or the addition of hydrogen to them.
2. Goldschmidt [1940] and DeVries [1905] also had considerable influence upon the controversial writings of Otta Schindewolf [1936, 1950] who believed that a single "mutation" might lead to a new species or family of organisms. Schindewolf however questioned on palaeontological grounds central gradualist assumptions. Four years after the publication of Goldschidt's book Simpson [1944] postulated the existence of quantum or non-gradualistic modes of evolution where taxon make their evolutionary transitions by passing through inadaptive phases of evolution in small isolated populations, after which the group moved into a new adaptive zone. The ingredients then for Gould and Eldredge's model of punctuated equilibria, had been laid by mid-century.

3. C. C. Li [1967] in a concise paper, has argued that Fisher's fundamental theorem is not applicable, or more precisely, cannot be mathematically justified in non-randomly mating populations. In particular Li shows that the average fitness would tend to decrease under selection.
4. Gene or point mutations occur through the alterations of nucleotide sequences within a "gene" (assuming that the gene is more than two nucleotides); chromosome mutations involve changes in the number of genes (deletion or duplication), in the location of genes (inversion or translocation) or changes in the number of chromosomes (fusion, fission, aneuploidy, haploidy, polyploidy)[Dobzhansky (et al), 1977, pp. 57–58].
5. In [Ayala and Valentine, 1979, p. 123], the following example is used to illustrate the applicability of the concept of Darwinian fitness: consider three genotypes AA, Aa and aa at a certain locus, such that individuals with AA and Aa produce one progeny, whilst those with aa produce (on average) 0.8 progeny. The Darwinian fitness of individuals with the highest reproductive efficiency is assigned a fitness value of 1. Hence the fitnesses of the three genotypes are respectively 1, 1, and 0.8. One may generalize from this example. For genotypes AA, Aa and aa, if A and a have frequencies p and q respectively and the Darwinian fitness w of AA and Aa are identical, but that of aa is less, then: $w(AA) = w(Aa) = 1$ and $w(aa) = 1 - s$. The frequency before selection according to the Hardy-Weinberg law will be $p^2\ (AA) + q^2\,(aa) + 2pq(Aa) = 1$. The frequency after selection is the fitness multiplied by the frequency before selection: $p^2\ (AA) + q^2\ (1-s)\ (aa) + 2pq(Aa) = 1$. A increases and a decreases by a frequency:
$$\Delta p = \frac{spq^2}{1 - sq^2}$$
6. Some appendicle remarks will be given however on the "species problem". The problem is: "what are species?" If we take species to be classes, and genera to be logical unions of classes, where classes are set-theoretical entities, then it follows that species cannot evolve, for abstract entities are not the sort of things which can evolve. The position of Ghiselin [1975, 1981] and Hull [1976, 1978] takes species as individuals, and thus directly avoids this problem. But it leads to others. By "individuals", we are referring to what philosophers call particulars. It is a matter of controversy that all particulars are located in space and time (excluding even the obvious transcendental exceptions); it is nevertheless an assumption Ghiselin accepts [Ghiselin, 1981, p. 271]. In addition, he maintains that organisms which belong to the same species have no ostensive properties in common which serve to individuate them from organisms of other species. This latter assumption is, as Kitts and Kitts [1979] argue, inconsistent with our theoretical practical use of the term, and we require better reasons than Ghiselin has given for a change in usage. The problems which Ghiselin's nominalism faces, is that in accepting species as individuals many basic properties of particulars must be abandoned. For example, the case of the extinction of a species S_1 at t_1 and its re-evolution from species S_2 at t_2 would mean that species are not, contrary to Ghiselin's assumption continuous in time. Nor are they continuous in space as we see, for example,

with the species *Homo sapiens*: some of its members for a short time might be even absent from the planet on space journeys. Further, two species can surely exist at precisely the same place at the same time. This does not mean that individual organisms exist precisely at the same place at the same time, and it is not necessary to accept this latter assumption to accept the former. This assumption also flies in the face of common assumptions about individuals.

I will follow Kitts and Kitts [1979] in treating species as natural kinds. This position needs further defence, especially in the light of strong criticisms [Sober, 1980] [Dupré, 1981], which also must be considered elsewhere.

7. Midgley [1979] argues that the concept of the "Selfish Gene" commits a category mistake, but she is certainly correct in my opinion. However so are the rejoinders by her Sociobiological opponents, especially Dawkins [1981]. The term "altruism" is a technical term Sociobiologists tell us. Fine, but unfortunately they have the tendency to slide from this technical term to the commonsense term when this suits them in the course of arguments. Wilson equates the cooperative behaviour of dolphins helping a wounded mate as a case of "altruism", [Wilson, 1975, pp. 476–477], but not *all* wounds affect the fitness of organisms, although many might, and the dolphin even though hurt might just as well have survived on its own. Thus an argument is needed for Wilson's assessment, not merely flat dogmatic assertion. Likewise in his discussion of human barter, he simply *assumes* that this is a case of *reciprocal altruism* [ibid., p. 551]. Alexander [1979, p. 271] tells us that *all* conflicts of "interest" are ultimately conflicts "over the differential reproduction of genetic units", a statement which is either outrightly false (intellectual conflicts seem as obscurely related to differential reproduction as the case of barter) or else it involves an equivocation on the term "interest", where the social sense is confused with the genetic sense.

Finally David Hull [1981] after criticizing philosophers for their misguided use of ordinary language claims that "altruistic genes", "increase the likelihood that some other gene, and not it, is replicated in greater relative frequency in later generations" [ibid., p. 308]. But the relationship between this concept and Wilson's "atruism" is not obvious. Altruistic genes in Hull's sense would by definition be rapidly eliminated from the gene pool *even if* kin selection was operating, since by hypothesis, the genes of the kin in question which are "altruistic" in Hull's sense would also be eliminated in time: if not, then they are not Hullian – altruistic genes. Such genes are doomed to certain death in the sample space of evolutionary time. Hull in response to this criticism has argued (personal communication 2/6/82) that my difficulties hinge around a confusion between (1) the question of identity or numerical sameness, concerning the likelihood that one allele rather than another is passed on, and (2) the percentage of replicates that two organisms have in common. Now whilst agreeing that this is an important distinction it does not affect my argument, since we can give a Hullian definition of both altruistic-genes (alleles) and replicators, and the argument simply proceeds as above.

8. *The Saga of the Flies Mating on the Cow Pats.* Maynard Smith [1978 (2), p. 265] states with regard to Parker's [1970] cow pat example: "if other males leave a pat quickly it will pay him to stay on, because he would be certain of mating any females which do come. If other males stay on, it would pay him to move". Now we must start analysis somewhere, so let us start on the pats, since the question is, after all, about departure.

   Suppose n flies are on their pats (n is a positive interger) monitoring each other. F_1 will stay if $F_2, F_3, \ldots F_n$ leave and leave if $F_2, F_3, \ldots F_n$ stay; F_2 will stay if $F_1, F_3, \ldots F_n$ leave and leave if $F_1, F_3, \ldots F_n$ stays; ... F_n will stay if $F_1, F_2, \ldots F_{n-1}$ leave, and leave if $F_1, F_2, \ldots F_{n-1}$ stay. To simply consider this only two flies F_1 and F_2. Then the following are the case:

   (1) Stays $(F_1) \rightarrow$ Goes (F_2)

   (2) Goes $(F_1) \rightarrow$ Stays (F_2)

   (3) Stays $(F_2) \rightarrow$ Goes (F_1)

   (4) Goes $(F_2) \rightarrow$ Stays (F_1)

   Now it is also the case that either:

   (5) Stays (F_1) & Stays (F_2)

   (6) Goes (F_1) & Goes (F_2)

   (7) Stays (F_1) & Goes (F_2)

   (8) Goes (F_1) & Stays (F_2)

   Since for any fly F_k: $\vdash$ Stays $(F_k) \vee$ Goes (F_k).

   (a) *Consider* (*5*). If (5), then Stays (F_1) & Stays (F_2). But from (1) if Stays $(F_1) \rightarrow$ Goes (F_2), which is inconsistent with (5), eliminating this possibility.

   (b) *Consider* (*6*). If (6), then Goes (F_1) & Goes (F_2). But from (2) if Goes $(F_1) \rightarrow$ Stays (F_2), which is inconsistent with (6), eliminating this possibility.

   (c) *Consider* (*7*). If (7), then Stays (F_1) & Goes (F_2). If Stays (F_1) then Goes (F_2), but for Goes (F_2) in turn Stays (F_1), but in turn for Stays (F_1), Goes (F_2), ... *ad infinitum*.

   (d) *Consider* (*8*). If (8), then Goes (F_1) and Stays (F_2). But if Goes (F_1), then Stays (F_2). But in turn for Stays (F_2), then in turn Goes (F_2), which in turn requires Stays (F_1), ... *ad infinitum*.

   Thus, (a) and (b) are impossible and (c) and (d) lead us into a vicious infinite spiral similar to that facing Grice's analysis of communicative intent [Giddens, 1976, p. 88]. It follows then the Maynard Smith's cited assumption is incoherent.

9. D. S. Wilson [1978] has argued that a cultural organism which: (1) does not have any offspring and (2) does not increase the numbers of offspring of relatives, can still increase its biological fitness by "altering the selective

pressures operating on the population in favour of its own genotype" [ibid., p. 235]. To take his own example, if proficiency with a computer has a genetic component, and working with computers have resource acquisition advantage, then computer developers increase the prevalence of their genotype by their cultural innovation. This is clever, but fallacious. If computer design and operation was genetically based, to put it crudely as is common in Sociobiology, there are "computer genes" (more precisely there are heritable differences with regard to computer operation) unless (1) or (2) are false, then the computers will not be operated once computer operators die out (from (1)). But if "actual genes can be recruited from the population at large" [ibid., p. 236] the example is irrelevant, since we are now considering a model where we have computer-genes in individuals and all Wilson has drawn our attention to is a fact of genetic migration. This is no argument that the fitness of computer operators has increased; on the contrary, for the argument to work we must already presuppose the existence of the required genotypes to be recruited from the population.

10. c.f. also [Slotkin, 1947, 1949] [Goodenough, 1949].
11. The reason for this according to statistical theory, is that the magnitude of sampling errors are inversely related to the size of the sample, the smaller the sample, the greater the probability of error.
12. The standard error of the mean (S.E. ($\bar{x}$)) is defined as follows:

$$\text{S.E. }(\bar{x}) = \sqrt{\left(\frac{\sigma^2}{n}\right) \cdot \frac{N-n}{N-1}}$$

where "σ" is the standard deviation in the population; "N" is the number of units in the population and "n" is the number of units in the sample.

13. *Haldane's dilemma* is summarized with a minimum of mathematical fuss and bother in [Dobzhansky (et al), 1977, pp. 163–164]. Consider a population P consisting of individuals with either gene G_1 or G_2, such that the fitness of individuals with G_1 is greater than that of individuals with gene G_2. Natural selection will favour G_1. This however puts a "genetic cost" upon the population, since G_2 type individuals must be eliminated. If individuals with such a genotype in P have a high Darwinian fitness, for a new mutation to be established, an extensive genetic load must be overcome. This, Haldane and others have shown, is usually much greater than the number of individuals alive in P in one generation. Given neo-Darwinist prejudices of gradualism, evolution can only occur at a tremendous cost measured in genetic deaths. This process seems to have performed its own *reductio ad absurdum*.

    One way of avoiding this dilemma has been to accept *pan-neutralism*. Another strategy has been to reject the assumption of *genetic additivity* (see the next chapter) of the model, but the latter strategy seems to me unsatisfactory, since (whilst I cannot show this here), *Haldane's dilemma* could be restated even if the level of selection was not the gene.

14. On this point Hoyle [1982, p. 11] has cited the case of a bacterium *Micro-*

coccus radiophilus which repaired 10,000 separate breakages in its DNA structure from an X-ray dose millions of times greater than that which has ever existed on Earth.

15. In this recent book, Michael Ruse [1982] has this to say about the "probability argument" against neo-Darwinism:

> "The Darwinian's point is that the evolution of life occurs in this sort of way. Natural selection allows the successes, but "rubs out" the failures. Thus, selection creates complex order, without the need for a designing mind. All of the fancy arguments about a number of improbabilities, having to be swallowed in one gulp, are irrelevant. Selection makes the improbable, actual". [ibid., p. 308]

The argument must assume the Dobzhansky position of natural selection – that it is a creative force, otherwise the argument is an utter *petitio principii,* for the critic would like to know about the origins of that which natural selection "rubs out".

Chapter 5

1. It is a standing dogma of the current literature on theoretical reduction, especially in the philosophy of mind, that the ideas of "type-type reductions" and "token-token reductions" are intuitively clear. But this is not true: these concepts even used within their rightful domain of semiotics lead us into difficult issues, especially over the problem of universals. There are however some specific problems which befall token-token reductions as described by Kimbrough [1979, p. 403] for example.

   In token-token reductions "individuals described in the language of one science are identified with individuals described in the language of another science" [ibid., p. 403]. This characterization is flawed: (1) It doesn't exclude mutual reducibility; (2) worse, any competing mutually inconsistent theories about individual I_1 and its properties are thereby mutually reducible; (3) It does not exclude mutual reducibility of absurd theories: say of poetry and physics, mathematics and Sociology, and so on. Here I mean that token-token reductions might very well occur as a matter of accident purely because two sciences might well, in context, make references and descriptions of the required individuals, even though the science may be inappropriate objects for reduction in Kimbrough's sense.

2. Matters of explicating the genotype-phenotype distinction become more difficult when one attempts, as Dawkins [1978] has attempted, to extend the concept of a phenotype to include more than one individual. Logically the extension of the concept of a phenotype must also include the genotype of the other individual, and that individual, the genotype of the former individual. The genotype-phenotype distinction then collapses: there exists only an "extended phenotype".

   I should also note a matter which relates to a previous argument in chapter 4 in my example of the behaviour of *Pepsis marginata* and *Cyrtopholis portoricae*. It might be suggested that this problem can be solved by Dawkins' idea of the extended phenotype: *Cyrtopholis portoricae* is clearly programmed to preserve the genes of *Pepsis marginata*. Thus genes may be selected for phenotypic effects upon other bodies. However, it is unclear how this could be, given the neo-Darwinist account of gene action which Dawkins presupposes. It must be said that the conflict between *Pepsis marginata* and *Cyrtopholis portoricae* genes is not different from the conflict of genes in one body [ibid., p. 72]. But the conflict of "power", i.e. replicator power, is seen in modification of the phenotype expression of other genes, and Dawkins hasn't in my opinion satisfactorily explained even in the vaguest of terms at a molecular level, how such extended expression occurs – unless he wishes to introduce action-at-a-distance. Further, one must draw critical attention to the idea of the *replicator* and replicator selection. Replicators in his 1978 position are neither individuals, populations, societies or even gene-tokens: "the whole world is potentially part of the phenotypic expression of a gene" [ibid., p. 72]. But given our previous argument that the genotype-phenotype distinction has collapsed for Dawkins, we are left now with a super holism where the gene is also the phenotypic expression of the whole world. Either that, or we accept an essentially circular characterization of replicators as "that which is selected and replicated".

3. The Hardy-Weinberg law which was also cited in the previous chapter deserves statement. The law is that in a randomly mating population (i.e. where no mating discrimination exists) the hereditary process does not (by itself) alter the genotypic frequencies at a given locus. Let K and M be two variants of a sexually reproducing species such that K and M "genotypically" differ only insofar as K has at a locus alleles $A_1 A_1$ and M, alleles $A_2 A_2$, which do not mutate for a time period t_n. Assume for t_n no migratory patterns. Let the proportion of $A_1 A_1$ and $A_2 A_2$ individuals in the population be p and q respectively such that $p + q = 1$. Then by Mendel's laws, the possible genotypes of the next generations are $\{A_1 A_1, A_1 A_2, A_2 A_2\}$ with the following frequency:

$$(1) \quad p^2(A_1 A_1) + 2pq(A_1 A_2) + q^2(A_2 A_2) = 1.$$

which is simply proved through an application of combinatorial theory. For three alleles A_1, A_2, A_3 with respective frequencies at t_n of p, q and r such that $p + q + r = 1$, the following frequency is predicted by the Hardy-Weinberg law:

$$(2) \quad p^2(A_1 A_1) + q^2(A_2 A_2) + r^2(A_3 A_3) + 2pq(A_1 A_2)$$
$$+ 2pq(A_1 A_3) + 2pr(A_2 A_3) = 1.$$

In general for $A_1, A_2, \ldots A_n$ alleles with respective frequencies $p_1, p_2, \ldots,$ p_n such that $p_1, p_2, \ldots p_n = 1$, then granted the usual qualifications, in the next generation the genotypic frequency is:

$$(3) \quad p_1^2(A_1 A_1) + p_2^2(A_2 A_2) + \ldots + p_n^2(A_n A_n) + 2p_1 p_2(A_1 A_2)$$
$$+ 2p_1 p_3(A_1 A_3) + \ldots + 2p_1 p_n(A_1 A_n) + 2p_2 p_3(A_2 A_3) + \ldots$$
$$+ 2p_{n-1} p_n(A_{n-1} A_n) = 1$$

which can also be proved by combinatorics. The derivation, however, depends upon Mendel's laws, and if these are strictly speaking false, then the argument of the derivation is unsound.

4. The consistency problem of Mendelian genetics and Molecular biology is this: if Mendelian genetics was axiomatized, then the statement:

   (S) Mutation and crossing over cannot occur within the gene

   is deducible from the axioms and theorems by means of accepted rules of logical inference. If Molecular biology was axiomatized, then a statement:

   (~ S) Mutation and crossing over can occur within the gene (i.e. at any point on the DNA strand)

   is deducible such that "S & ~ S" is a necessarily false statement. Now on the logical-empiricist account of theoretical reduction, Mendelian genetics (i.e. its axioms and theorems) are deducible from Molecular biology and certain bridge laws, i.e.

   $$(1)\quad M_0 \,\&\, M_1 \,\&\ldots\&\, M_n \,\&\, B_0 \,\&\, B_1 \,\&\ldots\&\, B_m \underset{L}{\rightarrow} G_1 \,\&\, G_2 \,\&\ldots G_0$$

   where "M_0 & M_1 & ... & M_n" is the conjunction of all the axioms and theorems of Molecular biology; "B_0 & B_1 & ... & B_m" are the bridge laws "G_1 & G_2 & ... & G_0" of all axioms and theorems of Mendelian genetics, and "$\underset{L}{\rightarrow}$" is the sign of logical implication. But if (1) and (S) and (~ S), then:

   $$(2)\quad M_0 \,\&\, M_1 \,\&\ldots\&\, M_n \,\&\, B_0 \,\&\, B_1 \,\&\ldots\&\, B_m \underset{L}{\rightarrow} S \,\&\sim S \text{ as}$$

   $$\Vdash (p \underset{L}{\rightarrow} q) \,\&\, (q \underset{L}{\rightarrow} r) \underset{L}{\rightarrow} (p \underset{L}{\rightarrow} r)$$

   in most standard systems of inference. But this would demonstrate the inconsistency of Molecular biology!

5. Robert Cummins [1977] has argued that "programs which are couched purely in information processing terms may have an explanatory interest for a given range of behaviour which is independent of physiological explanations of the same range of behaviour" [ibid., p. 269]. Cummins also has concluded that since for a programme P to be represented in a system S it is sufficient that S executes P, we cannot explain why P is executed by again appealing to the fact that the programme is represented in the system [ibid., pp. 277–278] – without vicious circularity. Hence Cummins' initial conclusion supports the position advocated in the present work. This conclusion we should note follows on Cummins' own generously wide account of a programme which seems to equate it with anything containing "instructions". Information processing programmes are programmes for manipulating symbols. It is difficult to see how information processing programmes can be satisfactory explanations of the behaviour of organisms, given Cummins' own acceptance that "two completely incommensurable analyses of the same complex capacity may both be correct" [ibid., p. 282]. On a realist theory of science two *completely* incommensurable explanations of the *same* phenomena certainly are not *both* correct. Second, Cummins' circularity argument seems quite general in its scope, and

thus, contrary to his denials, applicable to information processing programmes as well.

6. Whilst I agree with Lehrman's position, Katz and Goffman [1981] would not: they claim that the topologies of virtually all patterns present in early development are conserved throughout ontogeny. Now it hardly seems credible to propose that there is *no* inheritance of morphology from parents to offspring: cells after all are not indeterminate blobs, they have a determinate shape and characteristic structure, as do organisms. It is however another matter to propose that topological preformationism is true, and no science of morphology is necessarily committed to this position for its cogency. It is thus informative to give an exposition and criticism of Katz and Goffman's argument.

   By a "pattern" they mean "a stable configuration of discrete elements [ibid., p. 444]. If $\bar{p}$ and $\bar{p}^*$ are patterns, then $\bar{p}$ is a precursor to $\bar{p}^*$ if there is a deterministic mapping m: $\bar{p} \rightarrow \bar{p}^*$ such that if m(e) is an element of $\bar{p}^*$, then e is an element of $\bar{p}$. Now $\bar{p}^*$ inherits its topology from $\bar{p}$, if and only if $m(e_1)$ and $m(e_2)$ are neighbouring elements in $\bar{p}$. They then develop a mathematical theory based upon U. Grenander's mathematics for patterns of discrete elements [Grenander, 1976, 1978], which they believe and which I don't believe, represent or model some of the properties of developing organisms.

   "Generators" are the discrete elements in patterns. Generators are mutually linked at linkage sites or bonds on the generators. Let P be a population of n generators, i.e. $P = \{g_1, g_2, \ldots g_n\}$. Each generator is a set of bonds, i.e. $g = \{b_1, b_2, \ldots\}$. Let b_i be a bond of generator g_a and b_j a bond of generator g_b, then b_i and b_j form a *link* if and only if "Rb_ib_j" is true, where "R" is the linkage requirement relation, this being the relation of identity $Rb_ib_j \leftrightarrow b_i = b_j$.

   The measure M(a, b) of the *relative linkage* of generator g_a with generator g_b is given by:

$$(1)\quad M(a, b) = \frac{N(g_a \cap g_b)}{N(g_a)}$$

Where "N(x)" designates the number of elements in a set x [ibid., p. 445]. Generators g_a and g_b are said to *communicate* (Cg_ag_b) if and only if, for a "threshold value" t such that $0 \leqslant t - 1$, if $M(a, b) > t$ and $M(b, a) > t$. A sequence:

$$(2)\quad Cg_ag_{a+1}, Cg_bg_{b+1}, \ldots Cg_kg_{k+1}$$

for a sequence of generators g_a, g_{a+1}, g_b, g_{b+1}, ... g_k, g_{k+1} is called an *intercommunication chain*, and any pair of $\langle g_a, g_{a+1}\rangle$, $\langle g_a, g_{b+1}\rangle$, ... $\langle g_k, g_{k+1}\rangle$ is said to intercommunicate. For a given t, the intercommunicative relationship is reflexive, transitive and symmetric and hence establishes an equivalence relationship on P.

   Cells are disjoint classes which partition P, such "that all and only members of a given cell are in intercommunication with each other" [ibid.,

p. 446]. The length L of an intercommunication chain of k generators is: $\bar{L}(g_1, g_k) = k - 1$. $\bar{L}(g_1) = 0$ and if g_a^* and g_b^* are not intercommunicative, then $\bar{L}(g_a^*, g_b^*) = \infty$. P constitutes a metric space $(P, \bar{L}, t)$ since these postulates are satisfied:

$$(3)\ (1)\ \{[\bar{L}(g_a, g_b) \geqslant 0]\ \&\ [(\bar{L}(g_a, g_b) = 0) \leftrightarrow (g_a = g_b)]\}$$

$$(2)\ \{\bar{L}(g_a, g_b) = \bar{L}(g_b, g_a)\}$$

$$(3)\ \bar{L}(g_a, g_b) + \bar{L}(g_b, g_c) \geqslant \bar{L}(g_a, g_c).\}$$

As $t \to 0$, P tends towards a single "cell", indeed all of P itself. As $t \to 1$, P tends towards a state where each "cell", "contains only one distinguishable generator" [ibid., p. 446]. As $t \to 1$, each "cell" is a subset of one of the "cell"s precedingly covered, which they allege is similar to mitosis, and which I allege is not: in the latter we speak of a concrete object, not a set (and by the above definitions we cannot make the usual escape by claiming that we are talking about the elements of a set). More on this shortly.

Let C_a and C_b be two "cells" with generators g_a in C_a and g_b in C_b, then C_a communicates with C_b if and only if the relative linkage $M(a, t)_{max} > T$, where $0 \leqslant T \leqslant 1$. If C_a communicates with C_b, C_a and C_b are said to be adjacent. Organisms are represented by a graph, the cells being the points on the graph with adjacent cells connected. A pattern p^* of cells at $t > T$ for a given T, inherits its complete topology from a precursor pattern $\bar{p}$ of generators at $t = T$, as of the mapping $m: \bar{p} \to \bar{p}^*$. In particular, for adjacent cells C_a and C_b at $t > T$, there are generators g_a and g_b adjacent at $t = T$, such that $m(g_a) = C_a$ and $m(g_b) = C_b$ satisfying the previously stated definition.

An interesting mathematical model, but what has it to do with biology? True, biological cells are likewise generated from previously existing cells, but such a statement is uncontroversial. Yet the cells of Katz and Goffman are set – theoretical entities, and that they model biological cells is asserted, not argued for [ibid., p. 450]. Generators, for example, are identified with "certain cytoplasmic factors" which exist in a particular configuration within the zygote, and are parcelled out among the cells during development. In particular "the configuration of the generators in a given cell is a skeletal representation of the cytoplasmic organization of that cell" [ibid., p. 450]. Unfortunately no biological model of this structure has been given.

Next criticism. The term "pattern" in the language of biology is usually taken to mean a stable configuration of concrete elements, since biology is an empirical science. Generators are however sets of bonds. Now these generators are said to be mutually linked "attached by linkages formed among the available linkage sites or "bonds" of the generators" [ibid., p. 445]. Yet generators are also identified as sets of bonds: if "bonds" are a property *of* generators, then generators are not, under pains of vicious infinite regression, sets of bonds. We are also told that a bond b_i of generator g_a forms a link with bond b_j of generator g_b if and only if the linkage requirement "Rb_ib_j" is satisfied. But then we are told that "R" is the relation such that $b_i = b_j$, which when satisfied forms a link between generators g_a and g_b [ibid., p. 445]. Thus the only recognized linkages are identity relations, and since each generator is a set of bonds, if linkage occurs, each generator will be a singleton set. This hardly models cytoplasmic reality since cellular linkages are hardly identity relations.

Finally, a pattern $\bar{p}$ is said to be a precursor to pattern $\bar{p}^*$ if there is a deterministic mapping m: $\bar{p} \rightarrow \bar{p}^*$ such that if m(e) is an element of $\bar{p}^*$, then e is an element of $\bar{p}$. Consider the morphology of an organism, such as the human limb structure. Consider an element m(e) of $\bar{p}^*$ such as both of my hands. Then if m(e) is an element of $\bar{p}^*$, then e is an element of $\bar{p}$. Now one topological relationship which exists between my hands, is that which Nerlich [1976, p. 29] following Kant calls *enantiomorphism*. As far as I know, nothing like this occurs in the cell, and for example *only* "left handed" or levo amino acids occur in living cells [Crick, 1966, p. 60]. This is advanced as a test case: is the "enantiomorphism" topologically preformed? If so "where"?

Chapter 6

1. I am grateful to Dr. Lesley Rogers, Department of Pharmacology, Monash University for her comments upon the initial draft of this chapter, and Allan Patience, Sociology, Flinders University for reading and commenting upon this chapter at all stages of production.
2. See for example [Wittman, 1973] [Radicalesbians, 1973].
3. For a further sympathetic discussion of Dörner's work cf. [Goy and McEwen, 1980, pp. 64–73].
4. "Reproduction", in a narrow sense, I take here to refer to those processes by which a numerically distinct organism is produced from another organism. In a wider sense ("Reproduction_2"), the term refers to a change (most usually an increase) in the number of cells. Obviously not all the cells of an organism (e.g. human red blood cells, neurons etc.) are organisms, so they are incapable of "Reproduction", only "Reproduction_2". The context in the text should make the sense in which I am using these respective terms clear.
5. If the respectible gametes were approximately the same size and of the same number, the categories of "male" and "female" could not be defined, even though sexual reproduction (meiosis followed by syngamy) still occurred.
6. There is one challenge to this definition, insofar as "maleness" and "femaleness" have been ascribed to the conjunction activities of *E. Coli*. But the mere transfer of nucleoid material is not a sufficient condition for sexual reproduction – the genetic material needs to fuse as part of the process of forming a new individual, numerically distinct from the parent. Bacterial conjunction activities can be explained without use of sexual categories at all.

If we took sex to be merely "a process which changes the relationship between different elements (genes or linkage groups) of the genome, with or without the introduction of completely novel genetic material" [Bell, 1982, p. 20], then it is impossible to exclude counter-intuitive cases of mutational alterations of the genome as being sexual, such as when precisely half the genetic material of the genome is destroyed by a mutational agent, mimicing (and "mocking") meiosis. Reference is needed in any adequate definition of sex, to the process of the formation of a new individual.

7. McEwen [1976, p. 48 and p. 57] speaks of "the paradox that a 'male' hormone, testosterone, is converted into a 'female' hormone, estradiol, which then exerts strong male effects . . ." [ibid., p. 57]. Indeed, this would constitute good reason for concluding that estradiol is an androgynous hormone. The characterization given in the text, making reference to secondary sexual effects, does not rule out the logical possibility of some hormones being classified for some purposes as androgens, and the same hormones being classified for other theoretical purposes as estrogens. (NECER) seems to assume, that the androgen/estrogen dichotomy is both rigid and absolute, rather than vague and relative. The acceptance of the latter position strikes deeply to the heart of positions such as Goldberg's.
8. To produce explanations devoid of explanatory power is a general tendency in Goldberg's book. He also argues, in the latter section, that as there are more "famous" men than women, the "natural ability" of men must be greater than that of women. "Natural ability", whilst not defined as the ability to obtain fame (which would be circular) is at least measured by the attainment of fame. But such a criterion is inadequate, since people with natural ability are not always given the opportunity by society to pursue fame. (I will assume here that the feminist tradition has established that women have not had such opportunities (at least as freely as men)).
9. We exclude by this clause cases of brain transplants, which are irrelevant to our argument. We can clearly imagine transplanting a women's brain, which has by necessity a female brain organization (by (NECER)'s assumptions), into a male body. And it may well be that the (hopefully) functional individual after the transplant will still sexually prefer males. But even in (NECER)'s terms, this preference is *not* caused by the alleged hormonal abnormalities which allegedly cause homosexuality in men, because we are, by hypothesis transplanting a *women's brain*. Thus this *Gedaken experimente* does nothing by the way of supporting (NECER).
10. Even if logical fatalism or philosophical determinism were true, all this would show is that if a non-sexist society was created, then it was "inevitable" that it was created. Determinist arguments fail to show that some particular action cannot be done (unless, trivially, it is physically or logically impossible), only that it cannot be done *freely*.
11. This can be readily established by a Feyerabend-style argument: (1) It is generally admitted that in time contemporary scientific theories will be seen to be strictly false, but if (2) scientists always undertake theory choice by choosing the most parsimonious theories, then it follows that (3) it is not necessary that parsimonious theories are true.

Chapter 7

1. I shall give copious quotations from the writings of my opponents. I have noticed with this chapter, more than any other here, that the Sociobiologists who have been kind enough to review this material respond with *ad hominem* arguments which are not even amusing. One such critic, after abuse writes:

   > Optimality models are applied to specific behaviours, such as foraging, not to the sum of an animal's behaviours, still less to all the behaviours of a group, e.g. social insects or human society.

   The quotations which follow shall rebut this interpretation. If Ecological Sociobiology is to be of sociological and philosophical significance, a stronger claim than this must be made. This claim, it is argued in the text is made.
2. This is an impression gained also from criticisms of this chapter from readers. The point which Maynard Smith makes therefore deserves emphasis.
3. Niven (personal communication 10/12/82) states "The defining equations don't work for this Very Peculiar Animal [*Homo sapiens*]".
4. And more accurately c.f. [Oakes, 1981, p. 21 note 49].
5. This term is not Harris', but is taken from Oakes [1981, p. 4].
6. This contradiction can be avoided, but not the way Oakes [1981, p. 7] suggests. For Oakes, the Cultural Materialist must maintain that whilst emic explanations are never reasonable, they are permitted in situations where no etic explanations are available. But if Harris accepted this, then he would be committed to a large scale irrationalism in the light of Oakes' demonstration that Harris' major victories as a Cultural Materialist are emic victories [ibid., pp. 8–111]. What needs to be done is to modify the principle of emic mystification such that it asserts that all etic explanations are more reasonable than emic explanations. In doing this however, the principle of emic mystification becomes straightforwardly false, because we can easily cite nonsensical etic explanations or etic explanations which contain internal inconsistencies.
7. It is also interesting to note, that Harris nowhere attempts to operationalize the claim, that when infra-structural variables are "ripe" the appropriate thoughts and theories will occur. For example, what are the appropriate conditions of the mode of reproduction and mode of production for philosophical scepticism to be advocated? One might have some success explaining this phenomena politically and sociologically, but I don't see what "ecosystems" and "contraception" has to do with this matter. Yet it is precisely these claims which Cultural Materialists are committed to. These are theoretically absurd claims and illustrate the absurdity of the position.
8. In a later text, Hardin [1977] argues that any non-authoritarian "commons" system of environmental and resource management will result in ecological disaster that can only best be resolved by a private enterprise social system. The "Tragedy of the Commons" argument is, stated briefly, that in an area of grazing the benefits of adding additional animals to the herd belong to the individual whilst the costs of overgrazing are shared by all parties. Thus each and every rational herdsman will conclude that it is reasonable

to increasingly add more animals to the land, thus resulting in overgrazing. Thus a common-alism mode of resource management is self-defeating, for this example can be readily generalized.

The central difficulty with this argument is that it may just as easily be taken to constitute a *reductio ad absurdum* of Hardin's premises than of common-alism. Since the costs of overgrazing are by hypothesis, shared by all parties, unrestricted addition of livestock, or in general, unrestricted use and abuse of physical resources is irrational *precisely because* it leads to the frustration of the interests of everyone. Hardin's implicit assumption that all people will act egoistically is of course inconsistent with the interdependence of interests in a common-alism situation; from the point of view of common-alism, so much the worse for egoism.

9. For yet another repeat of the Hardin-style argument c.f. [Pascal, 1978, 1980] and for a criticism [Gray and Wolfe, 1980].
10. I have greatly benefited from talks with my friend and colleague Dr. Riaz Hassan and from [Hassan, 1979].

Chapter 8

1. Midgley [1978, p. 59] says even more strongly in this regard, that to have an "aggression instinct" means that

> in his [i.e. human beings] repertory of natural tendencies there is one to attack other members of his species sometimes, without being taught to, *without needing to as a means to another end, and without provocation* (italics added).

This statement is absurd, and I will attempt to demonstrate this in the course of the argument.

2. The "number of apostles or planets" argument which was cited earlier is this: the statement

(1) Number Planets = 9

is a true statement. But it is also true that:

(2) $\Box(9 > 7)$.

But it is also true that:

(3) $\sim\Box$(Number Planets > 7)

as there could have been any number of planets. Yet by applying indiscernibility to substitute the identity of (1) in (2) we obtain:

(4) $\Box$(Number of Planets > 7)

which is inconsistent with (3). In our previous citation we offered no explicit response to this issue. The answer I maintain now to this objection is that premise (3) is false, whilst (4) is true. A true *de re* statement is being confused with the false *de dicto* statement:

(5) "The number of planets is greater than seven" is a necessarily true proposition.

The argument from the "mathematical cyclist" is this: consider the statements:

(6) All cyclists are essentially two-legged but not essentially rational.

(7) All mathematicians are essentially rational but not essentially two-legged.

Consider an S such that S is both a mathematician and two-legged. Then it follows, (some believe) that S both is and is not essentially rational and essentially two-legged. The reply to this is that once more an insensitivity to the *de re/de dicto* distinction is made, and that (6) and (7) are being confused with the true *de dicto* claims (8) and (9):

(8) "All cyclists are two-legged" is a necessarily true statement and "All cyclists are rational" is not.

(9) "All mathematicians are rational" is a necessarily true statement and "All mathematicians are two-legged" is not [Plantinga, 1974, pp. 23–26].

A third argument, considered by Loux [1978, p. 167] can also be answerd by the above strategy. He considers the following analysis of *de re* modality by *de dicto* modality:

(10) An object exemplifies an attribute essentially = if the proposition that it exemplifies that attribute is necessarily true.

The problem with this analysis becomes evident when one considers a possible world w_1 where the object does not exist; since necessary truths are true in all possible worlds, the proposition that an object exemplifies an attribute will be false in w_1 and hence from (10) it follows that no contingent being can exemplify essential properties. This argument, apart from making explicit use of what Routley [1980 (1)] calls the "reference theory of truth", is a position which is problematic in that it makes an explicit equation of *de re* and *de dicto* modality. Reject this equation and the analysis collapses and the problem for the essentialist dissolves.

3. In [Brody, 1980], Brody argues that a satisfactory definition of essential properties is this:

(B_1) a has property p essentially = df a has p and a has always had p, there is no possible past in which a exists without p, and there is no moment of time at which a has had p and at which there is a possible future in which a exists without p [ibid., p. 123].

There are a number of points which could be raised here – especially with regards to cross-temporal identification [Hooker, 1976, pp. 264–265] [Linsky, 1982]. One point which I will make here is this: given Brody's acceptance of a Leibnizian account of identity, (B_1) trivializes once more the thesis of essentialism. Clearly (B_1) is restricted to entities which have a temporal existence and it does not appear to be immediately applicable to such atemporal entities as numbers. So, considering a temporal entity a and if there was a temporal entity b which has a property p*, such that a did not have p* at some moment of time, but they shared all other properties at all other times, then $a \neq b$, p* being then an essential property of b. But from the Leibnizian account of identity, if b lacked any of its permissible properties at some later time, we would say that a change in identity has occurred. This is true of all the properties of b. If so, then by (B_1) one cannot now exclude all the properties of b from being essential properties.

4. According to Routley [1980 (1), p. 113] the admission of contingent identity destroys Aristotelian essentialism. This claim is incorrect: essentialism can readily be constructed with the abandonment of full indiscernibility. The real difficulty remains, of course, developing a unified account of identity.
5. Lehrman [1953, p. 345] has also noted the inquiry-limiting nature of such concepts:

> The problem for the investigator who wishes to make a causal analysis of behavior is: How did this behavior come about? The use of 'explanatory' categories such as 'innate' and 'genetically fixed' obscures the necessity of investigating developmental *processes* in order to gain insight into the actual mechanisms of behavior and their interrelationships.

We shall follow up this point in the text very shortly.

6. The problems involved for the nativist in the very explication of this thesis are well known. This is especially so if the nativist is to argue that the "knowledge" which mature speakers of a native language have is most plausibly accounted for by postulating that they have "tacit knowledge" [Graves et al., 1973]. That is, a child manages to develop "linguistic competence" through a minimal exposure to what are essentially grammatically ill-formed sentence tokens so that the child must in turn possess innately, items of "tacit knowledge". Philip Kitcher [1978] has argued that if knowledge is taken in the traditional sense of "knowledge that", in volving justification, then I.H. is incoherent, but to free "knowledge" from the concept of "justification" gives us no reason for speaking of "innate knowledge" at all. Such problems have in fact, surprisingly enough, been recognized explicitly by Chomsky [1969, p. 524].

> In the past, I have tried to avoid the problem of explicating the notion "knowledge of language" by using an invented technical term, namely, the term "competence", in place of "knowledge". However, the term "competence" suggests "ability", "skill" and so on, through a chain of associations that lead direct to much new confusion. I do not think that the concepts of ordinary language suffice for the purpose at hand; they must either be sharpened, perhaps somewhat arbitrarily, or replaced by a new technical terminology. Either approach has a familiar disutility.

If this is so, the presently stated I.H. thesis collapses, for it cannot at present even satisfactorily formulated, and Chomsky thereby refutes himself.

7. In fact matters are conceptually worse. One would have to maintain now that language (in fact all possible humanly usable languages) is innate, so that "language acquisition" strictly speaking is impossible. The child then is like the slave in Plato's *Meno*, although even this comparison is problematic. The slave's knowledge "unfolds" under prompting by Socrates, and we assume would not have "unfolded" without this or similar actions. But, to continue the metaphor, what does the "unfolding" for the child, if the linguistic social environment is as Chomsky explicitly states, full of grammatical distortion. Why, even given I.H., should a language be acquired at

all? The question has not been addressed satisfactorily by Chomskyites, and therefore tells against their position. We press this objection further, shortly in the text.

8. For a further critique of I.H. cf [Margolis, 1978, ch. 7] [Stich, 1978].
9. On the problem of the sorites cf [Cargile, 1969] [Grim, 1982].
10. At the risk of stating the obvious (if there ever could be such a risk!) the reason why the historicity assumption is challenged by Creationism in any of its forms is that the historicity assumption is an assumption about the nature or essence of human beings. It is a metaphysical hypothesis about how human beings came to be what they are. In embracing Creationism we must embrace some substantial metaphysical assumptions about human being's place in the "scheme of things", as well as about the "scheme of things" themselves, if we are not to abandon the attempt to provide some sort of *causal explanation* of the phenomena of the world (as Russell's hypothesis about the creation of the world five minutes ago (as well as various forms of Deism) apparently does). These assumptions are usually "teleological" ones about the role, point and purpose of human beings in the world, especially with respect to the "Creative" source.

    With regard to the social sciences, historicity involves, as Anthony Giddens [1979, 1981 (1), 1981 (2)] has cogently argued, a critical rethinking of the concepts of *social system* and *social structure* with respect to time–space relations (although I have and detail later, objections to Gidden's account of both system and structure). This has implications which affect many aspects of received socioscientific theory, such as for example, the received view on socialization [Giddens, 1979, pp. 128–130]. Insofar as "human beings" are taken to be social beings, the implications of such a position will have substantial impacts upon adopted views of the nature of human social nature. This leaves totally unexplicated what precisely such impacts in fact are. We can only decide this after having thought through the question of "social being and time". This work here, unfortunately produces no detailed response to the question, it merely adds to the strength of Giddens' voice in affirming the importance of such a question, and matters of inquiry.
11. The philosophical sceptic's claim that none of the ontological claims of any metaphysical theory are justified, is not itself an ontological claim and is therefore not self refuting.
12. This problem of circularity also befalls attempts to refine Quinean style ontological commitment criteria. According, for example to Church [1958, p. 1014]: "The assertion of (Ex)Mx carries ontological commitment to entities x such that M". Unpacking this statement however gives us only this: "x's" exist having property M, if and only if "x" designates those x's and "Mx" designates the x's having property M, and the x's exist". Since the term "exists" occurs in the definiens, the account is straight-forwardly circular.
13. This is not to say that from some perspective we can formulate *sufficient* conditions for existence. One which I accept is this: if x's exist, then an essentialistic description of x's (i.e. a "real definition") is consistent. Routley

[1980 (1), p. 703] argues that this account fails as in some paraconsistent set theories the Russell class $\{x: x \notin x\}$ is taken to *exist.* But if mathematics is *ontologically neutral* this claim is absurd. Consequently we can only speak about whether an alleged object is admissibly defined or whether it is not.

14. It is a point tangential to the textual discussion, but not to the general interests of this work, to note Margolis' reasons for believing that a physicalist account of persons *must* fail. His argument is based upon concessions made by Feigl and Sellars with regards to the "irreducibility" of intentionality: this "irreducibility" being conceptual in nature, similar to the "irreducibility" of the "ought" to the "is" [Margolis, 1978, p. 15]. The essential properties of persons are intentional properties, but given the thesis of the conceptual irreducibility of intentionality "it does not follow that *the intentional properties of sentience or of sentient organisms* can be physicalistically construed" [ibid., p. 14]. Linguistic ability is irreducible in the sense of there being no purely "physical explanation" (i.e. an explanation involving the concepts, laws and theories adequate for explanations of inorganic and organic processes) of the linguistic capacities of human beings.

    Margolis' argument will unfortunately be unconvincing to the physicalist, for from the fact of the conceptual irreducibility of the intentional to the physical, nothing follows without further premises about the *explanatory irreducibility* of such phenomena. A pocket calculator may well be a completely macro-physical deterministic system, even though, mathematical entities are not physicalistically reducible in an ontological sense. For the physicalist, human beings are simply no more than complex "machines", made of meat rather than iron.

15. I would like to add at this point that not all perceptual experiences are "propositional perceptual experiences" analyzable with reference to (propositional) belief-concepts [Pappas, 1977]. If I maintained this, then I would have difficulties in saying what, if anything, constituted sentient mentality – and surely organic sensations and some perceptual capacities constitute sentient mentality. Further if I adopted an account of "experience", which took experiences to be so general as to be the "activities" of an entity, then I would be led into contradiction, for my categories of agents and mentalities then collapse. Defining the term "experience" is extremely difficult. Bunge [1979, p. 162] takes the experiences of an entity to be the set of all that is has learned up to a specific time. However this explication means that any organisms without what Bunge has called "plastic neural systems" are incapable of experiences, and indeed non-biosystems, such as "Star Wars" type robots, *would* be incapable of experiences. This seems to be to give some reason, but not conclusive reasons, nor even reasons as strong as one would wish, for taking the concept of experience to be indefinable with reference to the concept of learning. (Do we not also say, that we "learn by experience", and is not such talk an important part of our at least folk-psychological discourse?) The concept here is taken as a primitive.

16. The equation of intentionality purely with intensionality has underlined a number of cybernetic-mechanistic arguments seeking to show that simple

"mechanical systems" may be ascribed intentional properties. Whilst I would not deny intentionality to some artificial entities, Boden [1972, p. 128] has not established, to take one example, that an automatic potato locator is an intentional system. The behaviour of the machine certainly can be described in *intensional terms,* but whilst all *intentional* discourse is *intensional,* not all *intensional* discourse is *intentional.* Specifying what intentionality in fact consists in, remains an outstanding problem, another one not addressed here.

17. It is interesting to note that in recent years a re-establishment of alchemy has been undertaken by Kervran [1972], based upon the hypothesis that plants can transmutate elements! As such, if Kervran was right, mechanism would suffer a major blow, as indeed would contemporary physics. I do not intend to debate this issue here as I have not done repeat experiments myself. The point remains that once physicalism and mechanism are abandoned, such radical cases of downward causation are at least *logically possible. If* Kervran was right we would have a case of biology supplying evidence which refuted a major assumption within chemistry and physics! Further, contemporary mechanism would be false even for plants.

18. Some further remarks can be made here in this note about Giddens' concepts of generative rules and resources, which are tangential nevertheless to the central argument of the text. *Resources* refer to the attributes which agents can make recourse to in striving to achieve objects of their desires and interests in social interaction. Thus material status and authority are resources. Rules however, are properties of collectivities, not of specific individuals [Giddens, 1977, p. 132]. Generative rules are either *semantic rules* or *moral rules.* "Semantic rules" not only refer to rules of syntax of language, but also to the taken-for-granted rules of everyday life which constitute mutually shared knowledge. Moral rules are rules which generate moral evaluations of right and wrong, goodness and badness: in short, norms for the evaluation of action.

To know a rule is to know what is appropriate in the specific circumstances to which the rule applies. Accepting Winch's test for the existence of a rule, namely that it is cognitively meaningful to ask whether there is a right or a wrong way of doing that activity, Giddens distinguishes two senses in which there may be a right or a wrong way of doing things. First, in the appropriate identification of the act, second in the moral assessments of acts in specific circumstances [ibid., p. 133].

The problem with this explication of structure lies in the concept of a rule. Both "appropriation" and "making a mistake" involves recourse to *norms of appropriation.* Inappropriate behaviour in either a football game or any other social activity, makes sense only by recourse to *norms.* But as Ullman-Margalit [1977, p. 12] has pointed out, the concepts of "norms" and "rules" have the same sense and reference. Therefore Giddens' account of structure is circular.

BIBLIOGRAPHY

Abramson, N., Information Theory and Coding, (McGraw Hill, New York, 1963).

Achinstein, P., Concepts of Science, (The John Hopkins University Press, Baltimore and London, 1968).

Achinstein, P., "Critical Notice of R. L. Causey, Unity of Science" Nôus, Vol. XV, No. 1, March, 1981, pp. 67–75.

Adams, E. W., "The Foundations of Rigid Body Mechanics and the Derivation of Its Laws from Particle Mechanics" in L. Henkin, P. Suppes and A. Tarski (eds.), The Axiomatic Method, (North-Holland, Amsterdam, 1959), pp. 250–265.

Adby, P. R. and Dempster, M. A. H., Introduction to Optimization Methods, (Chapman and Hall, London, 1974).

Ager, T. A., Aronson, J. L. and Weingard, R., "Are Bridge Laws Really Necessary?", Nôus, Vol. 8, 1974, pp. 119–134.

Alexander, R. D., "The Evolution of Social Behaviour" Annual Review of Ecology and Systematics, Vol. 5, 1974, pp. 325–383.

Alexander, R. D. and Sherman, P. W., "Local Mate Competition and Parental Investment Patterns in the Social Insects", Science, Vol. 196, 1977, pp. 494–500.

Alexander, R. D., Darwinism and Human Affairs, (University of Washington Press, Seattle and London, 1979).

Allen, R. J., "The Idea of a Value Free Social Science", Journal of Value Inquiry, Vol. IX, No. 2, Summer, 1975, pp. 95–117.

Amin, S., Imperialism and Unequal Development, (Harvester Press, Sussex, 1977).

Anderson, P. W., "More is Different", Science, Vol. 177, 4 August, 1972, pp. 393–396.

Andreskil, S., Social Sciences As Sorcery, (Andre Deutsch, London, 1972).

Andrewartha, H. G. and Birch, L. C., The Distribution and Abundance of Animals, (University of Chicago Press, Chicago, 1954).

Andrewartha, H. G., Introduction to the Study of Animal Populations, 2nd edition, (Chapman and Hall, London, 1971).

Anonymous, "Too Soon for the Rehabilitation of Lamarck", Nature, Vol. 289, February, 1981, pp. 631–632.

Anscombe, G. E., "The First Person" in S. Guttenplan (ed.), Mind and Language, (Clarendon Press, Oxford, 1975), pp. 45–65.

Apter, M. J. and Wolpert, L., "Cybernetics and Development I. Information Theory", Journal of Theoretical Biology, Vol. 8, 1965, pp. 244–257.

Armstrong, D. M., A Materialist Theory of Mind, (Routledge and Kegan Paul, London, 1968).

Armstrong, D. M., Belief, Truth and Knowledge, (Cambridge University Press, Cambridge, 1973).

Armstrong, D. M. (1), "Naturalism, Materialism and First Philosophy", Philosophia, Vol. 8, No. 2–3, November, 1978, pp. 261–276.

Armstrong, D. M. (2), Nominalism and Realism: Universals and Scientific Realism Vol. I, (Cambridge University Press, Cambridge, 1978).

Armstrong, D. M. (3), A Theory of Universals: Universals and Scientific Realism Vol. III, (Cambridge University Press, Cambridge, 1978).

Armstrong, D. M., "Review of Margolis' Persons and Minds", Philosophy of the Social Sciences, Vol. 10, 1980, pp. 227–239.

Arrow, K. J., Social Choice and Individual Values, 2nd edition, (Yale University Press, New Haven, 1963).

Arruda, A. I., "A Survey of Paraconsistent Logic" in A. I. Arruda, R. Chuaqui and N. C. A. da Costa (eds.), Mathematical Logic in Latin America, (North-Holland, Amsterdam, 1980), pp. 1–41.

Aune, B., "Rorty on Language and the World", Journal of Philosophy, Vol. 69, 1972, pp. 665–667.

Ayala, F. J., "Evolution of Fitness I. Improvements in the Productivity and Size of Irradiated Populations of *Drosophila serrata* and *Drosophila birchii*", Genetics, Vol. 53, 1966, pp. 833–895.

Ayala, F. J., "Evolution of Fitness V. Rate of Evolution in Irradiated Populations of *Drosophila*", Proceedings of the National Academy of Sciences, (U.S.A.), Vol. 63, No. 3, 1969, pp. 790–793.

Ayala, F. J. and Valentine, J. W., Evolving: The Theory and Processes of Organic Evolution, (Benjamin Cummings, Menlo Park, California, 1979).

Baldwin, J. D. and Baldwin, J. I., "Sociobiology or Balanced Biosocial Theory", Pacific Sociological Review, Vol. 23, No. 1, January, 1980, pp. 3–27.

Baltimore, D., "RNA-dependent DNA Polymerase in Virons of RNA Tumor Virus", Nature, Vol. 226, 1970, pp. 1209–1211.

Bantz, D., Does Physics Explain Chemistry? (Ph.D. thesis, University of Chicago, Illinois, June, 1977).

Baran, P., The Political Economy of Growth, (Penguin, Harmondsworth, 1957).

Barash, D. P., Sociobiology and Behavior, (Heinemann, London, 1977).

Barash, D. P., Holmes, W. G. and Greene, P. J., "Exact Versus Probabilistic

Coefficients of Relationship: Some Implications for Sociobiology", American Naturalist, Vol. 112, No. 984, March–April, 1978, pp. 355–363.

Barash, D. P., Sociobiology: The Whispering Within, (Fontana, Collins, Glasgow, 1981).

Barker, A. D., "An Approach to the Theory of Natural Selection", Philosophy, Vol. 44, 1969, pp. 271–290.

Barkow, J. H., "Culture and Sociobiology", American Anthropologist, Vol. 80, 1978, pp. 5–20.

Barratt Brown, M., The Economics of Imperialism, (Penguin, Harmondworth, 1974).

Beach, F. A., "Hormonal Factors Controlling the Differentiation, Development, and Display of Copulatory Behavior in the Ramstergig and Related Species", in E. Tobach, L. Aronson and E. Shaw (eds.), The Biopsychology of Development (Academic Press, New York, 1971), pp. 249–296.

Bealer, G., "An Inconsistency in Functionalism", Synthese, Vol. 38, 1978, pp. 333–372.

Becker, G., The Economic Approach to Human Behavior, (University of Chicago Press, Chicago, 1976).

Becker, L. C., "Human Being: The Boundaries of the Concept", Philosophy and Public Affairs, Vol. 4, 1975, pp. 334–359.

Bell, G., The Masterpiece of Nature: The Evolution and Genetics of Sexuality, (Croom Helm, London and Canberra, 1982).

Bem, S. L., "Probing the Promise of Androgyny" in A. G. Kaplan and J. P. Bean (eds.), Beyond Sex-Role Stereotypes: Readings Towards a Psychology of Androgyny, (Little Brown and Company, Boston and Toronto), 1976, pp. 48–62.

Benzer, S., "The Fine Structure of the Gene", Scientific American, Vol. 206, 1962, pp. 70–84.

Berger, G., "Realism and Complex Entities", Philosophical Studies, Vol. 30, 1976, pp. 95–103.

Berger, P. and Luckmann, T., The Social Construction of Reality, (Allan Lane, London, 1967).

Berlinski, D., On Systems Analysis, (Massachusetts Institute of Technology Press, Cambridge, Massachusetts, 1976).

Bernstein, R. J., "The Challenge of Scientific Materialism", International Philosophical Quarterly, Vol. VIII, 1968, pp. 252–275.

Bernstein, R. J., The Restructuring of Social and Political Theory, (Methuen, London, 1976).

Bestor, T. W., "Dualism and Bodily Movements", Inquiry, Vol. 19, 1976, pp. 1–26.

Beth, E., Naturphilosphie, (Noorduyn, Gorinchem, 1948).

Beth, E., "Towards an Up-to-Date Philosophy of the Natural Sciences", Methodos, Vol. 1. 1949, pp. 178–185.

Beth, E., "Semantics of Physical Theories" in H. Freudenthal (ed.), The Concept and Role of the Model in Mathematics and Natural and Social Sciences, (D. Reidel, Dordrecht, 1961), pp. 48–51.

Bhaskar, R. (1), A Realist Theory of Science, (Harvester Press, Sussex, 1978).

Bhaskar, R. (2), "On the Possibility of Scientific Knowledge and the Limits of Naturalism", Journal for the Theory of Social Behaviour, Vol. 8, 1978, pp. 1–28.

Bhaskar, R., The Possibility of Naturalism, (Harvester Press, Sussex, 1979).

Bjorken, J. D. and Drell, S. D., Relativistic Quantum Mechanics, (McGraw Hill, New York, 1964).

Bjorken, J. D. and Drell, S. D., Relativistic Quantum Fields, (McGraw Hill, New York, 1965).

Block, N., "Troubles with Functionalism" in C. W. Savage (ed.), Minnesota Studies in the Philosophy of Science, Vol. IX, (University of Minnesota Press, Minneapolis, 1978), pp. 261–325.

Bloomfield, T. M., "About Skinner: Notes on the Theory and Practice of 'Radical Behaviourism'", Philosophy of the Social Sciences, Vol. 6, 1976, pp. 75–82.

Blum, J. M., Pseudoscience and Mental Ability: The Origins and Fallacies of the IQ Controversy, (Monthly Review Press, New York and London, 1978).

Blumenfield, J. B., "Action and Intention", Philosophia, Vol. 9, No. 3–4, July, 1981, pp. 299–315.

Boden, M., Purposive Explanations in Psychology, (Harvard University Press, Cambridge, Massachusetts, 1972).

Boden, M., Artificial Intelligence and Natural Man, (Harvester Press, Sussex, 1977).

Bodmer, W. F., "The Williams Allan Memorial Award Address: Gene Clusters, Genome Organization, and Complex Phenotypes: When the Sequence is Known, What Will it Mean?", American Journal of Human Genetics, Vol. 33, 1981, pp. 664–682.

Böhm, D., Causality and Chance in Modern Physics, (Routledge and Kegan Paul, London, 1957).

Böhm, D., "Quantum Theory as an Indication of a New Order in Physics. Part A. The Development of New Orders as Shown Through the History of Physics", Foundations of Physics, Vol. 1, No. 4, 1971, pp. 359–381.

Böhm, D., 'Quantum Theory as an Indication of a New Order in Physics. B. Implicate and Explicate Order in Physical Law", Foundations of Physics, Vol. 3, No. 2, 1973, pp. 139–168.

Böhm, D., Wholeness and the Implicate Order, (Routledge and Kegan Paul, London, Boston and Henley, 1980).

Bonevac, D. A., Ontological Reduction and Abstract Entities, (Ph.D. thesis, University of Pittsburgh, 1980).

Boolas, G. S. and Jeffrey, R. C., Computability and Logic, 2nd edition, (Cambridge University Press, Cambridge, 1980).

Bottomore, T. (ed.), Crisis and Contention in Sociology, (Sage Studies in International Sociology, Sage Publications, London, 1975).

Bouchard, T. J. and McGee, M. G., "Sex Differences in Human Spatial Ability: Not an X-linked Recessive Gene Effect", Social Biology, Vol. 24, 1977, pp. 332–335.

Boucher, D. H., "The Gospel According to Sociobiology", Perspectives in Biology and Medicine, Vol. 25, Autumn, 1981, pp. 63–65.

Boyd, R., "Realism, Underdetermination, and Causal Theory of Evidence", Nôus, Vol. 7, 1973, pp. 1–12.

Boyle, J. M., Grisez, G. and Tollefsen, O., Free Choice: A Self-Referential Argument, (University of Notre Dame Press, Notre Dame, 1976).

Bradbury, J. W., "Foraging, Social Disperson and Mating Systems" in G. W. Barlow and J. Silverberg (eds.), Sociobiology: Beyond Nature/Nurture? Reports, Definitions and Debates, AAAS Selected Symposium, 35 (Westview Press, Boulder, Colorado, 1980), pp. 189–207.

Brady, J. M., The Theory of Computer Science: A Programming Approach, (Chapman and Hall, London, 1977).

Brandon, R. N., "Adaptation and Evolutionary Theory", Studies in History and Philosophy of Science, Vol. 9, 1978, pp. 181–206.

Brandon, R. N., "A Structural Description of Evolutionary Theory" in P. D. Asquith and R. N. Giere (eds.), Philosophy of Science Association, (Philosophy of Science Association, East Lansing, Michigan, 1980), Vol. 2, pp. 427–439.

Bridges, C. B., "Non-disjunction of the Sex Chromosomes of Drosophila", The Journal of Experimental Zoology, Vol. 15, 1913, pp. 587–605.

Bridges, C. B., "Direct Proof Through Non-disjunction that the Sex-linked Genes of Drosophila are Borne by the X-chromosomes", Science, Vol. 15, 1914, pp. 107–109.

Brillouin, L., Science and Information Theory, (Academic Press, New York, 1962).

Briscoe, A. M., "Hormones and Gender" in E. Tobach and B. Rosoff (eds.), Genes and Gender I, (Gordian Press, New York, 1978), pp. 31–50.

Briskman, L., "Historical Relativism and Bootstrap Rationality", Monist, Vol. 60, 1977, pp. 509–539.

Brody, B. A., "Why Settle for Anything less than Good Old Fashioned Aristotelian Essentialism?", Nôus, Vol. 7, 1973, pp. 351–365.

Brody, B. A., "Fetal Humanity and the Theory of Essentialism" in R. Baker and F. Elliston (eds.), Philosophy of Sex, (Prometheus Books, Buffalo, New York, 1975), pp. 338–355.

Brody, B. A., Identity and Essence, (Princeton University Press, Princeton, New Jersey, 1980).

Browning, T. O., Animal Populations, (Hutchinson, London, 1963).

Brouwer, L. E. J., "On the Significance of the Principle of Excluded Middle in

Mathematics, Especially Function Theory" in J. Van Heijenoort (ed.), From Frege to Gödel: A Source Book in Mathematical Logic 1879–1931, (Harvard University Press, Cambridge, Massachusetts, 1967), pp. 334–345.

Buffery, A. W. H. and Gray, J. A., "Sex Differences in the Development of Spatial and Linguistic Skills" in C. Ounsted and D. C. Taylor (eds.), Gender Differences: Their Ontogeny and Significance, (Williams and Wilkins, Baltimore, 1972), pp. 123–157.

Bunge, M. (1), Scientific Research I: The Search for System, (Springer-Verlag, Heidelberg, Berlin, New York, 1967).

Bunge, M. (2), Foundations of Physics, (Springer-Verlag, Heidelberg, Berlin, New York, 1967).

Bunge, M., Philosophy of Physics, (D. Reidel, Dordrecht, 1973).

Bunge, M., Sense and Reference; Semantics 1: Treatise on Basic Philosophy, (D. Reidel, Dordrecht, 1974).

Bunge, M., "A Critical Examination of Dialectics" in C. Perelman (ed.), Dialectics, (Martinus Nijhoff, The Hague, 1975), pp. 63–77.

Bunge, M., Ontology 1; The Furniture of the World: Treatise on Basic Philosophy, (D. Reidel, Dordrecht, 1977).

Bunge, M., Ontology 2; A World of Systems: Treatise on Basic Philosophy, (D. Reidel, Dordrecht, 1979).

Bunge, M., The Mind-Body Problem: A Psychobiological Approach, (Pergamon Press, Oxford, 1980).

Bunge, M., Scientific Materialism, (D. Reidel, Dordrecht, 1981).

Burian, R. M., "A Methodological Critique of Sociobiology" in A. L. Caplan (ed.), The Sociobiology Debate: Readings on Ethical and Scientific Issues, (Harper and Row, New York, 1978), pp. 376–395.

Burkes, A. W., "Computer Science and Philosophy" in P. D. Asquith and H. E. Kyburg (eds.), Current Research in Philosophy of Science, (Philosophy of Science Association, East Lansing, Michigan, 1979), pp. 399–420.

Bush, E., "Rorty Revisited", Philosophical Studies, Vol. 25, 1974, pp. 33–42.

Cannon, W., "The Bases of Darwin's Achievement: A Revaluation", Victorian Studies, Vol. 5, 1961, pp. 109–134.

Capra, F., The Tao of Physics, (Bantam Books, New York, 1977).

Cargile, J., "The Sorites Paradox", British Journal for the Philosophy of Science, Vol. 20, 1969, pp. 193–202.

Carnap, R., Introduction to Symbolic Logic and Its Applications, (Dover Publications, New York, 1958).

Carnap, R., Philosophical Foundations of Physics, (Basic Books, New York, 1966).

Carter, W. R., "On Incorrigibility and Eliminative Materialism", Philosophical Studies, Vol. 28, 1975, pp. 113–121.

Cartwright, N., "Do Token-Token Identity Theories Show Why We Don't Need Reductionism?", Philosophical Studies, Vol. 36, 1979, pp. 85–90.

Causey, R., Unity of Science, (D. Reidel, Dordrecht, 1977).

Chase, A., The Legacy of Malthus: The Social Costs of the New Scientific Racism, (Alfred A. Knopf, New York, 1977).

Chew, G. F., "'Bootstrap': A Scientific Idea?", Science, Vol. 161, 23 August, 1968, pp. 762–765.

Chipman, L., "A Hole in Quine's Holism", Philosophical Papers, Vol. 3, 1974, pp. 46–47.

Chisholm, R. M., "Freedom and Action" in K. Lehrer (ed.), Freedom and Determinism, (Random House, New York, 1966), pp. 11–44.

Chisholm, R. M., "Identity Through Possible Worlds: Some Questions", Nôus, Vol. 1, 1967, pp. 1–8.

Chisholm, R. M., "Reflections on Human Agency", Idealistic Studies, Vol. 1, 1971, pp. 36–46.

Chisholm, R. M., The Problem of the Criterion, (The Aquinas Lecture 1973, Marquette University Press, Milwaukee, 1973).

Chisholm, R. M. (1), Person and Object: A Metaphysical Study, (George Allen and Unwin, London, 1976).

Chisholm, R. M. (2), "The Agent as Cause" in M. Brand and D. Walton (eds.), Action Theory, (D. Reidel, Dordrecht, 1976), pp. 199–211.

Chomsky, N., "Knowledge of Language", The Times Literary Supplement, 15-5-1969, pp. 523–525.

Chomsky, N., "The Case Against B. F. Skinner", New York Review of Books, Vol. XVII, No. 11, 30 December, 1971, pp. 18–24.

Chomsky, N., "Psychology and Ideology", Cognition, Vol. 1, No. 1, 1972, pp. 11–46.

Chomsky, N., "Recent Contributions to the Theory of Innate Ideas" in S. P. Stich (ed.), Innate Ideas, (University of California Press, Berkeley and Los Angeles, 1975), pp. 121–131.

Church, A., "Ontological Commitment", Journal of Philosophy, Vol. LV, 1958, pp. 1008–1014.

Churchland, P. M., "Eliminative Materialism and the Propositional Attitudes", Journal of Philosophy, Vol. LXXVIII, No. 2, February, 1981, pp. 67–90.

Clark, A., Psychological Models and Neural Mechanisms: An Examination of Reductionism in Psychology, (Clarendon Press, Oxford, 1980).

Clarke, B., "The Causes of Biological Diversity", Scientific American, Vol. 233, No. 2, 1975, pp. 50–60.

Clarke, P. G. H., "Chance, Repetition, and Error in the Development of Normal Nervous System", Perspectives in Biology and Medicine, Vol. 25, 1981, pp. 2–19.

Coates, S., "Sex Differences in Field Dependence-Independence between the Ages of 3 and 6", Perceptual and Motor Skills, Vol. 39, 1974, pp. 1307–1310.

Cochran, A. A., "Relationships between Quantum Physics and Biology", Foundations of Physics, Vol. 1, No. 3, 1971, pp. 235–250.

Cohen, J., "Maternal Constraints on Development" in D. R. Newth and M. Balls (eds.), Maternal Effects in Development: The Fourth Symposium of the British Society for Developmental Biology, (Cambridge University Press, Cambridge, 1979), pp. 1–28.

Cohen, Y., "The Disappearance of the Incest Taboo", Human Nature, Vol. 1, No. 7, 1978, pp. 72–78.

Colinvaux, P. A., "The Human Breeding Strategy", Nature, Vol. 261, 3 June, 1976, pp. 356–357.

Collins, R. L., "Origins of the Sense of Asymmetry: Mendelian and Non-Mendelian Models of Inheritance", New York Academy of Sciences, Annals. Vol. 299, 1977, pp. 283–305.

Cook, N. D., "The Case for Reverse Translation", Journal of Theoretical Biology, Vol. 64, 1977, pp. 113–135.

Cornman, J. W., "On the Elimination of 'Sensations' and Sensations", Review of Metaphysics, Vol. XXII, 1968, pp. 15–35.

Cornman, J. W., Materialism and Sensations, (Yale University Press, New Haven and London, 1971).

Cornman, J. W., "Craig's Theorem, Ramsey Sentences and Scientific Instrumentalism", Synthese, Vol. 25, 1972, pp. 82–128.

Cornman, J. W., "Can Eddington's 'Two' Tables be Identical?", Australasian Journal of Philosophy, Vol. 52, No. 1, May, 1974, pp. 22–38.

Cowley, F., "The Identity of the Person and His Body", Journal of Philosophy, Vol. 68, 1971, pp. 678–683.

Crick, F. H. C., Molecules and Men, (University of Washington Press, Seattle, 1966).

Crick, F. H. C. and Orgel, L., "Directed Panspermia", Icarus, Vol. 10, 1973, pp. 341–346.

Crick, F. H. C., "Split Genes and RNA Splicing", Science, Vol. 204, 1979, pp. 264–271.

Crook, J. H., The Evolution of Human Consciousness, (Clarendon Press, Oxford, 1980).

Crow, J. F. and Kimura, M., An Introduction to Population Genetics Theory, (Harper and Row, New York, 1970).

Crow, J. F., "Genes that Violate Mendel's Rules", Scientific American, Vol. 240, February, 1979, pp. 104–113.

Cummins, R., "Programs in the Explanation of Behavior", Philosophy of Science, Vol. 44, 1977, pp. 269–287.

Curtiss, S., A Psycholinguistic Study of a Modern-Day "Wild Child", (Academic Press, New York, 1977).

Danielli, J., Danielli, M., McGuiness, D., Goodwin, B. C., Pribram, K., Rosen, R. and Wheeler, H., "Constructional Biology", Journal of Social and Biological Structures, Vol. 5, 1982, pp. 11–14.

Darwin, C., The Descent of Man, 2nd edition, (Murray, London, 1889).

Davern, C. I. (ed.), Genetics: Readings from Scientific American, (W. H. Freeman & Co. San Francisco 1981).

Davidson, D., "Actions, Reasons and Causes", Journal of Philosophy, Vol. 60, 1963, pp. 685–700.

Davidson, D., "Mental Events" in L. Foster and J. W. Swanson (eds.), Experience and Theory, (University of Massachusetts Press, Cambridge, Massachusetts, 1970), pp. 79–101.

Davidson, D., "Thought and Talk" in S. Guttenplan (ed.), Mind and Language, (Clarendon Press, Oxford, 1975), pp. 7–23.

Dawkins, R., The Selfish Gene, (Oxford University Press, Oxford, 1976).

Dawkins, R., "Replicator Selection and the Extended Phenotype", Zeitschrift für Tierpsychologie, Vol. 47, 1978, pp. 61–76.

Dawkins, R., "Good Strategy or Evolutionary Stable Strategy?" in G. W. Barlow and J. Silverberg (eds.), Sociobiology: Beyond Nature/Nurture? Reports, Definitions and Debates, AAAS Selected Symposium, 35 (Westview Press, Boulder, Colorado, 1980), pp. 331–367.

Dawkins, R., "In Defence of Selfish Genes", Philosophy, Vol. 56, 1981, pp. 556–573.

DeFries, J. C., Johnson, R. C., Kuse, A. R., McCleam, G. E., Polovina, J., Vandenberg, S. G. and Wilson, J. R., "Family Resemblance for Specific Cognitive Abilities", Behavior Genetics, Vol. 9, 1979, pp. 23–43.

D'Espagnat, B., Conceptual Foundations of Quantum Mechanics, (W. A. Benjamin, California, 1971).

D'Espagnat, B., "Quantum Logic and Non-separability" in J. Mehra (ed.), The Physicist's Conception of Nature, (D. Reidel, Dordrecht, 1973), pp. 714–735.

D'Espagnat, B., Conceptual Foundations of Quantum Mechanics, 2nd revised edition, (W. A. Benjamin, California, 1976).

De Vries, H., Species and Varieties, Their Origin by Mutation, (Open Court, Chicago, 1905).

Dewan, E. M., "Consciousness as an Emergent Causal Agent in the Context of Control System Theory" in G. G. Globus, G. Maxwell, and I. Savodnik (eds.), Consciousness and the Brain: A Scientific and Philosophical Inquiry, (Plenum Press, New York and London, 1976), pp. 181–198.

Dobzhansky, T., Genetics and the Origin of the Species, (Columbia University Press, New York, 1951).

Dobzhansky, T., "A Review of Some Fundamental Concepts and Problems of Population Genetics", Cold Spring Harbor Symposia on Quantitative Biology, Vol. 20, 1955, pp. 1–15.

Dobzhansky, T., "On Some Fundamental Concepts of Darwinian Biology", Evolutionary Biology, Vol. 2, 1968, pp. 1–34.

Dobzhansky, T., "Chance and Creativity in Evolution" in F. Ayala and T. Dobzhansky (eds.), Studies in the Philosophy of Biology: Reduction and Related Problems, (Macmillan, London, 1974), pp. 307–338.

Dobzhansky, T., "Natural Selection in Mankind" in A. H. Halsey (ed.), Heredity and Environment, (Methuen, London, 1977), pp. 40–61.

Dobzhansky, T., Ayala, F. J., Stebbins, G. L. and Valentine, J. W., Evolution, (W. H. Freeman and Co., San Francisco, 1977).

Doore, G., "Functionalism and Absent Qualia", Australasian Journal of Philosophy, Vol. 59, No. 4, December, 1981, pp. 387–402.

Dörner, G., Rohde, W., Stahl, F., Krell, L. and Masius, W., "Neuroendocrine Conditioned Predisposition for Homosexuality in Men", Archives for Sexual Behavior, Vol. 4, 1975, pp. 1–8.

Dörner, G., Hormones and Brain Differentiation, (Elsevier Scientific Publishing Company, Amsterdam, Oxford, New York, 1976).

Dörner, G., "Hormone Dependent Differentiation, Maturation and Function of the Brain and Sexual Behaviour", Endokrinologie, Vol. 69, 1977, pp. 306–320.

Dörner, G., "Hormones, Brain Development and Fundamental Processes of Life" in G. Dörner and M. Kawakami (eds.), Hormones and Brain Development, (Elsevier, North Holland and Biomedical Press, 1978), pp. 13–25.

Douglas, J. and Miller, J., "Record Breaking Women", Science News, Vol. 112, 1977, p. 173.

Dresden, M., "Reflections on 'Fundamentality and Compexity'" in C. P. Enz and J. Mehra (eds.), Physical Reality and Mathematical Description, (D. Reidel, Dordrecht, 1974), pp. 133–166.

Dreyfus, H. L., "Holism and Hermeneutics", Review of Metaphysics, Vol. 34, September, 1980, pp. 3–23.

Driesch, H., History and Theory of Vitalism, (Macmillan, London, 1914).

Driesch, H., Mind and Body, (Methuen, London, 1927).

Driesch, H., Science and the Philosophy of the Organism, 2nd edition, (A & C. Black, London, 1929).

Dupré, J., "Natural Kinds and Biological Taxa", Philosophical Review, Vol. XC, No. 1, January, 1981, pp. 66–90.

Earman, J., "The Universality of Laws", Philosophy of Science, Vol. 45, 1978, pp. 173–181.

Eberhart, J. A., Herbert, I., Keverne, E. B. and Miller, R. E., "Some Hormonal Aspect of Primate Social Behaviour", Endocrinology 1980, (Australian Academy of Science, Canberra, 1980), pp. 622–625.

Eccles, J. C., The Human Psyche, (The Gifford Lectures, University of Edinburgh, 1978–1979), (Springer-Verlag, New York, Heidelberg, Berlin, 1980).

Eddington, A. S., The Nature of the Physical World, (University of Michigan Press, Ann Arbor, 1958).

Edge, H. L., "Richard Rorty on Identity", Journal of Value Inquiry, Vol. VIII, 1974, pp. 196–203.

Efron, E., "The Conditioned Reflex: A Meaningless Concept", Perspectives in Biology and Medicine, Vol. 9, Summer, 1966, pp. 488–514.

Ehrhardt, A. A. and Baker, S. W., "Fetal Androgens, Human Central Nervous System Differentiation and Behavior Sex Differences" in R. C. Friedman, R. M. Richart, and R. L. Vande Wiele (eds.), Sex Differences in Behavior, (John Wiley, New York, 1974), pp. 33–51.

Ehrman, L. and Parsons, P. A., The Genetics of Behavior, (Sinauer Associates Inc., Sunderland, Massachusetts, 1976).

Eigen, M., "Selforganization of Matter and the Evolution of Biological Macromolecules", Die Naturwissenschaften, Vol. 58, 1971, pp. 465–523.

Eiseley, L., Coming of the Giant Wasps, (Audobon, 1975), Cited by Watson [1980 (2)].

Eisenstadt, S. N. and Curelaru, M., The Form of Sociology – Paradigms and Crises, (John Wiley and Sons, New York, 1976).

Elgin, C. Z., "Interdeterminacy, Underdetermination, and the Anomalism of the Mental", Synthese, Vol. 45, 1980, pp. 233–255.

Ellis, B., "The Existence of Forces", Studies in the History and Philosophy of Science, Vol. 7, 1976, pp. 171–185.

Ellis, B., Rational Belief Systems, (Basil Blackwell, Oxford, 1979).

Ellis, L., "The Decline and Fall of Sociology 1975–2000", The American Sociologist, Vol. 12, No. 2, May, 1977, pp. 56–66; Rejoinder pp. 197–199.

Emlen, S. T., "Ecological Determinism and Sociobiology" in G. W. Barlow and J. Silverberg (eds.), Sociobiology: Beyond Nature/Nurture? Reports, Definitions and Debates, AAAS Selected Symposium, 35 (Westview Press, Boulder, Colorado, 1980), pp. 125–150.

Esposito, J. L., "Systems, Holons, and Persons: A Critique of Systems Philosophy", International Philosophical Quarterly, Vol. XVI, No. 2, June, 1976, pp. 219–236.

Essen, L., "Einstein's Special Theory of Relativity", Proceedings of the Royal Institute of Great Britain, Vol. 45, 1972, pp. 141–160

Evans, B. and Waites, B., IQ and Mental Testing: An Unnatural Science and Its Social History, (Macmillan, London and Basingstoke, 1981).

Everitt, N., "A Problem for the Eliminative Materialist", Mind, Vol. XC, 1981, pp. 428–434.

Fairweather, H., "Sex Differences in Cognition", Cognition, Vol. 4, 1976, pp. 231–280.

Falconer, D. S., Introduction to Quantitative Genetics, (Oliver and Boyd, Edinburgh, 1960).

Fay, B. and Moon, J. D., "What Would an Adequate Philosophy of Social Science Look Like?", Philosophy of Social Science, Vol. 7, 1977, pp. 209–227.

Feigl, H., "Unity of Science and Unitary Science" in H. Feigl and M. Brodbeck (eds.), Readings in the Philosophy of Science, (Appleton-Century-Crofts, Inc., New York, 1953), pp. 382–384.

Feigl, H., "Physicalism, Unity of Science and the Foundations of Psychology", in P. A. Schilpp (ed.), The Philosophy of Rudolf Carnap, (Open Court, La Salle, Illinois, 1963), pp. 227–267.

Feinberg, G., "Physics and the Thales Problem", Journal of Philosophy, Vol. 63, 1966, pp. 5–17.

Feyerabend, P. K., "Explanation, Reduction and Empricism" in H. Feigl and G. Maxwell (eds.), Minnesota Studies in the Philosophy of Science, (University of Minnesota Press, Minneapolis, 1962), Vol. 3, pp. 28–97.

Feyerabend, P. K., "Materialism and the Mind-Body Problem", Review of Metaphysics, Vol. 17, September, 1963, pp. 49–66.

Feyerabend, P. K., "The Structure of Science", British Journal for the Philosophy of Science, Vol. 17, 1966, pp. 237–249.

Feyerabend, P. K., Against Method, (New Left Books, London, 1975).

Feyerabend, P. K., "Changing Patterns of Reconstruction", British Journal for the Philosophy of Science, Vol. 28, 1977, pp. 351–382.

Finkelstein, D., "The Physics of Logic" in R. G. Colodny (ed.), Paradigms and Paradoxes: The Philosophical Challenge of the Quantum Domain, (University of Pittsburgh Press, Pittsburgh, 1972), pp. 47–66.

Fisher, L. E. and Werner, O., "Explaining Explanation: Tension in American Anthropology", Journal of Anthropological Research, Vol. 34, 1978, pp. 194–218.

Fisher, R. A., The Genetical Theory of Natural Selection, (Clarendon Press, Oxford, 1930).

Fisk, M., "Idealism, Truth and Practice", Monist, Vol. 59, 1976, pp. 373–391.

Flew, A., "'The Concept of Evolution': A Comment", Philosophy, Vol. 41, 1966, pp. 70–75.

Flew, A., A Rational Animal and Other Philosophical Essays on the Nature of Mind, (Clarendon Press, Oxford, 1978).

Fodor, J. A., "Special Sciences (OR: The Disunity of Science as a Working Hypothesis)", Synthese, Vol. 28, 1974, pp. 97–115.

Fodor, J. A., Representations: Philosophical Essays on the Foundations of Cognitive Science, (Harvester Press, Sussex, 1981).

Frank, A. G., Capitalism and Underdevelopment in Latin America, (Penguin, Harmondsworth, 1971).

Frankfurt, H. G., "The Problem of Action", The American Philosophical Quarterly, Vol. 15, No. 2, April, 1978, pp. 157–162.

Frazzetta, T. H., "From Hopeful Monster to Bolyerine Snake", American Naturalist, Vol. 104, 1970, pp. 55–72.

Freeman, D., "The Evolutionary Theories of Charles Darwin and Herbert Spencer", Current Anthropology, Vol. 15, No. 3, September, 1974, pp. 211–221.

Friedrichs, R. W., A Sociology of Sociology, (The Free Press, Macmillan, New York, 1970).

Freidrichs, R. W., "Dialectical Sociology: Towards a Resolution of the Current

'Crisis' in Western Sociology", British Journal of Sociology, Vol. 23, 1972, pp. 263–274.

Fuji, T., "Inherited Disorders in the Regulation of Serum Calcium in Rats Raised from Parathyroidectomised Mothers", Nature, Vol. 273, 1978, pp. 236–238.

Gaa, J., "The Replacement of Scientific Theories: Reduction and Explication", Philosophy of Science, Vol. 42, 1975, pp. 349–372.

Gal-Or, B., Cosmology, Physics and Philosophy, (Springer-Verlag, New York, Inc., 1981).

Gatlin, L. L., Information Theory and the Living System, (Columbia University Press, New York and London, 1972).

Gendron, B., "On the Relation of Neurological and Psychological Theories: A Critique of the Hardware Thesis" in R. C. Buck and R. S. Cohen (eds.), Boston Studies in the Philosophy of Science 1970, (D. Reidel, Dordrecht, 1970), pp. 483–495.

George, S., How the Other Half Dies, (Penguin Books, Harmondsworth, 1976).

Gerall, A. A., "Hormonal Factors Influencing Masculine Behavior of Female Guinea Pigs", Journal of Comparative and Physiological Psychology, Vol. 62, 1966, pp. 365–369.

Gettier, E. L., "Is Justified True Belief Knowledge?", Analysis, Vol. 23, No. 6, 1963, pp. 121–123.

Ghiselin, M. T., The Economy of Nature and the Evolution of Sex, (University of California Press, Berkeley, 1974).

Ghiselin, M. T., "A Radical Solution to the Species Problem", Systematic Zoology, Vol. 23, 1975, pp. 536–544.

Ghiselin, M. T., "The Economy of the Body", American Economic Review, Vol. 68, 1978, pp. 233–237.

Ghiselin, M. T., "Categories, Life, and Thinking", The Behavioral and Brain Sciences, Vol. 4, 1981, pp. 269–313.

Giannoni, C., "Quine, Grünbaum and the Duhem Thesis" in S. G. Harding (ed.), Can Theories Be Refuted? Essays on the Duhem-Quine Thesis, (D. Reidel, Dordrecht, 1976), pp. 162–175.

Giddens, A., New Rules of Sociological Method, (Hutchinson, London, 1976).

Giddens, A., Studies in Social and Political Theory, (Hutchinson, London, 1977).

Giddens, A., Central Problems in Social Theory: Action, Structure and Contradiction in Social Analysis, (Macmillan, London and Basingstoke, 1979).

Giddens, A. (1), "Agency, Institution, and Time-Space Analysis" in K. Knorr-Cetina and A. V. Cicourel (eds.), Advances in Social Theory and Methodology, (Routledge and Kegan Paul, London, Boston and Henley, 1981), pp. 161–174.

Giddens, A. (2), A Contemporary Critique of Historical Materialism Vol. 1, Power, Property and the State, (Macmillan, London and Basingstoke, 1981).

Gill, F. B. and Wolf, L. L., "Economics of Feeding: Territorality in the Golden-Winged Sunbird", Ecology, Vol. 56, 1975, pp. 333–345.

Girill, T. R., "Are Micro-Entities Picturable?", Philosophy of Science, Vol. 43, 1976, pp. 570–574.

Globus, G. G., "Mind, Structure, and Contradiction" in G. G. Globus, G. Maxwell and I. Savodnik (eds.), Consciousness and the Brain: A Scientific and Philosophical Inquiry, (Plenum Press, New York and London, 1976), pp. 271–293.

Glucksmann, A., "A Ventriloquist Structuralism", New Left Review, Vol. 72, 1972, pp. 68–92.

Glucksmann, M., Structuralist Analysis in Contemporary Social Thought: A Comparison of the Theories of Claude Lévi-Strauss and Louis Althusser, (Routledge and Kegan Paul, London, 1974).

Glymour, C., "On Some Patterns of Reduction", Philosophy of Science, Vol. 37, 1970, pp. 340–353.

Godbey, J. W., "Disjunctive Predicates and the Reduction of Psychology", Mind, Vol. LXXXVII, No. 347, July, 1978, pp. 433–435.

Goguen, J., "Fuzzy Sets and the Social Nature of Truth" in M. M. Gupla, R. K. Ragade and R. R. Yager (eds.), Advances in Fuzzy Set Theory and Applications, (North Holland, Amsterdam, 1979), pp. 49–67.

Goldberg, B., "A Problem with Anomalous Monism", Philosophical Studies, Vol. 32, 1977, pp. 175–180.

Goldberg, S., The Inevitability of Patriarchy, (Temple Smith, London, 1977).

Goldschmidt, R. B., "Different Philosophies of Genetics", Science, Vol. 119, 21 May, 1954, pp. 703–710.

Goldschmidt, R. B., The Material Basis of Evolution, (Yale University Press, New Haven, 1940).

Goodenough, U., Genetics, 2nd edition, (Holt, Rinehart and Winston, New York, 1978).

Goodenough, W. H., "Comments on the Question of Incestuous Marriages in Old Iran", American Anthropologist, Vol. 51, April–June, 1949, pp. 326–328.

Goodman, N., The Structure of Appearance, 2nd edition, (The Bobbs-Merrill Company, Indianapolis, 1966).

Goodwin, B. C., Analytical Physiology of Cells and Developing Organisms, (Academic Press, London, 1976).

Goodwin, B. C., "On Morphogenetic Fields", Theoria to Theory, Vol. 13, 1979, pp. 109–114.

Goodwin, B. C. and Trainor, L. E. H., "A Field Description of the Cleavage Process in Embryogenesis", Journal of Theoretical Biology, Vol. 86, 1980, pp. 757–770.

Goosens, W. K., "Reduction by Molecular Genetics", Philosophy of Science, Vol. 45, 1978, 73–95.

Gould, M. and Kern-Daniels, R., "Towards a Sociological Theory of Gender and Sex", The American Sociologist, Vol. 12, November, 1977, pp. 182–189.

Gould, S. J., "Is Uniformitarianism Necessary?", American Journal of Science, Vol. 263, March, 1965, pp. 223–228.

Gould, S. J. and Eldredge, N., "Punctuated Equilibria: An Alternative to Phyletic Gradualism" in T. J. M. Schopf (ed.), Modes in Paleobiology, (Freeman, Cooper and Co., San Francisco, 1972), pp. 82–115.

Gould, S. J. and Eldredge, N., "Punctuated Equilibria: The Tempo and Mode of Evolution Reconsidered", Paleobiology, Vol. 3, 1977, pp. 115–151.

Gould, S. J. (1), "Evolution's Erratic Pace", Natural History, Vol. 86, May, 1977, pp. 12–16.

Gould, S. J. (2), "The Return of Hopeful Monsters", Natural History, Vol. 86, June–July, 1977, pp. 22–30.

Gould, S. J. (3), Ontogeny and Phylogeny, (Harvard University Press, Cambridge, Massachusetts, 1977).

Gould, S. J. and Lewontin, R. C., "The Spandrels of San Marco and the Panglossian Paradigm: A Critique of the Adaptationist Programme", Proceedings of the Royal Society of London, Series B., Vol. 205, 1979, pp. 581–598.

Gouldner, A. W., The Coming Crisis of Western Sociology, (Heinemann, London, 1971).

Gouldner, A. W., The Future of the Intellectuals and the Rise of the New Class, (Seabury Press, New York, 1979).

Goy, R. W. and McEwen, B. S., Sexual Differentiation of the Brain, (Massachusetts Institute of Technology Press, Cambridge, Massachusetts, 1980).

Granett, S. and Villarejo, M., "Regulation of Gene Expression in *Escherichia coli* by the Local Anaesthetic Procaine", Journal of Molecular Biology, Vol. 160, 1982, pp. 363–367.

Graves, C., Katz, J. J., Nishiyama, Y, Soames, S., Stecker, R. and Tovey, P., "Tacit Knowledge", Journal of Philosophy, Vol. 70, 1973, pp. 318–330.

Gray, J. P. and Wolfe, L., "The Loving Parent Meets the Selfish Gene", Inquiry, Vol. 23, 1980, pp. 233–242.

Grenander, U., Pattern Synthesis, (Springer-Verlag, New York, 1976).

Grenander, U., Pattern Analysis, (Springer-Verlag, New York, 1978).

Grene, M., "Notes on Maynard Smith's 'Status of neo-Darwinism'" in C. H. Waddington (ed.), Towards a Theoretical Biology, (Edinburgh University Press, Edinburgh, 1969), Vol. 2, pp. 97–98.

Grene, M., "Reducibility: Another Side Issue?" in M. Grene (ed.), Interpretations of Life and Mind, (Routledge and Kegan Paul, London, 1971), pp. 14–37.

Grene, M., The Understanding of Nature: Essays in the Philosophy of Biology, (D. Reidel, Dordrecht, 1974).

Grene, M., "The Paradoxes of Historicity", Review of Metaphysics, Vol. XXXII, September, 1978, pp. 15–36.

Griffin, D. R. (ed.), Animal Mind – Human Mind, Dahlem Workshop, 1981, (Springer-Verlag, Berlin, Heidelberg, New York, 1982).

Griffin, N., Relative Identity, (Clarendon Press, Oxford, 1977).
Griffin, N. and Harton, M., "Sceptical Arguments", Philosophical Quarterly, Vol. 31, No. 122, January, 1981, pp. 17–30.
Grim, P., "What Won't Escape Sorites Arguments", Analysis, Vol. 42, No. 192, January, 1982, pp. 38–43.
Gruner, R., "On Evolution and Its Relation to Natural Selection", Dialogue, Vol. XVI, No. 4, 1977, pp. 708–714.
Guilford, J., The Nature of Human Intelligence, (McGraw Hill, New York, 1967).
Guleserian, T. (1), "Contemporary Materialism and Epistemological Values", International Philosophical Quarterly, Vol. XI, 1971, pp. 403–426.
Guleserian, T. (2), "On Two Aspects of Eliminative Materialism", Philosophy of Science, Vol. XXXVIII, 1971, pp. 282–289.
Guttman, R., "Genetic Analysis of Analytic Spatial Ability: Raven's Progressive Matrices", Behavior Genetics, Vol. 4, 1974, pp. 273–284.
Habermas, J., Towards a Rational Society, (Heinemann, London, 1971).
Habermas, J. Knowledge and Human Interests, (Heinemann, London, 1972).
Hacking, I., "Is the End in Sight for Epistemology?", Journal of Philosophy, Vol. LXXVII, No. 10, October, 1980, pp. 579–588.
Haken, H., Synergetics: An Introduction; Nonequilibrium Phase Transitions and Self-Organization in Physics, Chemistry and Biology, (Springer-Verlag, Berlin, Heidelberg, New York, 1978).
Haldane, J. B. S., "The Cost of Natural Selection", Journal of Genetics, Vol. 55, 1957, pp. 511–524.
Hamilton, W. D., "The Evolution of Altruistic Behavior", American Naturalist, Vol. 97, 1963, pp. 354–356.
Hamilton, W. D. (1), "The Genetical Evolution of Social Behaviour I", Journal of Theoretical Biology, Vol. 7, 1964, pp. 1–16.
Hamilton, W. D. (2), "The Genetical Evolution of Social Behaviour II", Journal of Theoretical Biology, Vol. 7, 1964, pp. 17–52.
Hamilton, W. D., "Geometry for the Selfish Herd", Journal of Theoretical Biology, Vol. 31, 1971, pp. 295–311.
Hamilton, W. C., "Altruism and Related Phenomena, Mainly in Social Insects", Annual Review of Ecology and Systematics, Vol. 3, 1972, pp. 193–232.
Hardin, G., "Editorial", Science, Vol. 172, 1971, p. 1297.
Hardin, G., "Living on a Lifeboat", BioScience, Vol. 24, 1974, pp. 561–568.
Hardin, G., The Limits of Altruism, (Indiana University Press, Bloomington, 1977).
Harding, S. G. (ed.), Can Theories Be Refuted? Essays on the Duhem-Quine Thesis, (D. Reidel, Dordrecht, 1976).
Hardy, G. H., "Mendelian Proportions in a Mixed Population", Science, Vol. 28, 1908, pp. 49–50.
Harman, G. H., "The Inference to the Best Explanation", Philosophical Review, Vol. 74, 1965, pp. 88–95.

Harré, R., The Principles of Scientific Thinking, (Macmillan, London, 1970).
Harré, R. and Madden, E. H., Causal Powers, (Basil Blackwell, Oxford, 1975).
Harré, R., "Philosophical Aspects of the Macro-Micro Problem" in K. Knorr-Cetina and A. V. Cicourel (eds.), Advances in Social Theory and Methodology, (Routledge and Kegan Paul, London and Henley, 1981), pp. 141–160.
Harris, E. E., The Foundations of Metaphysics in Science, (George Allen and Unwin, London, 1965).
Harris, G. W., "Hormonal Differentiation of the Developing Central Nervous System with Respect to Patterns of Endocrine Function", Philosophical Transactions of the Royal Society of London, Series B, Vol. 259, 1970, pp. 165–177.
Harris, M., The Rise of Anthropological Theory: A History of Theories of Culture, (Routledge and Kegan Paul, London, 1968).
Harris, M., Cows, Pigs, Wars and Witches, (Random House, New York, 1974).
Harris, M., Cannibals and Kings, (Random House, New York, 1977).
Harris, M., Cultural Materialism: The Struggle for a Science of Culture, (Random House, New York, 1979).
Harris, M., Culture, People, Nature, (Harper and Row, New York, 1980).
Harrison, J. W. H., "Induced Changes in the Pigmentation of the Pupae of the Butterfly *Pieris napi L*, and their inheritance", Proceedings of the Royal Society of London, Series B, Vol. 102, 1928, pp. 347–353.
Hartnack, J., "On Thinking", Mind, Vol. 81, 1972, pp. 543–552.
Hartung, J., "On Natural Selection and the Inheritance of Wealth", Current Anthropology, Vol. 17, No. 4, December, 1976, pp. 607–613.
Hassan, R., "Famine: A Sociological Interpretation", Unpublished Manuscript, Flinders University, 1979.
Hempel, C. G., Aspects of Scientific Explanation, (Free Press, New York, 1965).
Hempel, C. G., "The Theoretician's Dilemma" in Aspects of Scientific Explanation, (Free Press, New York, 1965), pp. 173–226.
Hesse, M., "In Defense of Objectivity", Proceedings of the British Academy, Vol. LVII, (Oxford University Press, Oxford, 1972), pp. 275–292.
Himmelfarb, G., Darwin and the Darwinian Revolution, (Doubleday, New York, 1962).
Hindess, B. and Hirst, P. Q., Pre-Capitalist Modes of Production, (Routledge and Kegan Paul, London, 1975).
Hindess, B., Philosophy and Methodology in the Social Sciences, (Harvester Press, Sussex, 1977).
Ho, M. W. and Saunders, P. T., "On the Increase in Compexity in Evolution", Journal of Theoretical Biology, Vol. 63, 1976, pp. 375–384.
Ho, M. W. and Saunders, P. T., "Beyond neo-Darwinism – An Epigenetic Approach to Evolution", Journal of Theoretical Biology, Vol. 78, 1979, pp. 573–591.
Ho, M. W. and Saunders, P. T., "Adaptation and Natural Selection: Mechanism

and Teleology". Paper delivered at the International Conference of the Dialectics of Biology and Society in the Production of Mind, University of Padua, Italy, 23–26 March, 1980.

Holland, J. H., "Studies of the Spontaneous Emergence of Self-Replicating Systems Using Cellular Automata and Formal Grammars" in A. Lindenmayer and G. Rozenberg (eds.), Automata, Language, Development, (North Holland, Amsterdam, 1976), pp. 385–404.

Holman, E. L., "Is the Physical World Colourless?", Australasian Journal of Philosophy, Vol. 57, No. 4, December, 1979, pp. 295–304.

Honderich, T., "The Argument for Anomalous Monism", Analysis, Vol. 42, January, 1982, pp. 59–64.

Hooker, C. A., "Review of Ronald M. Yoshida Reduction in the Physical Sciences", Dialogue, Vol. 18, 1979, pp. 81–99.

Hooker, C. A., "Towards a General Theory of Reduction: Pt. I, Historical and Scientific Setting", Dialogue, Vol. 20, March, 1981, pp. 38–60.

Hooker, C. A., "Towards a General Theory of Reduction: Pt. II, Identity in Reduction", Dialogue, Vol. 20, June, 1981, pp. 201–236.

Hooker, C. A., "Towards a General Theory of Reduction: Pt. III, Cross-Categorical Reduction", Dialogue, Vol. 20, September, 1981, pp. 496–529.

Hooker, J., "Brody on Essentialism", Philosophical Studies, Vol. 29, 1976, pp. 263–269.

Horner, J. K., "Putnam's Complaint", Auslegung, Vol. 3, 1976, pp. 166–173.

Howard, J. C., "A Tropical Volute Shell and the Icarus Syndrome", Nature, Vol. 290, April, 1981, pp. 441–442.

Hoyle, F. and Wickramasinghe, C., Lifecloud, (Dent, London, 1978).

Hoyle, F., "Evolution from Space", Omni Lecture 12th January, 1982.

Hubbard, R. and Lowe, M., Genes and Gender II: Pitfalls in Research on Sex and Gender, (Gordian Press, New York, 1979).

Hubbard, R., Henifin, M. S. and Fried, B. (eds.), Women Looking at Biology Looking at Women: A Collection of Feminist Critiques, (Schenkman Publishing Co., Cambridge, Massachusetts, 1979).

Hull, D. L., Philosophy of Biological Science, (Prentice Hall Inc., Englewood Cliffs, New Jersey, 1974).

Hull, D. L., "Are Species Really Individuals?", Systematic Zoology, Vol. 25, 1976, pp. 174–191.

Hull, D. L., "A Matter of Individuality", Philosophy of Science, Vol. 45, 1978, pp. 174–191.

Hull, D. L., "Review of Beast and Man: The Roots of Human Nature by Mary Midgley", Philosophical Review, Vol, XC, 2 April, 1981, pp. 307–310.

Hunt, E. K. and Schwartz, J. G. (eds.), A Critique of Economic Theory, (Penguin, Harmondsworth, 1973).

Hyde, J. S. and Rosenberg, B. G., Half the Human Experience: The Psychology of Women, (Heath, Lexington, Massachusetts, 1976).

Jacob, F. and Monod, J., "On the Regulation of Gene Activity", Cold Spring Harbor Symposia on Quantitative Biology, Vol. 26, 1961, pp. 193–209.
Jacob, F., The Logic of Life: A History of Heredity, translation B. E. Spillman, (Pantheon Books, New York, 1973).
Jammer, M., The Philosophy of Quantum Mechanics: The Interpretations of Quantum Mechanics in Historical Perspective, (John Wiley and Son, New York, 1974).
Jenni, D. A. and Jenni, M. A., "Carrying Behavior in Humans: Analysis of Sex Differences", Science, Vol. 194, November, 1976, pp. 859–860.
Johannsen, W., Elemente der exakten Erblichkeitslehre, (Gustav Fischer, Jena, 1909).
Jollos, V., "Experimentelle Protisten Studien I. Untersuchungen über Variabilität und Vererbung bei Infusorien", Archiv für Protistenkunde, Vol. 43, 1921, pp. 1–222.
Jones, D. H., "Emergent Properties, Persons, and the Mind-Body Problem", Southern Journal of Philosophy, Vol. 10, Winter, 1972, pp. 423–433.
Kaffman, M., "Sexual Standards and the Behavior of Kibbutz Adolescent", American Journal of Orthopsychiatry, Vol. 47, No. 2, 1977, pp. 207–217.
Kalke, W., "What is Wrong with Fodor and Putnam's Functionalism?", Noûs, Vol. 3, 1969, pp. 83–93.
Katz, B. D., "Davidson on the Identity Theory", Canadian Journal of Philosophy, Vol. VII, No. 1, March, 1977, pp. 81–90.
Katz, J. J., "Innate Ideas" in S. P. Stich (ed.), Innate Ideas, (University of California Press, Berkeley and Los Angeles, 1975, pp. 145–163.
Katz, J. J., Language and Other Abstract Entities, (Rowman and Littlefield, Totowa, New Jersey, 1981).
Katz, M. J. and Goffman, W., "Preformation of Ontogenetic Patterns", Philosophy of Science, Vol. 48, 1981, pp. 438–453.
Kauffman, S. A., "Metabolic Stability and Epigenesis in Randomly Constructed Genetic Nets", Journal of Theoretical Biology, Vol. 22, 1969, pp. 437–467.
Keat, R., "A Critical Examination of B. F. Skinner's Objections to Mentalism", Behaviorism, Vol. 1, 1972, pp. 53–70.
Keat, R. and Urry, J., Social Theory and Science, (Routledge and Kegan Paul, London, 1975).
Kemeny, J. G. and Oppenheim, P., "On Reduction", Philosophical Studies, Vol. VII, No. 1–2, January–February, 1956, pp. 6–19.
Kerr, R. A., "Origin of Life: New Ingredients Suggested", Science, Vol. 210, 1980, pp. 42–43.
Kervran, C. L., Biological Transmutations, (Swan House Publishing Co., Binghamton, New York, 1972).
Keuth, H., "Objective Knowledge Out of Ignorance: Popper on Body, Mind and the Third World", Theory and Decision, Vol. 5, 1974, pp. 391–412.

Khin Zaw, S., "The Case for a Cognitive Biology II", The Aristotelian Society, (Supplementary), Vol. LIV, 1980, pp. 51–71.

Kim, J., "Rorty on the Possibility of Philosophy", Journal of Philosophy, Vol. LXXVII, No. 10, October, 1980, pp. 588–597.

Kimbrough, S. O., "On the Reduction of Genetics to Molecular Biology", Philosophy of Science, Vol. 46, 1979, pp. 389–406.

Kimbrough, S. O., "The Concepts of Fitness and Selection in Evolutionary Biology", Journal of Social and Biological Structures, Vol. 3, 1980, pp. 149–170.

King, J. L. and Jukes, T. H., "Non-Darwinian Evolution", Science, Vol. 164, 1969, pp. 788–798.

King, M-C. and Wilson, A. C., "Evolution at Two Levels in Humans and Chimpanzees", Science, Vol. 188, No. 4184, 11 April, 1975, pp. 107–116.

Kitchener, R. F., "Are There Molar Psychological Laws?", Philosophy of the Social Sciences, Vol. 6, 1976, pp. 143–154.

Kitchener, R. F., "Epigenesis: The Role of Biological Models in Developmental Psychology", Human Development, Vol. 21, 1978, pp. 141–160.

Kitcher, P., "The Nativist's Dilemma", The Philosophical Quarterly, Vol. 28, No. 110, January, 1978, pp. 1–16.

Kitts, D. B. and Kitts, D. J., "Biological Species as Natural Kinds", Philosophy of Science, Vol. 46, 1979, pp. 613–622.

Koertge, N., "Towards a New Theory of Scientific Inquiry" in G. Radnitzky and G. Anderson (eds.), Progress and Rationality in Science, (D. Reidel, Dordrecht, 1978), pp. 253–278.

Kojima, K-I., "Is There a Constant Fitness Value for a Given Genotype? No!", Evolution, Vol. 25, 1971, pp. 281–285.

Korner, A., "Hormone Control of Protein Synthesis", Proceedings of the Royal Society of London, Series B, Vol. 176, 1970, pp. 287–290.

Krebs, J. R., Ericksen, J. T., Webber, M. I., and Charnov, E. L., "Optimal Prey Selection in the Great Tit", (*Parus major*), Animal Behavior, Vol. 25, 1977, pp. 30–38.

Krebs, J. R., "Optimal Foraging: Decision Rules for Predators" in J. R. Krebs and N. B. Davies, Behavioural Ecology: An Evolutionary Approach, (Basil Blackwell, Oxford, 1978), pp. 23–63.

Kuhn, T. S., The Structure of Scientific Resolutions, 2nd edition, (University of Chicago Press, Chicago, 1970).

Kuipers, T. A. F., "The Reduction of Phenomenological to Kinetic Thermostatics", Philosophy of Science, Vol. 49, 1982, pp. 107–119.

Kyburg, H. E., "Propensities and Probabilities", British Journal for the Philosophy of Science, Vol. 25, No. 4, 1974, pp. 358–374.

Lack, D., Population Studies of Birds, (Oxford University Press, Oxford, 1966).

Laszlo, E., System, Structure and Experience, (Gordon and Breach, New York, 1969).

Laszlo, E. (1), Introduction to Systems Philosophy, (Gordon and Breach, New York, 1972).
Laszlo, E. (2), The Systems View of the World, (George Braziller, New York, 1972).
Laudan, E., Progress and Its Problems: Towards a Theory of Scientific Growth, (Routledge and Kegan Paul, London, 1977).
Laudan, L., "Historical Methodologies: An Overview and Manifesto" in P. D. Asquith and H. E. Kyburg (eds.), Current Research in Philosophy of Science, (Philosophy of Science Association, East Lansing, Michigan, 1979), pp. 40–54.
Laudan, L., "A Refutation of Convergent Realism" in U. J. Jensen and R. Harré (eds.), The Philosophy of Evolution, (Harvester Press, Sussex, 1980), pp. 232–268.
Laudan, L., "A Confutation of Convergent Realism", Philosophy of Science, Vol. 48, 1981, pp. 19–49.
Lehrer, K., "When Rational Disagreement is Impossible", Nôus, Vol. 10, 1976, pp. 327–332.
Lehrer, K., "Self Profile: 2. Freedom and Determinism" in R. J. Bogdan (ed.), Keith Lehrer, (D. Reidel, Dordrecht, 1981), pp. 13–39.
Lehrman, D. S., "A Critique of Konrad Lorenz's Theory of Instinctive Behavior", The Quarterly Review of Biology, Vol. 28, No. 4, December, 1953, pp. 337–363.
Lehrman, D. S., "Semantical and Conceptual Issues in the Nature-Nurture Problem" in L. R. Aronson, E. Tobach, D. S. Lehrman and J. S. Rosenblatt (eds.), Development and Evolution of Behavior: Essays in Memory of T. C. Schneirla, (W. H. Freeman and Co., San Francisco, 1970), pp. 17–52.
Lenin, V. I., Materialism and Empirico-Criticism, (Progress Publishers, Moscow, 1970).
Leonard, H. S. and Goodman, N., "The Calculus of Individuals and Its Uses", Journal of Symbolic Logic, Vol. 5, No. 2, June, 1940, pp. 45–55.
Lerner, R. M., "Nature, Nurture, and Dynamic Interactionism", Human Development, Vol. 21, 1978, pp. 1–20.
Levin, J. C., Lloyd, C. W., Lobotsky, I. and Friedrich, E. H., "The Effects of Epinephrine on Testosterone Production", Acta Endocrinologica, (Copenhagen), Vol. 55, 1967, pp. 184–192.
Levin, M. E., Metaphysics and the Mind-Body Problem, (Clarendon Press, Oxford, 1979).
Levins, R., Evolution in Changing Environments, (Monographs in Population Biology, No. 2), (Princeton University Press, Princeton, 1968).
Levins, R. and Lewontin, R., "Dialectics and Reductionism in Ecology", Synthese, Vol. 43, 1980, pp. 47–78.
Levy, J., "Lateral Specialization of the Human Brain: Behavioral Manifestation and Possible Evolutionary Basis" in J. A. Kiger (ed.), The Biology of Behavior, (Oregon University Press, Corvalis, 1972), pp. 109–127.

Levy, J., "Psychobiological Implications of Bilateral Asymmetry" in S. Dimond and J. G. Beaumont (eds.), Hemisphere Function in the Human Brain, (Paul Elek, London, 1974), pp. 121–183.

Levy-Agresti, J. and Sperry, R. W., "Differential Perceptual Capacities in Major and Minor Hemispheres", Proceedings of the National Academy of Science, Vol. 61, 1968, p. 1151. (Paper presented at the Fall meeting, National Academy of Sciences, California Institute of Technology, Pasadena.)

Lewontin, R. C., "The Adaptations of Populations to Varying Environments", Cold Spring Harbor Symposia on Quantitative Biology, Vol. 22, 1957, pp. 395–408.

Lewontin, R. C., The Genetic Basis of Evolutionary Change, (Columbia University Press, New York and London, 1974).

Lewontin, R. C., "Sociobiology – A Caricature of Darwinism" in P. D. Asquith and F. Suppe (eds.), Philosophy of Science Association, (Philosophy of Science Association, East Lansing, Michigan, 1977), Vol. 2, pp. 22–31).

Lewontin, R. C., "Adaptation", Scientific American, Vol. 239, 1978, pp. 157–169.

Lewontin, R. C., "Sociobiology as an Adaptationist Program", Behavioral Science, Vol. 24, 1979, pp. 5–14.

Lewontin, R. C., "On Constraints and Adaptation", The Behavioral and Brain Sciences, Vol. 4, 1981, pp. 244–245.

Li, C. C., "Fundamental Theorem of Natural Selection", Nature, Vol. 214, 29 April, 1967, pp. 505–506.

Linsky, B., "Critical Notice: Baruch Brody Identity and Essence", Canadian Journal of Philosophy, Vol. XII, No. 2, June, 1981, pp. 391–407.

Linsky, L. (ed.), Reference and Modality, (Oxford University Press, Oxford, 1979; reprinted from 1971 with corrections).

Livingstone, F. B., "Genetics, Ecology, and the Origins of Incest and Exogamy", Current Anthropology, Vol. 10, 1969, pp. 45–62.

Livingstone, F. B., "Cultural Causes of Genetic Change" in G. W. Barlow and J. Silverberg (eds.), Sociobiology: Beyond Nature/Nurture? Reports, Definitions and Debates, AAAS Selected Symposium, 35 (Westview Press, Boulder, Colorado, 1980), pp. 307–329.

Loehlin, J. C., Sharon, S. and Jacoby, R., "In Pursuit of the 'Spatial Gene': A Family Study", Behavior Genetics, Vol. 8, 1978, pp. 229–241.

Long, D. C., "The Philosophical Concept of a Human Body", The Philosophical Review, Vol. 73, 1964, pp. 321–337.

Long, D. C., "The Bodies of Persons", Journal of Philosophy, Vol. LXXI, No. 10, May, 1974, pp. 291–301.

Loux, M. J., Substance and Attribute: A Study in Ontology, (D. Reidel, Dordrecht, 1978).

Løvtrup, S., Epigenetics: A Treatise on Theoretical Biology, (John Wiley and Sons, London, 1974).

Løvtrup, S., "On the Falsifiability of neo-Darwinism", Evolutionary Theory, Vol. 1, December, 1976, pp. 267–283.
Luce, R. D. and Raiffa, H., Games and Decisions: Introduction and Critical Survey, (John Wiley and Sons, New York, 1958).
Lukes, S., "The Underdetermination of Theory by Data", The Aristotelian Society, (Supplement), Vol. LII, 1978, pp. 93–107.
Luria, A. R., "Scientific Perspectives and Philosophical Dead Ends in Modern Linguistics", Cognition, Vol. 3, No. 4, 1975, pp. 377–385.
Lycan, W. C. and Pappas, G. S., "What is Eliminative Materialism?", Australasian Journal of Philosophy, Vol. 50, No. 2, August, 1972, pp. 149–159.
MacBeth, N., Darwin Retried: An Appeal to Reason, (Gambit, Boston, 1971).
Maccoby, E. E. (ed.), The Development of Sex Differences, (Stanford University Press, Stanford, California, 1966).
Maccoby, E. E. and Jacklin, C. N., The Psychology of Sex Differences, (Stanford University Press, Stanford, California, 1974).
Maelzer, D. A., "A Discussion of Components of Environment in Ecology", Journal of Theoretical Biology, Vol. 8, 1965, pp. 141–162.
Magdoff, H., The Age of Imperialism, (Monthly Review Press, New York, 1970).
Malcolm, N., "Thoughtless Brutes", Proceedings and Addresses of the American Philosophical Association, Vol. 46, 1972–73, pp. 5–20.
Mampilly, P., "Agricultural Credit in Asia", Report Prepared for SSAII, Asian Development Bank, Manila, 1976.
Mannheim, K., Ideology and Utopia, (Routledge and Kegan Paul, London, 1960).
Manser, A. R., "The Concept of Evolution", Philosophy, Vol. 40, 1965, pp. 18–34.
Marchal, J. H., "On the Concept of a System", Philosophy of Science, Vol. 42, 1975, pp. 448–468.
Margolis, J. (1), "Behaviorism and Alien Languages", Philosophia, Vol. 3, No. 4, October, 1973, pp. 413–427.
Margolis, J. (2), Knowledge and Existence, (Oxford University Press, New York, 1973).
Margolis, J., Persons and Minds, (D. Reidel, Dordrecht, 1978).
Marshall, J., "Some Problems and Paradoxes Associated with Recent Accounts of Hemispheric Specialization", Neuropsychologia, Vol. 11, 1973, pp. 463–470.
Martin, C. P., Psychology, Evolution and Sex, (Charles C. Thomas, Springfield, Illinois, 1956).
Mates, B., Skeptical Essays, (University of Chicago Press, Chicago, 1981).
Massey, G. J., "Reflections on the Unity of Science", Annals of the Japan Association for Philosophy of Science, Vol. 4, No. 3, March, 1973, pp. 203–212.
Maynard Smith, J., "The Status of neo-Darwinism" in C. H. Waddington (ed.),

Towards a Theoretical Biology, (Edinburgh University Press, Edinburgh, 1969), Vol. 2, pp. 82–89.

Maynard Smith, J., "Time in the Evolutionary Process", Studium Generale, Vol. 23, 1970, pp. 266–272.

Maynard Smith, J., "What Use is Sex?", Journal of Theoretical Biology, Vol. 30, 1971, pp. 319–335.

Maynard Smith, J., "Group Selection", The Quarterly Review of Biology, Vol. 51, 1976, pp. 277–283.

Maynard Smith, J. (1), "Optimization Theory in Evolution", Annual Review of Ecology and Systematics, Vol. 9, 1978, pp. 31–56.

Maynard Smith, J. (2), "Evolution and the Theory of Games", American Scientist, Vol. 64, No. 1, January–February, 1976, pp. 41–45. Reprinted in T. H. Clutton-Brock and P. H. Harvey (eds.), Readings in Sociobiology, (W. H. Freeman and Co., San Francisco, 1978), pp. 258–270. References to latter.

Maynard Smith, J. (3), The Evolution of Sex, (Cambridge University Press, Cambridge, 1978).

Maynard Smith, J. (4), "The Concepts of Sociobiology" in G. S. Stent (ed.), Morality as a Biological Phenomena, (Dahlem Konferenzen, Berlin, 1978), pp. 23–34.

Maynard Smith, J., "Commentary on Webster and Goodwin's 'The Origin of Species: A Structuralist Approach'", Journal of Social and Biological Structures, Vol. 5, 1982, pp. 49–68.

Mayr, E., "Behavior Programs and Evolutionary Strategies", American Scientist, Vol. 62, 1974, pp. 650–659.

Mays, L. L., Genetics: A Molecular Approach, (Macmillan, New York, 1981).

McCann, H., "Volition and Basic Action", The Philosophical Review, Vol. 83, 1974, pp. 451–473.

McCann, H., "Trying, Paralysis and Volition", Review of Metaphysics, Vol. 28, 1975, pp. 423–442.

McCauley, R. N., "Hypothetical Identities and Ontological Economizing: Comments on Causey's Program for the Unity of Science", Philosophy of Science, Vol. 48, 1981, pp. 218–227.

McClare, C. W. F., "Chemical Machines, Maxwell's Demon and Living Organisms", Journal of Theoretical Biology, Vol. 30, 1971, pp. 1–34.

McClare, C. W. F., "A Quantum Mechanical Muscle Model", Nature, Vol. 240, 10 November, 1972, pp. 88–90.

McCullagh, C. B., "Causal Theories of Action", Philosophical Studies, Vol. 27, 1975, pp. 201–209.

McDougall, W., "An Experiment For the Testing of the Hypothesis of Lamarck", British Journal of Psychology, Vol. 17, 1927, pp. 267–304.

McDougall, W., "Second Report on a Lamarckian Experiment", British Journal of Psychology, Vol. 20, 1930, pp. 201–218.

McDougall, W., "Fourth Report on a Lamarckian Experiment", British Journal of Psychology, Vol. 28, 1938, pp. 321–345.

McEwen, B. S., "Interactions between Hormones and Nerve Tissues", Scientific American, Vol. 235, No. 1, July, 1976, pp. 48–58.

McMullin, E., "Persons in the Universe", Zygon, Vol. 15, No. 1, March, 1980, pp. 69–89.

McMurty, J., "Making Sense of Economic Determinism", Canadian Journal of Philosophy, Vol. III, No. 2, December, 1973, pp. 249–261.

McShane, P., Randomness, Statistics and Emergence, (Gill and Macmillan, Ireland, 1970).

Mehra, J., The Quantum Principle: Its Interpretation and Epistemology, (D. Reidel, Dordrecht, 1974).

Mekler, L. B., "Mechanisms of Biological Memory", Nature, Vol. 215, 1967, pp. 481–484.

Melden, A. I., Free Action, (Routledge and Kegan Paul, London, 1961).

Mellor, D. H., "Physics and Furniture", American Philosophical Quarterly, (Monograph Series), No. 3, Oxford, 1969. Studies in the Philosophy of Science, pp. 171–187.

Mellor, D. H., "Materialism and Phenomenal Qualities", Aristotelian Society, (Supplement), Vol. XLVI, 1973, pp. 107–119.

Mendel, G., "Versuche Ueber Pflanzenhydriden", Verhandlugen des Naturforschunden Vereines in Bruenn, Vol. 4, 1866, pp. 3–47. Reprinted in C. J. Davern (ed.), Genetics: Readings from Scientific American, (W. H. Freeman and Company, San Francisco, 1981), pp. 8–17.

Mendelson, E., Introduction to Mathematical Logic, (D. Van Nostrand Company Inc., Princeton, New Jersey, 1964).

Mercer, E. H., The Foundations of Biological Theory, (John Wiley and Sons, New York, 1981).

Mermin, N. D., "Quantum Mysteries for Anyone", Journal of Philosophy, Vol. LXXVIII, No. 7, July, 1981, pp. 397–408.

Meyer, R. K. and Routley, R., "Extensional Reduction I", Monist, Vol. 60, 1977, pp. 355–369.

Meyer-Bahlburg, H. F. L., "Sex Hormones and Male Homosexuality in Comparative Perspective", Archives of Sexual Behavior, Vol. 6, 1977, pp. 297–325.

Middleton, R., "Brother-Sister and Father-Daughter Marriage in Ancient Egypt", American Sociological Review, Vol. 27, No. 5, October, 1962, pp. 603–611.

Midgley, M., Beast and Man: The Roots of Human Nature, (Harvester Press, Sussex, 1978).

Midgley, M., "Gene-juggling", Philosophy, Vol. 54, 1979, pp. 439–458.

Miller, S. J., "A Production of Amino Acids Under Possible Primitive Earth Conditions", Science, Vol. 117, 1953, p. 528.

Miller, S. J. and Orgel, L. E., The Origins of Life on Earth (Prentice-Hall, Englewood Cliffs, New Jersey, 1974).

Mills, S. K. and Beatty, J. H., "The Propensity Interpretation of Fitness", Philosophy of Science, Vol. 46, 1979, pp. 263–286.

Mitchison, N. A., "Inheritance: The Immunological Challenge", Nature, Vol. 288, 1980, p. 306.

Money, J. and Ehrhardt, A. A., "Progestin-induced Hermaphroditism: IQ and Psychosexual Identity in a Study of Ten Girls", Journal of Sex Research, Vol. 3, 1967, pp. 83–100.

Money, J., Ehrhardt, A. A. and Epstein, R., "Fetal Androgens and Female Gender Identity in the Early Treated Androgentical Syndrome", John Hopkins Medical Journal, Vol. 122, 1968, pp. 160–167.

Money, J. and Ehrhardt, A. A., Man and Woman: Boy and Girl, (John Hopkins University Press, Baltimore, 1972).

Money, J. and Tucker, P., Sexual Signatures: On Being a Man or a Woman, (Little Brown, Boston and Toronto, 1975).

Money, J. and Daléry, J., "Iatrogenic Homosexuality", Journal of Homosexuality, Vol. 1, No. 4, 1976, pp. 357–371.

Monod, J., Chance and Necessity: An Essay on the Natural Philosophy of Modern Biology, translation A. Wainhouse, (Collins, St. James Place, London, 1972).

Moorhead, P. S. and Kaplan, M. M. (eds.), Mathematical Challenges to the Neo-Darwinist Interpretation of Evolution, (Wistar Institute Press, Philadelphia, 1967).

Morris, H. M. (ed.), Scientific Creationism, (Creation Life Publishers, San Diego, California, 1974).

Morrison, D. E. and Henkel, R. E. (eds.), The Significance Test Controversy: A Reader, (Aldine, Chicago, 1970).

Munson, R., "Is Biology a Provincial Science?", Philosophy of Science, Vol. 42, 1975, pp. 428–447.

Muntz, M. K., Existence and Logic, (New York University Press, New York, 1974).

Musgrave, A. E., "Falsification and Its Critics" in P. Suppes, L. Henkin, A. Joja and G. R. G. Moisil (eds.), Logic, Methodology and Philosophy of Science IV, (North Holland, Amsterdam, 1973), pp. 393–406.

Nagel, E., The Structure of Science: Problems in the Logic of Scientific Explanation, (Routledge and Kegan Paul, London, 1961).

Nanney, D. L., "Molecules and Morphologies: The Perpetuation of Pattern in the Ciliated Protozoa", Journal of Protozoology, Vol. 24, 1977, pp. 27–35.

Needham, R., "Remarks on the Analysis of Kinship and Marriage" in R. Needham (ed.), Rethinking Kinship and Marriage, (Tavistock, London, 1971), pp. 1–35.

Nelson, A. F., "Emergentism Reconsidered and an Alternative", Southern Journal of Philosophy, Vol. 7, 1969, pp. 187–192.

Nerlich, G., The Shape of Space, (Cambridge University Press, Cambridge, 1976).

Newton-Smith, W., "The Underdetermination of Theory by Data", The Aristotelian Society, (Supplement), Vol. LII, 1978, pp. 71–91.

Nickles, T., "Two Concepts of Intertheoretic Reduction", Journal of Philosophy, Vol. 70, April, 1973, pp. 181–201.

Nickles, T., "Davidson on Explanation", Philosophical Studies, Vol. 31, 1977, pp. 141–145.

Nielsen, K., "Impediments to Radical Egalitarianism", American Philosophical Quarterly, Vol. 18, No. 2, April, 1981, pp. 121–129.

Niven, B. S., "The Formal Definition of the Environment of an Animal", Australasian Journal of Ecology, Vol. 5, 1980, pp. 37–46.

Niven, B. S., "Formalization of the Basic Concepts of Animal Ecology", Erkenntnis, Vol. 17, 1982, pp. 307–320.

Noren, S. J., "The Picturability of Micro-Entities", Philosophy of Science, Vol. 40, 1973, pp. 234–241.

Noren, S. J., "Cornman on the Colour of Micro-Entities", Australasian Journal of Philosophy, Vol. 53, No. 1, May, 1975, pp. 65–67.

Noyles, P., "The Abandonment of Simultaneity", Theoria to Theory, Vol. 9, 1975, pp. 23–32.

Oakes, G., "The Epistemological Foundations of Cultural Materialism", Dialectical Anthropology, Vol. 6, No. 1, August, 1981, pp. 1–21.

O'Connor, S. M., Vietze, P. M., Hopkins, J. B. and Altemeier, W. A., "Postpartum Extended Maternal-Infant Contact: Subsequent Mothering and Child Health", Pediatric Research, Vol. 11, No. 4, 1977, p. 380.

O'Dwyer, M. F., Kent, J. E. and Brown, R. D., Valency, (Holt, Rinehart and Winston, Sydney, 1978).

Oparin, A. I., The Origin of Life, (Dover Books, New York, 1953).

Oppenheim, P. and Putnam, H., "Unity of Science as a Working Hypothesis" in H. Feigl and M. Scriven (eds.), Minnesota Studies in the Philosophy of Science, (University of Minnesota Press, Minneapolis, 1958), Vol. 2, pp. 3–36.

Orlove, M. J., "A Model of Kin Selection Not Invoking Coefficients of Relationship", Journal of Theoretical Biology, Vol. 49, 1975, pp. 289–310.

Otte, R., "A Critique of Suppes' Theory of Probabilistic Causality", Synthese, Vol. 48, 1981, pp. 167–189.

Pappas, G., "Perception Without Belief", Ratio, vol. XIX No. 1, June, 1977, pp. 142–161.

Parker, G. A., "The Reproductive Behaviour and the Nature of Sexual Selection in Scatophaga Stercoraria LII", Journal of Animal Ecology, Vol. 39, 1970, pp. 205–228.

Parker, S., "The Precultural Basis of the Incest Taboo: Towards a Biosocial Theory", American Anthropologist, Vol. 76, 1976, pp. 285–305.

Parkin, F., Marxism and Class Theory: A Bourgeois Critique, (Tavistock, London, 1979).

Parsons, T. and Bales, R. F., Family, Socialization and Interaction Process, (Free Press, New York, 1955).

Parsons, T., The Structure of Social Action, (Free Press, New York, 1964).

Parsons, T., "Essentialism and Quantified Modal Logic", The Philosophical Review, Vol. LXXVIII, 1 January, 1969, pp. 35–52. Reprinted in L. Linsky (ed.), Reference and Modality, (Oxford University Press, Oxford, 1979; Reprinted from 1971 with corrections), pp. 73–87.

Pascal, L., "Human Tragedy and Natural Selection", Inquiry, Vol. 21, 1978, pp. 443–460.

Pascal, L., "Rejoinder to Gray and Wolfe", Inquiry, Vol. 23, 1980, pp. 242–251.

Patt, D. I. and Patt, G. R., An Introduction to Modern Genetics, (Addison-Wesley, Reading, Massachusetts, 1975).

Pearson, K., "Variation in Man and Woman" in The Chances of Death, Vol. 1, (Edward Arnold, London, 1897), (Cited by [Shields, 1975]).

Pearson, K. and Weldon, W, F. R., "Inheritance in *Phaseolus vulgaris*", Biometrika, Vol. 2, 1903, pp. 499–503.

Peters, R. H., "Tautology in Evolution and Ecology", American Naturalist, Vol. 110, 1976, pp. 1–12.

Petersen, A. F., "Biological Evolution or Anti-Chaos: OR the Problem of Reduction in Biology and Psychology", Danish Yearbook of Philosophy, Vol. 12, 1975, pp. 65–92.

Petrunkevitch, A., "Tarantula Versus Tarantula-Hawk: A Study in Instinct", Journal of Experimental Zoology, Vol. 45, 1926, pp. 367–397.

Petryszak, N. G. L., The Sociology of Human Nature, (Ph.D. thesis, Simon Frazer University, Vancouver, 1978).

Petryszak, N. G. L. (1), "Human Nature: Ideology and Social Theory", Archiv für Rechts-und-Sozial Philosophie, Vol. 66, 1980, pp. 1–23.

Petryszak, N. G. L. (2), "Sociological Theory and Human Nature", Pacific Sociological Review, Vol. 23, No. 2, April, 1980, pp. 131–150.

Phoenix, C. H., Coy, R. W. and Resko, J. A., "Psychosexual Differentiation as a Function of Androgenic Stimulation" in M. Diamond (ed.), Perspectives in Reproduction and Sexual Behavior, (Indiana University Press, Bloomington, Indiana, 1968), pp. 33–49.

Pitcher, G., A Theory of Perception, (Princeton University Press, Princeton, New Jersey, 1971).

Plantinga, A., The Nature of Necessity, (Clarendon Press, Oxford, 1974).

Platt, J. R., "Properties of Large Molecules that Go Beyond the Properties of their Chemical Sub-Groups", Journal of Theoretical Biology, Vol. 1, 1961, pp. 342–358.

Plomin, R. and Kuse, A. R., "Genetic Differences Between Humans and Chimps and Among Humans", American Psychologist, Vol. 34, February, 1979, pp. 188–190.

Plomin, R., DeFries, J. C. and McClearn, G. E., Behavioral Genetics: A Primer, (W. H. Freeman and Company, San Francisco, 1980).

Plotkin, H. C. and Odling-Smee, F. J., "A Multiple-Level Model of Evolution and Its Implications for Sociobiology", The Behavioral and Brain Sciences, Vol. 4, 1981, pp. 225–268.

Polten, E. P., Critique of the Psycho-Physical Identity Theory, (Mouton, The Hague, Paris, 1973).

Pompa, L., "Review Discussion: Philosophy Without Epistemology", Inquiry, Vol. 24, 1981, pp. 359–380.

Popper, K. R., "The Propensity Interpretation of the Calculus of Probability and Quantum Theory" in S. Körner (ed.), Observation and Interpretation, (Butterworth Scientific Publications, London, 1957), pp. 65–70.

Popper, K. R., "Time's Arrow and Entropy", Nature, Vol. 207, July, 1965, pp. 233–234.

Popper, K. R., "Scientific Reduction and the Essential Incompleteness of All Sciences" in F. J. Ayala and T. Dobzhansky (eds.), Studies in the Philosophy of Biology, (Macmillan, London, 1974), pp. 259–284.

Popper, K. R. and Eccles, J. C., The Self and Its Brain, (Springer International, Berlin, Heidelberg, London, New York, 1977).

Potter, K. H., "The Background of Skepticism, East and West", Journal of Indian Philosophy, Vol. 3, September–December, 1975, pp. 299–313.

Poulantzas, N., Political Power and Social Classes, (New Left Books, London, 1975).

Prezelecki, M., "A Set Theoretic Versus a Model Theoretic Approach to the Logical Structure of Physical Theories", Studia Logica, Vol. XXXIII, 1974, pp. 91–105.

Pribram, K., Languages of the Brain: Experimental Paradoxes and Principles of Neuropsychology, (Prentice Hall, Englewood Cliffs, New Jersey, 1971).

Prigogine, I. (Discussion), "Can Thermodynamics Explain Biological Order?", Impact of Science on Society, Vol. XXIII, No. 3, 1973, pp. 159–179.

Puccetti, R., Persons: A Study of Possible Moral Agents in the Universe, (Macmillan, London, 1968).

Puccetti, R., "Brain Bisection and Personal Identity", British Journal for the Philosophy of Science, Vol. 24, 1973, pp. 339–355.

Puccetti, R., "Bilateral Organization of Consciousness in Man", Annals of the New York Academy of Sciences, Vol. 299, 1977, pp. 448–458.

Putnam, H., "How Not to Talk About Meaning" in R. S. Cohen and M. W. Wartofsky (eds.), Boston Studies in the Philosophy of Science, (Humanities Press, New York, 1965), Vol. II, pp. 205–222.

Putnam, H., "The Innateness Hypothesis' and Explanatory Models in Linguistics", in S. P. Stich (ed.), Innate Ideas, (University of California Press, Berkeley and Los Angeles, 1975), pp. 121–131.

Putnam, H., Meaning and the Moral Sciences, (Routledge and Kegan Paul, London, 1978).

Pyke, G. H., Pulliam, H. R. and Charnov, E. L., "Optimal Foraging: A Selective Review of Theory and Tests", The Quarterly Review of Biology, Vol. 52, 1977, pp. 137–154.

Quadagno, J. S., "Paradigms in Evolutionary Theory: The Sociobiological Model of Natural Selection", American Sociological Review, Vol. 44, 1979, pp. 100–109.

Quine, W. V. O., Word and Object, (Massachusetts Institute of Technology Press, Cambridge, Massachusetts, 1960).

Quine, W. V. O., "On What There Is" in From a Logical Point of View, (Harper and Row, New York, 1961), pp. 1–19.

Quine, W. V. O., "Reference and Modality" in his From a Logical Point of View, (Harper and Row, New York, 1961), pp. 139–157.

Quine, W. V. O., The Ways of Paradox and Other Essays, (Random House, New York, 1966).

Quine, W. V. O., "Ontological Reduction and the World of Numbers", Journal of Philosophy, Vol. LXI, 26 March, 1964, pp. 209–216. Reprinted in The Ways of Paradox and Other Essays, (Random House, New York, 1966), pp. 199–214.

Quine, W. V. O., Ontological Relativity and Other Essays, (Columbia University Press, New York, 1969).

Quine, W. V. O., "On the Reasons for Indeterminacy of Translation", Journal of Philosophy, Vol. 67, 1970, pp. 178–183.

Quine, W. V. O., "On Empirically Equivalent Systems of the World", Erkenntnis, Vol. 9, 1975, pp. 313–328.

Quinton, A., The Nature of Things, (Routledge and Kegan Paul, London, 1973).

Raacke, I. D., "Protein Hormones and the Eucaryolic Genome: A General Theory of Hormone Action", Perspectives in Biology and Medicine, Vol. 21, Autumn, 1977, pp. 139–157.

Rabinow, P. and Sullivan, W. M., "The Interpretive Turn: Emergence of an Approach" in P. Rabinow and W. M. Sullivan (eds.), Interpretive Social Science: A Reader, (University of California Press, Berkeley, California, 1979), pp. 1–21.

Radicalesbians., "Women-Identified Women" in P. Brown (ed.), Radical Psychology, (Harper Colophon, New York, 1973), pp. 471–480.

Radnitzky, G., Contemporary Schools of Metascience, (Akademiförlaget, Göteborg, Sweden, 1970).

Radnitzky, G., "Progress and Rationality in Research" in M. D. Grmek, R. S. Cohen and G. Cimino (eds.), On Scientific Discovery, (D. Reidel, Dordrecht, 1980), pp. 43–102.

Ramsey, F. P., The Foundations of Mathematics, (Routledge and Kegan Paul, London, 1931), edited by R. B. Braithwaite.
Rantala, V., "The Old and New Logic of Metascience", Synthese, Vol. 39, 1978, pp. 233–247.
Rantala, V., "On the Logical Basis of the Structuralist Philosophy of Science", Erkenntnis, Vol. 15, 1980, pp. 269–286.
Rapoport, A., N-Person Game Theory: Concepts and Applications, (University of Michigan Press, Ann Arbor, 1970).
Rashevsky, N., Mathematical Biophysics, 3rd edition, (Dover, New York, 1961).
Rescher, N. (1), The Coherence Theory of Truth, (Clarendon Press, Oxford, 1973).
Rescher, N. (2), The Primacy of Practice: Essays Towards a Pragmatically Kantian Theory of Empirical Knowledge, (Basil Blackwell, Oxford, 1973).
Rescher, N. Methodological Pragmatism: A Systems-Theoretic Approach to the Theory of Knowledge, (Basil Blackwell, Oxford, 1977).
Rescher, N., "Philosophical Disagreement: An Essay Towards Orientational Pluralism in Metaphilosophy", The Review of Metaphysics, Vol. XXII, No. 2, December, 1978, pp. 217–251.
Rescher, N., Cognitive Systematization: A Systems Theoretic Approach to the Theory of Knowledge, (Basil Blackwell, Oxford, 1979).
Rescher, N. and Brandon, R., The Logic of Inconsistency: A Study in Non Standard Possible World Semantics and Ontology, (Rowman and Littlefield, Totowa, New Jersey, 1979).
Reznek, L., "A Note on Relativism", Philosophical Papers, Vol. VIII, No. 2, October, 1979, pp. 69–71.
Richardson, R. C., "Functionalism and Reductionism", Philosophy of Science, Vol. 46, 1979, pp. 533–558.
Richardson, R. C., "Disappearance and the Identity Theory", Canadian Journal of Philosophy, Vol. XI, No. 3, September, 1981, pp. 473–483.
Richardson, R. C., "How Not to Reduce a Functional Psychology", Philosophy of Science, Vol. 49, 1982, pp. 125–137.
Riedl, R., "A Systems-Analytical Approach to Macro-Evolutionary Phenomena", The Quarterly Review of Biology, Vol. 52, No. 4, 1977, pp. 351–370.
Riedl, R., Order in Living Organisms: A Systems Analysis of Evolution, (John Wiley and Sons, New York, 1978).
Riegl, K. F., "Towards a Dialectical Theory of Development", Human Development, Vol. 18, 1975, pp. 50–64.
Ripley, C., "A Theory of Volitions", American Philosophical Quarterly, Vol. II, No. 2, April, 1974, pp. 141–147.
Rogers, L. and Walsh, J., "Shortcomings of the Psycho-medical Research of John Money and Co-Workers into Sex Differences in Behavior: Social and Political Implications", Sex Roles, Vol. 8, No. 3, 1982, pp. 269–281.
Rorty, R. (1), "Mind-Body Identity, Privacy and Categories", The Review of

Metaphysics, Vol. XIX, 1965, pp. 24–54. Reprinted in C. V. Borst (ed.), The Mind-Brain Identity Theory, (Macmillan, St. Martin's Press, London, 1970), pp. 187–213. References to latter.

Rorty, R. (2), "Incorrigibility as the Mark of the Mental", Journal of Philosophy, Vol. 67, 1970, pp. 399–424.

Rorty, R., "The World Well Lost", Journal of Philosophy, Vol. 69, 1972, pp. 649–665.

Rorty, R., Philosophy and the Mirror of Nature, (Princeton University Press, Princeton, New Jersey, 1979).

Rorty, R., "A Reply to Dreyfus and Taylor", The Review of Metaphysics, Vol. 34, September, 1980, pp. 39–46.

Rorty, R., Taylor, C. and Dreyfus, H., "A Discussion", The Review of Metaphysics, Vol. 34, September, 1980, pp. 47–55.

Rorty, R., "Pragmatism and Philosophy", Lecture delivered University of Adelaide, 12-7-1982.

Rosaldo, M. Z., "The Use and Abuse of Anthropology: Reflections on Feminism and Cross Cultural Understanding", Signs: Journal of Women in Culture and Society, Vol. 5, No. 3, 1980, pp. 389–417.

Rose, R., Gordon, T. P. and Bernstein, I. S., "Plasma Testosterone Levels in the Male Rhesus: Influences of Sexual and Social Stimuli", Science, Vol. 178, 1972, pp. 643–645.

Rose, R. M., Bourne, P. and Peo, R., "Androgen Responses to Stress", Psychosomatic Medicine, Vol. 31, 1969, pp. 418–436.

Rose, S. and Rose, H., "'Do Not Adjust Your Mind, There is a Fault in Reality' – Ideology in Neurobiology", Cognition, Vol. 2, No. 4, 1973, pp. 479–502.

Rose, S., The Conscious Brain, (Weidenfeld and Nicholson, London, 1973).

Rosen, R., Optimality Principles in Biology, (Butterworths, London, 1967).

Rosen, R., "Morphogenesis" in R. Rosen (ed.), Foundations of Mathematical Biology Vol. II: Cellular Systems, (Academic Press, New York and London, 1972), pp. 1–77.

Rosenberg, A. (1), "Ruse's Treatment of the Evidence for Evolution: A Reconsideration" in P. D. Asquith and R. N. Giere (eds.), Philosophy of Science Association, Philosophy of Science Association, East Lansing, Michigan, 1980), Vol. 1, pp. 83–93.

Rosenberg, A. (2), Sociobiology and the Preemption of Social Science, (John Hopkins University Press, Baltimore and London, 1980).

Rosenberg, A., "On the Propensity Definition of Fitness", Philosophy of Science, Vol. 49, 1982, pp. 268–273.

Rossi, A. S., "A Biosocial Perspective on Parenting", Daedalus, Vol. 106, Spring, 1977, pp. 1–31.

Rotkin, K. F., "The Phallacy of Our Sexual Norm" in A. G. Kaplan and J. P. Bean (eds.), Beyond Sex-Role Stereotypes: Readings Towards a Psychology of Androgyny, (Little Brown and Co., Boston, 1976), pp. 154–162.

Routley, R., "Repairing Proofs of Arrow's General Impossibility Theorem and Enlarging the Scope of the Theorem", Notre Dame Journal of Formal Logic, Vol. XX, No. 4, October, 1979, pp. 879–890.

Routley, R. (1), Exploring Meinong's Jungle and Beyond: An Investigation of Noneism and the Theory of Items, (Australian National University, Canberra, 1980).

Routley, R. (2), "The Choice of Logical Foundations: Non-classical Choices and the Ultra Logical Choice", Studia Logica, Vol. XXXIX, 1980, pp. 78–98.

Routley, R. (3), "On the Impossibility of an Orthodox Social Theory and of an Orthodox Solution to Environmental Problems", Logique et Analyse, N. S. Vol. 23, Mar., 1980, pp. 145–166.

Rubin, D-H., Marxism and Materialism: A Study in Marxist Theory of Knowledge, (Harvester Press, Sussex, 1977).

Ruse, M., "Are There Laws in Biology?", Australasian Journal of Philosophy, Vol. 48, No. 2, August, 1970, pp. 234–246.

Ruse, M., Philosophy of Biology, (Hutchinson University Library, London, 1973).

Ruse, M., "Reduction in Genetics" in R. S. Cohen (et al)(eds.), Philosophy of Science Association, (D. Reidel, Dordrecht, 1976), pp. 633–651.

Ruse, M., Sociobiology: Sense or Nonsense?, (D. Reidel, Dordrecht, 1979).

Ruse, M., Is Science Sexist? And Other Problems in the Biomedical Sciences, (D. Reidel, Dordrecht, 1981).

Ruse, M., "Is Human Sociobiology A New Paradigm?", The Philosophical Forum, Vol. XIII, Nos. 2–3, Winter–Spring, 1981–82, pp. 119–143.

Ruse, M., Darwinism Defended: A Guide to the Evolution Controversies, (Addison-Wesley Publishing Company, Massachusctts, 1982).

Russell, E. S., Form and Function: A Contribution to the History of Animal Morphology, (John Murray, London, 1916).

Ryle, G., The Concept of Mind, (Hutchinson, London, 1949).

Ryle, G., Dilemmas, (Cambridge University Press, Cambridge, 1954).

Sadovsky, V. N., "Problems For a General Systems Theory as a Metatheory", Ratio, Vol. XVI, No. 1, June, 1974, pp. 33–50.

Sahlins, M., The Use and Abuse of Biology: An Anthropological Critique of Sociobiology, (Tavistock, London, 1977).

Salisbury, F. B., "Doubts About the Modern Synthetic Theory of Evolution", The American Biology Teacher, Vol. 33, September, 1971, pp. 335–338 and p. 354.

Salmon, W. (ed.), Zeno's Paradoxes, (The Bobbs-Merrill Company, Indianapolis, 1970).

Salt, D., "Physical Axiomatics: Freudenthal vs. Bunge", Foundations of Physics, Vol. 1, No. 4, 1971, pp. 307–313.

Sartre, J. P., Existentialism and Humanism, translation P. Mairet, (Eyre Methuen, London, 1958).

Saunders, P. T., An Introduction to Catastrophe Theory, (Cambridge University Press, Cambridge, 1980).

Savitt, S., "Rorty's Disappearance Theory", Philosophical Studies, Vol. 28, 1975, pp. 433–436.

Schaffner, K. F., "Approaches to Reduction", Philosophy of Science, Vol. 34, 1967, pp. 137–147.

Schaffner, K., "Reductionism in Biology: Prospects and Problems" in R. S. Cohen (et al)(eds.), Proceedings of the 1974 Biennial Meeting of the Philosophy of Science Association, (D. Reidel, Dordrecht, 1976), pp. 613–632.

Schievella, P. S., "Emergent Evolution and Reductionism", Scientia, Vol. 108, 1973, pp. 323–330.

Schindewolf, O., Paläontologie, Entwicklungslehre und Genetik, (Borntraeger, Berlin, 1936).

Schindewolf, O., Der Zeit Faktor in Geologie und Paläontologie, (Schweizerbart, Stuttgort, 1950).

Schlesinger, G. N., "What is Metaphysics?", American Philosophical Quarterly, Vol. 18, No. 3, July, 1981, pp. 229–235.

Schneider, D. M., "The Meaning of Incest", Journal of the Polynesian Society, Vol. 85, 1976, pp. 149–169.

Schulman, S. R., "Kin Selection, Reciprocal Altruism, and the Principle of Maximization: A Reply to Sahlins", The Quarterly Review of Biology, Vol. 53, September, 1978, pp. 283–286.

Schwartz, J. G. (ed.), The Subtle Anatomy of Capitalism, (Goodyear Publishing Company, California, 1977).

Scriven, T., "Preference, Rational Choice and Arrow's Theorem", Journal of Philosophy, Vol. LXXVIII, No. 12, December, 1981, pp. 778–785.

Secord, P. F., "Making Oneself Behave: A Critique of the Behavioral Paradigm and an Alternative Conceptualization" in T. Mischel (ed.), The Self: Psychological and Philosophical Issues, (Rowman and Littlefield, Totowa, New Jersey, 1977), pp. 250–273.

Seidler, M. J., "Problems of a Systems Epistemology", International Philosophical Quarterly, Vol. XIX, No. 1, March, 1979, pp. 29–60.

Sellars, W., "The Language of Theories" in H. Feigl and G. Maxwell (eds.), Current Issues in the Philosophy of Science, (Holt, Rinehart and Winston, New York, 1961), pp. 57–77.

Sellars, W. (1), "Philosophy and the Scientific Image of Man" in Science, Perception and Reality, (Routledge and Kegan Paul, London, 1963), pp. 1–40.

Sellars, W. (2), "Empricism and the Philosophy of Mind" in Science, Perception and Reality, (Routledge and Kegan Paul, London, 1963), pp. 127–196.

Sellars, W., "The Identity Approach to the Mind Body Problem", The Review of Metaphysics, Vol. 18, 1965, pp. 430–451.

Sen, A. K., Collective Choice and Social Welfare, (Holden-Day Inc., San Francisco, 1970).

Shannon, C. E. and Weaver, W., The Mathematical Theory of Communication, (The University of Illinois Press, Urbana, Illinois, 1949).
Sheldrake, R., "Three Approaches to Biology. Part I. The Mechanistic Theory of Life", Theoria to Theory, Vol. 14, 1980, pp. 125–144.
Sheldrake, R. (1), "Three Approaches to Biology. Part II. Vitalism", Theoria to Theory, Vol. 14, 1981, pp. 227–240.
Sheldrake, R. (2), "Three Approaches to Biology. Part III. Organicism", Theoria to Theory, Vol. 14, 1981, pp. 301–311.
Sheldrake, R. (3), A New Theory of Life: The Hypothesis of Formative Causation, (Blond and Briggs, London, 1981).
Shepher, J., "Mate Selection Among Second Generation Kibbutz Adolescents and Adults: Incest Avoidance and Negative Imprinting", Archives of Sexual Behaviour, Vol. 1, No. 4, 1971, pp. 293–307.
Shepher, J., Incest, The Biosocial View, (Harvard University Press, Cambridge, Massachusetts, 1979).
Sherman, J. A., Sex-Related Cognitive Differences: An Essay on Theory and Evidence, (Charles C. Thomas, Springfield, Illinois, 1978).
Shields, S. A., "Functionalism, Darwinism and the Psychology of Women", American Psychologist, Vol. 30, 1975, pp. 739–754.
Shimony, A. (1), "Is Observation Theory-Laden? A Problem in Naturalistic Epistemology", The Philosopher's Annual, Vol. 1, 1978, pp. 116–145.
Shimony, A. (2), "Metaphysical Problems in the Foundations of Quantum Mechanics", International Philosophical Quarterly, Vol. XVIII, No. 1, March, 1978, pp. 2–17.
Shirley, E. S., "Rorty's 'Disappearance' Version of the Identity Theory", Philosophical Studies, Vol. 25, 1974, pp. 73–75.
Shope, R. L., "Eliminating Mistakes About Eliminative Materialism", Philosophy of Science, Vol. 46, 1979, pp. 590–612.
Shrader-Frechette, K., "Atomism in Crisis: An Analysis of the Current High Energy Paradigm", Philosophy of Science, Vol. 44, 1977, pp. 409–440.
Simpson, G. G., Tempo and Mode in Evolution, (Columbia University Press, New York, 1944).
Skinner, B. F., "The Operational Analysis of Psychological Terms", Psychological Review, Vol. 52, No. 5, 1945, pp. 270–277.
Skinner, B. F., Contingencies of Reinforcement: A Theoretical Analysis, (Appleton-Century-Crofts, New York, 1969).
Skinner, B. F., Beyond Freedom and Dignity, (Bantam Books, New York, 1971; Penguin Books, Harmondsworth, 1976). Reference to the 1976 edition.
Skjervheim, H., "Objectivism and the Study of Man (Part I)", Inquiry, Vol. 17, 1974, pp. 213–239.
Skjervheim, H., "Objectivism and the Study of Man (Part II)", Inquiry, Vol. 17, 1974, pp. 265–302.

Sklar, L., "Types of Inter-Theoretic Reduction", British Journal for the Philosophy of Science, Vol. 18, 1967, pp. 109–124.

Slobodkin, L. B. and Rapoport, A., "An Optimal Strategy of Evolution", The Quarterly Review of Biology, Vol. 49, No. 3, 1974, pp. 181–200.

Sloman, A., The Computer Revolution in Philosophy: Philosophy, Science and Models of Mind, (Harvester Press, Sussex, 1978).

Slote, M., Reason and Scepticism, (George Allen and Unwin, Ltd., London, 1970).

Slotkin, J. S., "On a Possible Lack of Incest Regulations in Old Iran", American Anthropologist, Vol. 49, October–December, 1947, pp. 612–617.

Slotkin, J. S., "Reply to Goodenough", American Anthropologist, Vol. 51, July–September, 1949, pp. 531–532.

Smart, J. J. C. (1), Philosophy and Scientific Realism, (Routledge and Kegan Paul, London, 1963).

Smart, J. J. C. (2), "Materialism", Journal of Philosophy, Vol. LX, 1963, pp. 651–662.

Smart, J. J. C., "Conflicting Views About Explanation" in R. S. Cohen and M. W. Wartofsky (eds.), Boston Studies in the Philosophy of Science, Humanities Press, New York, 1965), Vol. 2, pp. 157–169.

Smart, J. J. C. (1), "The Content of Physicalism", The Philosophical Quarterly, Vol. 28, 1978, pp. 339–341.

Smart, J. J. C. (2), "Is Occam's Razor a Physical Thing?", Philosophy, Vol. 53, 1978, pp. 382–385.

Smith, J. W. (1), "The Problem of the Criterion of Truth", Aspect, Vol. 1, No. 2, September, 1982, pp. 5–19.

Smith, J. W. (2), "A Reply to Frankel's Criticism of Harré's Theory of Cauaslity", Philosophy of Science, Vol. 49, 1982, pp. 282–289.

Smith, J. W. (3), "A Critique of Some Critiques of the General Systems Weltanschauung". (Forthcoming.)

Smith, J. W. (4), "Tal Scriven on Preference, Rational Choice and Arrow's Theorem". (Forthcoming.)

Smith, J. W. (5), "Two Recent Self-Referential Arguments". (Forthcoming.)

Sneed, J. D., The Logical Structure of Mathematical Physics, (D. Reidel, Dordrecht, 1971).

Sober, E., "Evolution, Population Thinking and Essentialism", Philosophy of Science, Vol. 47, 1980, pp. 350–383.

Sociobiology Study Group of Science for the People, "Sociobiology – A New Biological Determinism" in The Ann Arbor Science for the People Editorial Collective, Biology as a Social Weapon, (Burgess Publishing Company, Minneapolis, Minnesota, 1977), pp. 131–149.

Sonneborn, T. M., "Gene Action in Development", Proceedings of the Royal Society of London Series B, Vol. 176, 1970, pp. 347–366.

Sorokin, P. A., Fads and Foibles in Modern Sociology and Related Sciences, (Greenwood Press, Chicago, 1956).

Sperry, R. W., "Mental Phenomena as Causal Determinants in Brain Function" in G. G. Globus, G. Maxwell and I. Savodnik (eds.), Consciousness and the Brain: A Scientific and Philosophical Inquiry, (Plenum Press, New York and London, 1976), pp. 163–177.

Sperry, R. W., Zaidel, E. and Zaidel, D., "Self-Recognition and Social Awareness in the Disconnected Minor Hemisphere", Neuropsychologia, Vol. 17, 1979, pp. 156–166.

Spetner, L. M., "Natural Selection Versus Genetic Uniqueness", Nature, Vol. 226, 6 June, 1970, pp. 948–949.

Stafford, R. E., "Sex Differences in Spatial Visualization as Evidence of Sex-linked Inheritance", Perception and Motor Skills, Vol. 13, 1961, p. 428.

Stanley, S. M., Macroevolution: Pattern and Process, (W. H. Freeman and Company, San Francisco, 1979).

Star, S. L. (1), "Sex Differences and the Dichotomization of the Brain: Methods, Limits and Problems in Research on Consciousness" in R. Hubbard and M. Lowe (eds.), Genes and Gender II: Pitfalls in Research on Sex and Gender, (Gordian Press, New York, 1979), pp. 113–130.

Star, S. L. (2), "The Politics of Right and Left: Sex Differences in Hemispheric Brain Asymmetry" in R. Hubbard, M. S. Henifin and B. Fried (eds.), Women Looking at Biology Looking at Women: A Collection of Feminist Critiques, (Schenkman Publishing Co., Cambridge, Massachusetts, 1979), pp. 61–74.

Stebbing, L. S., Philosophy and the Physicists, (Methuen, London, 1937).

Stebbins, G. L., Variation and Evolution in Plants, (Columbia University Press, New York, 1950).

Steele, E. J., Somantic Selection and Adaptive Evolution: On the Inheritance of Acquired Characteristics, (William-Wallace International Inc., Toronto, 1979).

Steele, E. J. and Gorczynski, R. M., "Inheritance of Acquired Immunological Tolerance to Foreign Histocompatibility Antigens in Mice", Proceedings of the National Academy of Science, Vol. 77, 1980, pp. 2871–2875.

Stegmüller, W., The Structure and Dynamics of Theories, (Springer-Verlag, Berlin, Heidelberg, New York, 1976).

Stegmüller, W., The Structuralist View of Theories, (Springer-Verlag, Berlin, Heidelberg, New York, 1979).

Stern, J. T., "The Meaning of 'Adaptation' and its Relation to the Phenomena of Natural Selection", Evolutionary Biology, Vol. 4, 1970, pp. 39–66.

Stich, S. P., "Introduction: The Idea of Innateness" in S. P. Stich (ed.), Innate Ideas, (University of California Press, Berkeley and Los Angeles, 1975), pp. 1–22.

Stich, S. P., "Emprìcism, Innateness, and Linguistic Universals", Philosophical Studies, Vol. 33, 1978, pp. 273–286.

Stich, S. P., "Do Animals Have Beliefs?", Australasian Journal of Philosophy, Vol. 57, 1979, pp. 15–28.

Stone, C. P., "A Note on 'Feminine' Behavior in Adult Male Rats", American Journal of Physiology, Vol. 68, 1924, pp. 39–41.

Strawson, P. F., Individuals, (Methuen, London, 1959).

Stryer, L., Biochemistry, (W. H. Freeman and Company, San Francisco, 1975).

Suppe, F., The Meaning and Use of Models in Mathematics and the Exact Sciences, (Ph.D. thesis, University of Michigan, 1967).

Suppe, F., "What's Wrong with the Received View of the Structure of Scientific Theories?", Philosophy of Science, Vol. 39, 1972, pp. 1–19.

Suppe, F., "Theories, Their Formulations, and the Operational Imperative", Synthese, Vol. 25, 1973, pp. 129–164.

Suppe, F., "Theories and Phenomena" in W. Leinfellner and E. Kohler (eds.), Developments of the Methodology of the Social Sciences, (D. Reidel, Dordrecht, 1974), pp. 45–92.

Suppe, F. (ed.), The Structure of Scientific Theories, 2nd edition, (University of Chicago Press, Urbana, 1977).

Suppe, F., "Theory Structure" in P. D. Asquith and H. E. Kyburg (eds.), Current Research in Philosophy of Science, (Philosophy of Science Association, East Lansing, Michigan, 1979), pp. 317–338.

Suppes, P., Introduction to Logic, (D. Van Nostrand Co., Princeton, 1957).

Suppes, P., "What is a Scientific Theory?" in S. Morgenbesser (ed.), Philosophy of Science Today, (Basic Books, New York, 1967), pp. 55–67.

Suppes, P., A Probabilistic Theory of Causality, Acta Philosophica Fennica, Fasc. XXIV, (North-Holland, Amsterdam, 1970).

Sussman, H. J., "Catastrophe Theory: A Preliminary Critical Study" in F. Suppe and P. D. Asquith (eds.), Philosophy of Science Association, (Philosophy of Science Association, East Lansing, Michigan, 1976), Vol. 1, pp. 256–286.

Sussman, H. J. and Zahler, R. S., "Catastrophe Theory as Applied to the Social and Biological Sciences: A Critique", Synthese, Vol. 37, 1978, pp. 117–216.

Swartz, N., "Emergence and Materialist Theories of Sentience", Philosophy Forum, Vol. 14, 1975, pp. 241–267.

Swinburne, R., "Are Mental Events Identical with Brain Events?", American Philosophical Quarterly, Vol. 19, No. 2, April, 1982, pp. 173–189.

Sztompka, P., Sociological Dilemmas: Towards a Dialectical Paradigm, (Academic Press, London, 1979).

Taylor, C., "Interpretation and the Sciences of Man", The Review of Metaphysics, Vol. 25, 1971, pp. 3–51.

Taylor, C., "Understanding Human Science", The Review of Metaphysics, Vol. 34, September, 1980, pp. 27–38.

Taylor, R., Action and Purpose, (Prentice Hall, Englewood Cliffs, 1966).

Taylor, R., Metaphysics, 2nd edition, (Prentice Hall, Englewood Cliffs, 1974).

Temin, H. M. and Mizutuni, S., "RNA-dependent DNA Polymerase in Virons of Rous Sarcomo Virus", Nature, Vol. 226, 1970, pp. 1211–1213.

Thagard, P. R. (1), "The Best Explanation: Criteria for Theory Choice", Journal of Philosophy, Vol. 75, 1978, pp. 76–92.

Thagard, P. R. (2), "Why is Astrology a Pseudoscience?" in P. D. Asquith and I. Hacking (eds.), Philosophy of Science Association, (Philosophy of Science Association, East Lansing, Michigan, 1978), Vol. 1, pp. 223–234.

Thalberg, I., Enigmas of Agency: Studies in the Philosophy of Human Action, (George Allen and Unwin, London, 1972).

Thalberg, I., "Agent Causality and Reasons for Acting", Philosophia, Vol. 7, No. 3–4, July, 1978, pp. 555–565.

Thom, R., "Structuralism and Biology" in C. H. Waddington (ed.), Towards a Theoretical Biology, (Edinburgh University Press, Edinburgh, 1972), Vol. 4, pp. 68–82.

Thom, R., Structural Stability and Morphogenesis: An Outline of a General Theory of Models, translation by D. H. Fowler, (W. A. Benjamin, Massachusetts, 1975).

Thomas, D., Naturalism and Social Science: A Post-Empiricist Philosophy of Social Science, (Cambridge University Press, Cambridge, 1979).

Thomason, J. M., "Ontological Relativity and the Inscrutability of Reference", Philosophical Studies, Vol. 22, June, 1971, pp. 50–56.

Thompson, D. W., On Growth and Form, Revised edition (Macmillan, New York, 1945).

Thompson, D. W., On Growth and Form, Abridged Edition, edited by J. T. Bonner (Cambridge University Press, Cambridge, 1961).

Thompson, V., "Does Sex Accelerate Evolution?", Evolutionary Theory, Vol. 1, 1976, pp. 131–156.

Tiger, L. and Fox, R., The Imperial Animal, (Secker and Warburg, London, 1972).

Tiger, L. and Shepher, R., Women in the Kibbutz, (Harcourt, Brace and Jovanovitch, New York, 1975).

Tilney-Bassett, R. A. E. and Abdel-Wahab, O. A. L., "Maternal Effects and Plastid Inheritance" in D. R. Newth and M. Balls (eds.), Maternal Effects in Development: The Fourth Symposium of the British Society for Developmental Biology, (Cambridge University Press, Cambridge, 1979), pp. 29–45.

Torrey, E. F., The Death of Psychiatry, (Penguin, New York, 1974).

Treisman, M., "The Evolution of Sexual Reproduction: A Model which Assumes Individual Selection", Journal of Theoretical Biology, Vol. 60, 1976, pp. 421–431.

Trigg, R. L., The Shaping of Man: Philosophical Aspects of Sociobiology, (Basil Blackwell, Oxford, 1982).

Trivers, R. L., "The Evolution of Reciprocal Altruism", The Quarterly Review of Biology, Vol. 46, 1971, pp. 35–57.

Trivers, R. L., "Parent-Offspring Conflict", American Zoologist, Vol. 14, 1974, pp. 249–264.

Trivers, R. L. and Hare, H., "Haplodiploidy and the Evolution of Social Insects", Science, Vol. 191, 1976, pp. 249–263.

Trivers, R. L., "Parental Investment and Sexual Selection" in B. Campbell (ed.), Sexual Selection and the Descent of Man 1871–1971, (Aldine, Chicago, 1972). Reprinted in T. H. Clutton-Brock and P. H. Harvey (eds.), Readings in Sociobiology, (W. H. Freeman and Company, San Francisco, 1978), pp. 52–97. References to latter.

Tuomela, R., "On the Structuralist Approach to the Dynamics of Theories", Synthese, Vol. 39, 1978, pp. 211–231.

Turner, L., Multinational Companies and the Third World, (Allen Lane, London, 1973).

Ullmann-Margalit, E., The Emergence of Norms, (Clarendon Press, Oxford, 1977).

Unger, P., "Why There Are No People", Midwest Studies in Philosophy, Vol. IV, 1979, pp. 177–222.

Van Den Berghe, P. L., Man in Society: A Biosocial View, (Elsevier, New York, 1975).

Van Den Berghe, P. L., "Bridging the Paradigms: Biology and the Social Sciences" in M. S. Gregory, A. Silvers and D. Sutch (eds.), Sociobiology and Human Nature: An Interdisciplinary Critique and Defense, (Jossey-Bass Publishers, San Francisco, 1978), pp. 33–52.

Vandermeer, J., "Ecological Determinism" in The Ann Arbor Science for the People Editorial Collective, Biology as a Social Weapon, (Burgess Publishing Company, Minneapolis, Minnesota, 1977), pp. 108–122.

Van Fraassen, B., "On the Extension of Beth's Semantics of Physical Theories", Philosophy of Science, Vol. 37, 1970, pp. 325–339.

Van Fraassen, B., "A Formal Approach to the Philosophy of Science" in R. Colodny (ed.), Paradigms and Paradoxes: The Philosophical Challenge of the Quantum Domain, (University of Pittsburgh Press, Pittsburgh, 1972, pp. 303–366.

Van Fraassen, B., "The Labyrinth of Quantum Logic" in R. S. Cohen and M. Wartofsky (eds.), Boston Studies in the Philosophy of Science, Vol. XIII, (D. Reidel, Dordrecht, 1974), pp. 224–254.

Vender, Z., Res Cogitans, (Cornell University Press, Ithaca, 1972).

Von Bertalanffy, L., General System Theory, (George Braziller, New York, 1968).

Von Bertalanffy, L., "Chance or Law?" in A. Koestler and J. R. Smythies (eds.), Beyond Reductionism: New Perspectives in the Life Sciences, (Macmillan, New York, 1969), pp. 56–84.

Waber, D. P., "Sexual Differences in Mental Abilities, Hemispheric Laterlization, and the Rate of Physical Growth at Adolescence", Developmental Psychology, Vol. 13, 1977, pp. 29–38.

Waddington, C. H., The Strategy of the Genes, (George Allen and Unwin, London, 1957).

Waddington, C. H., "The Basic Ideas of Biology" in C. H. Waddington (ed.), Towards a Theoretical Biology, (Edinburgh University Press, Edinburgh, 1968), Vol. 1, pp. 1–32.

Waddington, C. H., "Paradigm for an Evolutionary Process" in C. H. Waddington (ed.), Towards a Theoretical Biology, (Edinburgh University Press, Edinburgh, 1969), Vol. 2, pp. 106–124.

Waddington, C. H., The Evolution of an Evolutionist, (Cornell University Press, Ithaca, 1975).

Wahlsten, D., "A Critique of the Concepts of Heritability and Heredity in Behavioral Genetics" in J. R. Royce and L. P. Mos (eds.), Theoretical Advances in Behavior Genetics, (Sijthoff and Noordhoff, The Netherlands, 1979), pp. 425–481.

Washburn, S. L., "Human Behavior and the Behavior of Other Organisms", American Psychologist, Vol. 33, 1978, pp. 405–418.

Watson, J. D. and Crick, F. H. C., "Molecular Structure of Nucleic Acids: A Structure for Deoxyribose Nucleic Acid", Nature, Vol. 171, 1953, pp. 737–738.

Watson, J. D., Molecular Biology of the Gene, 3rd edition, (W. A. Benjamin Inc., Menlo Park, California, 1977).

Watson, L. (1), Supernature: A Natural History of the Supernatural, (Coronet Books, Hodder and Stoughton, Kent, 1980).

Watson, L. (2), Lifetide: A Biology of the Unconscious, (Coronet Books, Hodder, and Stoughton, Kent, 1980).

Webster, G. and Goodwin, B. C., "History and Structure in Biology", Perspectives in Biology and Medicine, Vol. 25, Autumn, 1981, pp. 39–62.

Webster, G. and Goodwin, B. C., "The Origins of Species: A Structuralist Approach", Journal of Social and Biological Structures, Vol. 5, 1982, pp. 15–47.

Weinrich, J. D., "The Author Replies", Behavioral Ecology and Sociobiology, Vol. 3, 1978, pp. 95–98.

Werner, E. E. and Hall, D. J., "Optimal Foraging and Size Selection of Prey by the Bluegill Sunfish, (*Lepomis machrochirus*)", Ecology, Vol. 55, 1974, pp. 1042–1052.

West Eberhard, M. J., "The Evolution of Social Behavior by Kin Selection", The Quarterly Review of Biology, Vol. 50, No. 1, March, 1975, pp. 1–33.

Wheelwright, E. E. and Stilwell, F. J. B. (eds.), Readings in Political Economy, (Australian and New Zealand Book Company, Sydney, 1976), Vol. 2.

White, A., Handler, P., Smith, E. L., Hill, R. L. and Lehman, S. R., Principles of Biochemistry, McGraw Hill, New York, 1978).

Whitrow, G. J., What is Time?, (Thames and Hudson, London, 1972).

Whyte, L. L., Internal Factors in Evolution, (George Braziller, New York, 1965).

Wigner, E. P., "The Probability of the Existence of a Self-Reproducing Unit" in The Logic of Personal Knowledge, (Routledge and Kegan Paul, London,

1961), pp. 231–238.
Wigner, E. P., Symmetries and Reflections, (Indiana University Press, Indiana, 1967).
Wigner, E. P., "Physics and the Explanation of Life", Foundations of Physics, Vol. 1, No. 1, 1970, pp. 35–45.
Wilkes, K. V., Physicalism, (Routledge and Kegan Paul, London, 1978).
Williams, B. O. A., Problems of the Self, (Cambridge University Press, Cambridge, 1973).
Williams, G. C., Adaptation and Natural Selection, (Princeton University Press, Princeton, 1966).
Williams, G. C., "Kin Selection and the Paradox of Sexuality" in G. W. Barlow and J. Silverberg (eds.), Sociobiology: Beyond Nature/Nurture? Reports, Definitions and Debates, AAAS Selected Symposium, 35 (Westview Press, Boulder, Colorado, 1980), pp. 371–384.
Williams, M. B., "Deducing the Consequences of Evolution: A Mathematical Model", Journal of Theoretical Biology, Vol. 29, 1970, pp. 343–385.
Williams, M. B., "Falsifiable Predictions of Evolutionary Theory", Philosophy of Science, Vol. 40, 1973, pp. 518–537.
Williams, T., "Family Resemblance in Abilities: The Wechsler Scales", Behavior Genetics, Vol. 5, 1975, pp. 405–409.
Wilson, D. S., "A Cultural Route to Biological Fitness", Evolutionary Theory, Vol. 3, March, 1978, pp. 235–236.
Wilson, E., The Mental as Physical, (Routledge and Kegan Paul, London, 1979).
Wilson, E. O., The Insect Societies, (Belknap Press of Harvard University Press, Cambridge, Massachusetts, 1971).
Wilson, E. O., Sociobiology: The New Synthesis, (Belknap Press of Harvard University Press, Cambridge, Massachusetts, 1975).
Wilson, E. O. (1), "What is Sociobiology?" in M. S. Gregory, A. Silvers and D. Sutch (eds.), Sociobiology and Human Nature: An Interdisciplinary Critique and Defense, (Jossey-Bass Publishers, San Francisco, 1978), pp. 1–12.
Wilson, E. O. and Oster, G. F. (2), Caste and Ecology in the Social Insects, (Princeton University Press, Princeton, 1978).
Wilson, E. O., On Human Nature, (Bantam Books, New York, 1979).
Wilson, E. O., "The Relation of Science to Theology", Zygon, Vol. 15, No. 4, December, 1980, pp. 425–434.
Wilson, E. O. and Lumsden, C. J., Genes, Mind, and Culture: The Coevolutionary Process, (Harvard University Press, Cambridge, Massachusetts and London, 1981).
Wimsatt, W. C., "Reduction and Reductionism" in P. D. Asquith and H. E. Kyburg (eds.), Current Research in Philosophy of Science, (Philosophy of Science Association, East Lansing, 1979), pp. 352–377.
Winch, P., The Idea of a Social Science and Its Relation to Philosophy, (Routledge and Kegan Paul, London, 1958).

Wittgenstein, L., Philosophical Investigations, translation by G. E. M. Anscombe (Macmillan, New York, 1953).

Wittig, M. A., "Sex Differences in Intellectual Functioning: How Much of a Difference do Genes Make?", Sex Roles, Vol. 2, 1976, pp. 63–74.

Wittman, C., "Gay Liberation Manifesto" in P. Brown (ed.), Radical Psychology, (Harper Colophon, New York, 1973), pp. 453–470.

Wolf, A., "Childhood Association and Sexual Attraction: A Further Test of the Westermarck Hypothesis", American Psychologist, Vol. 72, 1970, pp. 503–515.

Wright, S., "Evolution in Mendelian Populations", Genetics, Vol. 16, 1931, pp. 97–159.

Wynne-Edwards, V. C., Animal Dispersion in Relation to Social Behaviour, (Oliver and Boyd, Edinburgh, 1962).

Yamazaki, T., "Measurement of Fitness at the Esterase-5 Locus In *Drosophila pseudoobscura*", Genetics, Vol. 67, 1971, pp. 579–603.

Yoshida, R. M., Reduction in the Physical Sciences, Philosophy in Canada, Vol. 4, (Dalhousie University Press, Dalhousie, 1977).

Zeeman, E. C., Catastrophe Theory: Selected Papers, 1972–1977, (Addison-Wesley Publishing Company, Massachusetts, 1977).

Zetterlund, B., Winberg, J., Lundgren, G., and Johansson, G., "Lead in Umbilical Cord Blood Correlated with the Blood Lead of the Mother in Areas with Low, Medium or High Atmospheric Pollution", Acta Paediatrica Scandanavia, Vol. 66, 1977, pp. 169–175.

Zwart, P. J., About Time, (North-Holland, Amsterdam, 1976).

Subject index

NIJHOFF INTERNATIONAL PHILOSOPHY SERIES

1. Rotenstreich N: Philosophy, History and Politics – Studies in Contemporary English Philosophy of History. 1976. ISBN 90-247-1743-4.
2. Srzednicki JTJ: Elements of Social and Political Philosophy. 1976. ISBN 90-247-1744-2.
3. Tatarkiewicz W: Analysis of Happiness. 1976. ISBN 90-247-1807-4.
4. Twardowski K: On the Content and Object of Presentations – A Psychological Investigation. Translated and with an Introduction by R. Grossman. 1977. ISBN 90-247-1726-7.
5. Tatarkiewicz W: A History of Six Ideas – An Essay in Aesthetics. 1980. ISBN 90-247-2233-0.
6. Noonan HW: Objects and Identity – An Examination of the Relative Identity Thesis and Its Consequences. 1980. ISBN 90-247-2292-6.
7. Crocker L: Positive Liberty – An Essay in Normative Political Philosophy. 1980. ISBN 90-247-2291-8.
8. Brentano F: The Theory of Categories. Translated by R.M. Chisholm and N. Guterman. 1981. ISBN 90-247-2302-7.
9. Marciszewski W (ed): Dictionary of Logic as Applied in the Study of Language – Concepts / Methods / Theories. 1981. ISBN 90-247-2123-7.
10. Ruzsa I: Modal Logic with Descriptions. 1981. ISBN 90-247-2473-2.
11. Hoffman P: The Anatomy of Idealism – Passivity and Activity in Kant, Hegel and Marx. 1982. ISBN 90-247-2708-1.
12. Gram, MS: Direct Realism – A Study of Perception. 1983. ISBN 90-247-2870-3.
13. Srzednicki, JTJ and Rickey, VF (eds): Leśniewski's Systems – Ontology and Mereology. ISBN 90-247-2879-7.
14. Smith, Joseph Wayne: Reductionism and Cultural Being – A Philosophical Critique of Sociobiological Reductionism and Physicalist Scientific Unificationism. 1984. ISBN 90-247-2884-3.
15. Zumbach C: The Transcendent Science – Kant's Conception of Biological Methodology. 1984. ISBN 90-247-2904-1.
16. Notturno, MA: Objectivity, Rationality and the Third Realm: Justification and the Grounds of Psychologism – A Study of Frege and Popper. 1984. ISBN 90-247-2956-4.